AFRICA'S EMERGING MAIZE REVOLUTION

AFRICA'S EMERGING MAIZE REVOLUTION

edited by
Derek Byerlee and Carl K. Eicher

LYNNE
RIENNER
PUBLISHERS

BOULDER
LONDON

Published in the United States of America in 1997 by
Lynne Rienner Publishers, Inc.
1800 30th Street, Boulder, Colorado 80301

and in the United Kingdom by
Lynne Rienner Publishers, Inc.
3 Henrietta Street, Covent Garden, London WC2E 8LU

© 1997 by Lynne Rienner Publishers, Inc. All rights reserved

Library of Congress Cataloging-in-Publication Data
Byerlee, Derek.
   Africa's emerging maize revolution / edited by Derek Byerlee and
Carl K. Eicher.
     p.   cm.
   Includes bibliographical references (p.   ) and index.
   ISBN 1-55587-776-1 (hardcover : alk. paper).
   ISBN 1-55587-754-0 (pbk. : alk. paper)
   1. Corn—Africa, Sub-Saharan—Congresses. 2. Corn—Economic
aspects—Africa, Sub-Saharan—Congresses. I. Eicher, Carl K.
SB191.M2B924 1997
338.1'7315'096—dc21   97-10662
                                                                         CIP

British Cataloguing in Publication Data
A Cataloguing in Publication record for this book
is available from the British Library.

Printed and bound in the United States of America

∞   The paper used in this publication meets the requirements
of the American National Standard for Permanence of
Paper for Printed Library Materials Z39.48-1984.

5 4 3 2 1

Contents

List of Tables and Figures		vii
Preface		xi

Part 1 The Maize Economy of Africa

1	Introduction: Africa's Food Crisis *Derek Byerlee and Carl K. Eicher*	3
2	Evolution of the African Maize Economy *Derek Byerlee and Paul W. Heisey*	9

Part 2 Country Studies

3	Zimbabwe's Emerging Maize Revolution *Carl K. Eicher and Bernard Kupfuma*	25
4	Zambia's Stop-and-Go Maize Revolution *Julie A. Howard and Catherine Mungoma*	45
5	Maize Technology and Productivity in Malawi *Melinda Smale and Paul W. Heisey*	63
6	Increasing Maize Production in Kenya: Technology, Institutions, and Policy *Rashid M. Hassan and Daniel D. Karanja*	81
7	Maize Technology Development in Ghana During Economic Decline and Recovery *Robert Tripp and Kofi Marfo*	95
8	Fostering Sustainable Increases in Maize Productivity in Nigeria *Joyotee Smith, Georg Weber, M.V. Manyong, and M. A. B. Fakorede*	107

Part 3 Technology, Institutions & Policy

9 The Technological Foundation of the Revolution 127
   Derek Byerlee and David Jewell

10 Maize Research Priorities: The Role of Consumer Preferences 145
    Lawrence Rubey, Richard W. Ward, and David Tschirley

11 Soil Fertility Management in Southern Africa 157
    *John D. T. Kumwenda, Stephen R. Waddington,
    Sieglinde S. Snapp, Richard B. Jones, and Malcolm J. Blackie*

12 Institutional Innovations in the Maize Seed Industry 173
    Joseph Rusike and Carl K. Eicher

13 Fertilizer Use and Maize Production 193
    Paul W. Heisey and Wilfred Mwangi

14 Maize Marketing and Pricing Policy in Eastern and
    Southern Africa 213
    *Thomas S. Jayne, Stephen Jones, Mulinge Mukumbu, and
    Share Jiriyengwa*

Part 4 Conclusions

15 Accelerating Maize Production: Synthesis 247
    Carl K. Eicher and Derek Byerlee

Acronyms and Abbreviations 263
References 265
Contributors 291
Index 295
About the Book 301

Tables and Figures

Tables

2.1	Africa: Major Trends and Distribution of Maize Production	13
3.1	Zimbabwe: Land Classification by Natural Region, 1995	27
3.2	Zimbabwe: Smallholders' Use of Inputs, Loans, Fertilizer Application Rates, and Average Maize Yields in Low- and High-Rainfall Areas, 1990–1991	33
3.3	Zimbabwe: Number of Scientists in the Department of Research and Specialist Services (DR&SS), 1980–1992	39
4.1	Zambia: Maize Technology and the Policy/Organizational Environment	47
5.1	Malawi: Farmers' Ranking of Yield, Processing, and Storage Characteristics in Maize Hybrids and Local Maize, 1991–1992	73
5.2	Malawi: Mean Yields of Maize Hybrids and Local Maize, With and Without Fertilizer, in "Normal" and Drought Seasons, 1989–1993	74
5.3	Malawi: Adoption of Maize Technology by Farm Size, 1985 and 1990	76
6.1	Kenya: Maize Varieties Released, 1961–1995	83
6.2	Kenya: Growth in Maize Area, Yield, and Production and Trends in Adoption of Improved Maize Seed and Fertilizer, 1963–1991	84
6.3	Kenya: Adoption of Improved Maize Seed and Fertilizer, Extension and Market Access, and Proportion of Area Planted to Maize by Zone, 1990–1993	86
6.4	Kenya: Percentage of Farmers Adopting Improved Maize Seed and Fertilizer and Having Access to Extension and Credit, by Zone and Farm Size, 1992–1993	86
6.5	Kenya: Gap Between Maize Yields Obtained in Research Trials and by Farmers and Disparity Between Farmers' Nutrient Use and Recommended Nutrient Levels, by Zone and Farm Size, 1992–1993	87

7.1	Ghana: Maize Area, Yield, and Production, 1950–1994	96
7.2	Ghana: Adoption of Maize Technology in Three Areas, Mid-1980s and 1990	102
8.1	Western Africa: Distinguishing Characteristics of Agricultural Systems in Humid and Subhumid Areas of Nine Countries	110
8.2	Northern Guinea Savanna, Nigeria: Importance of Maize in the Mid-1970s and 1989	115
8.3	Nigeria: Maize Production Technologies and an Assessment of their Adoption	119
8.4	Northern Guinea Savanna, Nigeria: Some Strategic Research Issues	122
9.1	Africa, Asia, and Latin America: A Comparison of Maize Research Resources and Varietal Releases, 1991	128
9.2	Sub-Saharan Africa: Maize Area Planted to Improved OPVs and Hybrids, 1990	130
9.3	Sub-Saharan Africa: Summary of Estimates of Returns to Investment in Maize Research	133
10.1	Zimbabwe: Percentage of Consumers by Income Quintile Switching from White Roller Meal to Yellow Roller Meal at a Specified Price Discount, 1993	152
12.1	Key Market Problems and Innovations During the Life Cycle of the Maize Seed Industry	175
12.2	Sub-Saharan Africa: Public and Private Investment Share of Commercial Maize Seed Sales and Seed Prices	177
12.3	Sub-Saharan Africa: Number of Companies Engaged in the Maize Seed Industry by Type of Company, 1992	178
12.4	Eastern and Southern Africa: Stages of the Life Cycle of the Maize Seed Industry in Six Countries, 1996	179
12.5	Eastern and Southern Africa: Public and Private Companies Distributing Maize Seed in Six Countries, 1996	188
13.1	Sub-Saharan Africa: Fertilizer Use by Crop, 1991–1992	196
13.2	Sub-Saharan Africa and Other Regions of the Developing World: Fertilizer Applied to Maize, 1989–1993 period	198
13.3	Sub-Saharan Africa and Other Regions of the Developing World: Nitrogen-Maize Price Ratios	200
13.4	Lilongwe, Malawi: Effect of Price Assumptions on the Profitability of Alternative Maize Technologies in 110 On-Farm Demonstrations, 1990 and 1991	201
14.1	Phase 1 of Maize Marketing Policy in Eastern and Southern Africa: Colonial Regime	217
14.2	Phase 2 of Maize Marketing Policy in Eastern and Southern Africa: Independence	221

14.3	Phase 3 of Maize Marketing Policy in Eastern and Southern Africa: Structural Adjustment	226
14.4	Eastern and Southern Africa: Real Maize Grain and Maize Meal Price Trends in Selected Countries, 1980–1994	231
14.5	Eastern and Southern Africa: Trends in Maize Production Per Capita, Area, Yield, Net Exports, and Fertilizer Use in Selected Countries	233

Figures

2.1	Africa: Net Imports of Wheat, Rice, and Maize	11
2.2	Africa, Selected Countries: Adjusted Population Density and Proportion of Arable Land Utilized, Mid-1980s	12
2.3	Areas of Sub-Saharan Africa Classified as Suitable for Maize Production	14
2.4	Africa, Selected Countries: Contribution of Maize to Total Calorie Consumption	17
2.5	Developing Countries in Africa and Other Regions: Mean Yields and Yield Variability	19
3.1	Zimbabwe: Average Maize Yields on Smallholder and Large-Scale Farms, 1965–1994	28
4.1	Zambia: Maize Area and Production, 1962–1995	46
4.2	Zambia: Sales of Fertilizer and Improved Maize Seed, 1962–1995	53
5.1	Malawi: Maize Yields, 1961–1995	64
5.2	Malawi: Maize Production and Utilization Per Capita, 1961–1995	65
5.3	Malawi: Diffusion of Hybrid Maize, 1981–1996	73
6.1	Kenya: Maize Area and Yield, 1961–1995	83
6.2	Kenya: Public Maize Research Expenditure, 1955–1988	85
6.3	Kenya: Sales of Improved Maize Seed, Early 1960s to Mid-1980s	88
8.1	Nigeria: Maize Area, Yield, and Production, 1969–1993	112
11.1	Malawi: Scenarios for Maize Surplus/Deficit	158
14.1	Zimbabwe: Per Capita Grain Production in African Communal Lands, 1914–1994	219

Preface

In 1966, Marvin Miracle published *Maize in Tropical Africa*, the first comprehensive study of the role of maize in food production and consumption in Africa. At that time, attention to the world food production problem was focused on Asia—most countries in Africa were self-sufficient or net exporters of food. Today, however, that focus has shifted to Africa, which is facing a food crisis similar to that of Asia in the early 1960s. Per capita food consumption in Africa has declined sharply over the past two decades, and many African countries are now dependent on food imports. Increased food production is critical to achieving food security, peace, and democracy in Africa in the twenty-first century.

It is also clear that food production strategies employed successfully in Asia cannot simply be transferred to Africa; the continent's food crisis demands fresh thinking based on regional experiences in promoting increased food production. To that end, the editors of this book have assembled a large body of research on one major staple food crop, maize, which has been at the forefront of efforts to increase food production in much of Africa in recent years. The book provides the first overview of maize in Africa since the publication of Miracle's book more than three decades ago.

The book had its genesis in the early 1970s, when the editors, then both at Michigan State University, collaborated with African scholars in some of the first detailed microlevel studies of African food production systems. In the late 1970s, Derek Byerlee moved to the International Maize and Wheat Improvement Center (CIMMYT) in Mexico to undertake multidisciplinary research on technologies for smallholder production systems. This work, carried out in collaboration with national agricultural research programs, led in the early 1990s to a detailed series of farm-level studies on technology needs and impacts in several African countries in which CIMMYT had been actively involved. Similar studies by the International Institute of Tropical Agriculture (IITA) were underway in Nigeria and other West African countries, where maize production had expanded rapidly from the late 1970s.

Meanwhile, Michigan State University, with support from the United States Agency for International Development (USAID), was involved in a number of collaborative country studies on food security, many of which featured work on institutional and policy issues in maize research, production, and marketing. In 1994, we recognized that the two groups of studies constituted a unique Africa-wide perspective on technological, institutional, and policy issues for a major food staple. We then approached the Rockefeller Foundation, which generously agreed to fund a workshop to bring together a set of 13 studies on the African maize economy. The debate among the 30 workshop participants (from Africa, MSU, CIMMYT, and IITA) are reflected in this book.

The book has four major objectives: to synthesize historical and contemporary experiences related to Africa's maize economy; to foster cross-country comparisons of major technological, institutional, and policy issues; to provide guidelines on strategies for increasing maize production; and to identify research gaps and priorities. The introductory chapter sets out these objectives in more detail and, together with Chapter 2, provides an overview of the importance of and major trends in the African maize economy. The body of the book consists of two major parts. The first provides six country case studies of the evolving maize economies of Zimbabwe, Zambia, Malawi, Kenya, Ghana, and Nigeria. The second part synthesizes major technological, institutional, and policy issues with chapters on research and extension, soil fertility, seed and fertilizer delivery systems, and marketing and price policy. In the last chapter we draw together the book's major findings. We owe much to the contributing authors for this last chapter, although they do not bear responsibility for the product.

This book is aimed at a broad audience of policymakers and others interested in Africa's food production problems, including donors, university professors, researchers, and students. The chapters have been written by both social scientists and maize specialists, but an effort has been made to ensure that they are accessible to the nonspecialist.

* * *

The editors and authors owe thanks to many who have contributed in some way to this book. First, we are most grateful for the extensive support received from CIMMYT, USAID, and the Rockefeller Foundation. In addition to funding research for several of the chapters, CIMMYT contributed generously to support for the workshop and for the editing and publication of the book. We particularly thank Larry Harrington, Prabhu Pingali, Delbert Hess, and David Jewell for facilitating this support. USAID, through its Food Security in Africa Cooperative Agreements with Michigan State University, supported research for several chapters. The Rockefeller Foundation, through Bob Herdt, Malcom Blackie, and John Lynam, not only provided generous financial support for research for some chapters and for the workshop, but has also been a continuing source of intellectual stimulation on the challenges to African food production. Although we owe a great debt to these sponsors, our conclusions should not be attributed to them nor to the World Bank, where Derek Byerlee is employed.

In addition, many people have assisted in some way on the individual chapters. The following is undoubtedly an incomplete list of those persons: Comfort Ateh, Tracy Atwood, Doyle Baker, Lewis Bangwe, Duncan Boughton, Munhamo Chisvo, Eric Crawford, A. A. Danyi, S. Dapaah, Alex Duncan, Richard Goldman, Patrick Kambewa, Zikani H. W. Kaunda, Yilma Kebede, Charles Machethe, Hermes L. Makina, K. O. Makinde, Mywish Maredia, the late Martin M. K. Mkandawire, Mac N. S. Msowaya, Watson Mwale, Bethel Nakaponda, Jim Oehmke, S. Oikeh, A. Opoku-Apau, B. Oyewole, Coty Pinckney, Al Schmid, Jim Shaffer, John Staatz, Timothy Thobane, S. Twumasi-Afriyie, Mike Weber, and Kim Witte.

Finally, this book would not have been possible without the untiring support of Pat Eisele, who organized the workshop and provided invaluable support in finalizing the chapters. We were also fortunate to have the services of a first-class editor—Kelly Cassaday. To both Pat and Kelly, we owe an extreme debt of gratitude.

Derek Byerlee
Carl K. Eicher

Part 1
The Maize Economy of Africa

1

Introduction: Africa's Food Crisis

Derek Byerlee & Carl K. Eicher

At the beginning of the independence movement in 1960, African agriculture was moving. Africa was self-sufficient in food and a leading agricultural exporter. In contrast, Asia was the epicenter of the world food crisis. But by the mid-1960s, Asia had launched the green revolution, which presently adds 50 million tons of grain to the world's food supply each year. Today, although Asia still struggles to increase household food security, it is Africans, not Asians, who bear the brunt of the world food problem.

The food crisis began shifting to Africa in the early 1970s as the continent's food balance sheet changed from positive to negative. Food production grew at half (1.5% per year) the rate of population growth (3.0% per year) from 1970 to 1985. Since then, the situation has continued to deteriorate. For the period 1988 to 1993, 33 African countries experienced a reduction in per capita food production (FAO 1994b). On the consumption side, sub-Saharan Africa is the only region of the world where the average caloric intake has declined over time. In 1990, it was estimated that 37% of Africa's population was undernourished. Even if economic growth resumes, sub-Saharan Africa will have 300 million undernourished people by the year 2000, nearly half of the world's total (Alexandratos 1995).

The problem of stagnating food production is reflected in growing dependence on food imports, rising poverty, and degradation of the natural resource base, all of which have become increasingly critical problems in Africa over the past two decades. Although many solutions have been proposed, they have often been at a level that is too general for this diverse continent, which comprises 47 countries, a wide range of ecologies, seven colonial histories, and an array of food staples.

Increased food production has a vital role to play in enhancing food security, peace, and democracy in Africa in the twenty-first century. Africa's population will grow by an additional 100 million over the coming six to seven years, and its 1995 population of 600 million is expected to double to 1.2 billion by 2020. Africa's food production gap demands fresh thinking and urgent attention by scientists and policymakers.

Two preconditions are essential for alleviating the downward spiral of poverty and malnutrition in Africa. First, in nearly all African countries, the key to economic growth is growth in agriculture. The bulk of the population depends on agriculture, and increases in agricultural household income generate further rounds of spending that stimulate economic growth by increasing demand for rural nonfarm products, as well as urban industrial products. Second, the key to renewed growth in the agricultural sector is rapid technical change in food production. Greater productivity, brought about by technical change in staple food crops, generates broad-based growth through increased rural incomes and reduced prices of food.

The success of Asia's green revolution encouraged many scientists, governments, and donor agencies to try to replicate Asia's model for technical change in Africa, but virtually all of these attempts have failed. There is abundant evidence that Asia's green revolution model cannot be automatically replicated in Africa (Eicher 1989). Instead, there is a need to turn inward, to study Africa's own experience in developing food production models compatible with its cultures, histories, environments, and population densities. This book contributes to the debate over the future course of African agriculture by focusing on a large body of recent research on one staple food crop, maize. Africa's emerging revolution in maize production represents one of the few rays of hope for stepping up food production over the coming 10 to 20 years.

Technology alone, however, will not provide the momentum for a maize revolution. Institutional change, rural infrastructure, and policy are critical to success, and this book focuses on these complex issues. During much of the 1980s, development assistance agencies promoted reforms to remove the large policy distortions that characterized most countries in Africa and that almost invariably discriminated against the agricultural sector. Policy reforms also encouraged privatization of product and input markets, in which governments had been dominant, and the elimination of input subsidies, especially on fertilizer. Much progress has been made, but policy reforms are still incomplete (Donovan 1995). We will return to some of the issues in the policy debate later in this chapter, but first we discuss our reasons for focusing on maize rather than on any other staple food crop.

Why Maize?

We have chosen to focus on maize in this book for several reasons. First, although maize is a relatively new crop in Africa, its production has expanded so rapidly since independence that it is the most important food crop for urban and rural consumers.[1] Maize is the dominant food staple throughout most of eastern and southern Africa, where its importance equals that of rice and wheat in much of Asia. Consumption of rice and wheat is increasing more rapidly in Africa, but these increases are supplied largely through imports, whereas maize is a home-grown food.

A second reason for focusing exclusively on maize is that the past two decades have yielded some compelling success stories for this crop, as the use of new seed and associated technologies has increased smallholder maize production. The diffusion of new technologies in Africa has been more widespread for maize than for other food crops, and Africa's maize experience can provide lessons for increasing food production more generally.

Third, as we look to the future, maize—with its high yield potential and ease of processing and marketing for urban consumers—has considerable potential to help reverse the downward spiral of food production in Africa (Blackie 1994b). More knowledge is available about production systems, processing, and marketing for maize than for any other food crop in Africa. Finally, maize is a politically important crop in many countries because it is the most important food staple. The chapters in this book bring together a large body of new evidence on the evolving maize economies of Africa, including country studies that summarize results of recent surveys of thousands of farmers.

Food Policy Debates

The six country studies bring fresh evidence to bear on policy debates surrounding future food production strategies, beginning with evidence related to several technical questions. To what extent should food production strategies emphasize high-potential areas and/or areas with good infrastructure, where the potential for quick payoffs is highest? Given the extremely critical nature of the food crisis in Africa, the country case studies make a strong case for seeking an increase in production in high-potential areas with adequate rainfall, where maize is often the leading food crop. Despite the continuing, gradual spread of maize into more marginal areas in Africa, most maize is still produced in medium- to high-potential areas, although the crop is subject to infrequent but sometimes severe drought. Maize also tends to be grown in more densely populated areas and areas with better infrastructure. Thus maize is concentrated in areas that, because of their ecological and geographic characteristics, have the potential to be major breadbaskets of Africa.

A related question that permeates much of the debate about strategies for technical change in Africa is whether the major agents for increasing food production should be external inputs (fertilizers, hybrid seed, and pesticides) or new, low-input production systems that emphasize farmer-saved seed (i.e., open-pollinated varieties) and internally generated sources of crop nutrients (Low 1993). Fertilizer use in Africa is still very low and, even more important, the rate of growth in fertilizer consumption in the 1980s was slow compared to Asia and Latin America, where fertilizer use has grown rapidly, even in rainfed areas. One school of thought vigorously maintains that the poor infrastructure in much of Africa, combined with the poverty of African farmers, precludes the use of such external, usually imported, inputs. Adherents of this view contend that other means, based on internally generated sources of nutrients, must be found to restore and maintain Africa's impoverished soils. Given that increased doses of chemical fertilizer have been an important source of growth in crop yields in both industrialized and developing countries since World War II, this question is obviously central to future strategies for African food production. If the model of change in agricultural productivity that has served much of the world in recent decades is not applicable in Africa, the implications for investment in research and extension in Africa to develop and diffuse new types of technologies are enormous.

The worldwide emphasis on protecting the environment has stimulated interest in natural resources, especially soil and water conservation. There is no doubt that critical problems of soil and water conservation exist in Africa. An important issue to resolve is the extent to which the historical emphasis on increasing agricultural productivity should be shifted to give more emphasis to conservation of land resources for succeeding generations.

Sustainable agricultural systems require more than a one-time boost to agricultural productivity. Institutional structures must be in place to provide a continual stream of new innovations and to adapt to an ever-changing environment. These structures include research and extension systems and seed and fertilizer distribution systems. The record of building research and extension systems in Africa has been disappointing. From a very small base at independence, these systems have grown rapidly in numbers of scientific staff, but real budgets have eroded steadily. Declining funds, combined with poor incentive systems and defective management, have provoked a crisis in African research and extension systems (Eicher 1989). Many African governments are grappling with the question of how to revitalize these systems. New management and incentive systems, in which the private sector, farmers, and nongovernmental organizations (NGOs) participate, are part of these ongoing institutional experiments. Some of these, such as the Training and Visit (T&V) extension system, have been widely applied in Africa. Given the weak state of African national research systems and the small size of many African countries, regional research bodies and networks and international centers assume greater importance in Africa relative to other regions. These institutions, however, have also been undergoing rapid evolution.

Finally, from the colonial period into the postindependence period, the heavy hand of government has pervaded almost all aspects of food crop production in Africa, including input supply, processing, and marketing. The period since 1980 has seen an effort not only to align local prices with international prices through reforms in exchange rates, price policy, and trade but also to minimize or eliminate government interventions in agricultural markets, allowing the private sector to take over. Under pressure from international lenders and donors, governments are withdrawing marketing board monopolies on food procurement and distribution, privatizing fertilizer distribution, removing fertilizer subsidies, and selling state seed monopolies. Although considerable progress has been made—especially in freeing trade and exchange rates—the process is still incomplete, and the debate continues over how to sequence and implement reforms. It is now recognized that it is not enough for governments to abruptly withdraw, for example, from fertilizer distribution and assume that the private sector will effectively take over (Tripp 1993). Likewise, the dismemberment of effective maize marketing boards in some countries has led to difficulty in stabilizing maize prices and administering a floor price to producers. Thus a critical but still unresolved debate concerns the appropriate role of government intervention in input and output markets in the climate of fiscal austerity and privatization of the 1990s (Smith 1995).

All of these issues are discussed to varying degrees in the chapters of this book. Through a focus on one crop from the farm level to the consumer, we hope to add significantly to the debate and identify appropriate policy interventions to increase food production.

Objectives and Outline

This book has four main objectives. First, we seek to synthesize historical and contemporary experience related to Africa's emerging maize-based revolution; second, to foster cross-country comparisons and exchange of information on technical, policy, and institutional issues in maize production, marketing, and processing; third, to provide guidelines to African policymakers and international agencies for developing environmentally sound strategies for increasing the production of maize; and fourth, to identify research gaps and priorities for technology generation, farmer support systems, and policy reforms to meet future food supply needs.

The book is divided into four parts. A general overview of maize production and consumption in Africa (Chapter 2) concludes part one. Part two presents case studies that synthesize information on the evolution of the maize economies of six countries (Chapters 3–8). The six countries were selected on the basis of the importance of maize in local diets and success in introducing improved maize production technologies. The chapters draw on in-depth studies of maize production, processing, and marketing completed within the past five years. All of the studies are based on extensive fieldwork, including surveys of thousands of farmers. The results of these studies provide a unique opportunity to draw implications for future food production strategies, policy reforms, and donor assistance.

The third part of the book examines technologies, institutions, and policies to increase food production in Africa in the twenty-first century. Chapter 9 reviews past achievements and future challenges related to the generation and diffusion of technology through research and extension systems. The next chapter discusses the need to take account of consumer preferences when setting priorities for maize research. Chapter 11 focuses on a critical issue in much of Africa's maize belt: the search for cost-effective and sustainable approaches to increasing soil fertility.

Government monopolies are rapidly being replaced by new forms of public and private partnerships and new indigenous and multinational seed companies in eastern and southern Africa. These encouraging institutional innovations are discussed in Chapter 12 on the maize seed industry. Chapter 13 examines the contentious issues surrounding the development of efficient fertilizer distribution systems. Chapter 14 pulls together the burgeoning literature on maize marketing, processing, and pricing to shed light on the complex issues of public grain marketing monopolies. In eastern and southern Africa, most of these monopolies were established during the world depression in the 1930s to regulate the pricing and distribution of maize. In the wake of the policy reforms that have occurred over the past decade, the appropriate roles of government and the private sector in marketing maize and other food crops are evolving and subject to vigorous debate.

In part four, we draw together the main findings from the country studies and technical chapters to identify the crucial decisions that must be taken to increase maize production in Africa in the next 20 years. Although these findings are based on the analysis of data for only one crop, we believe many of the conclusions and generalizations are relevant to the broader issues of increasing food production in Africa.

Note

1. In 1992, the per capita calorie consumption of food staples in sub-Saharan Africa was as follows: maize, 302; cassava, 299; sorghum, 162; rice, 162; millet, 123; wheat, 120; sweet potatoes, 107; and plantain, 57 (FAO, various years).

2
Evolution of the African Maize Economy

Derek Byerlee & Paul W. Heisey

Maize was a relatively minor food crop in Africa in 1900, but today it is the continent's most important food crop. This chapter first briefly traces the evolution of maize in Africa during the twentieth century and updates the historical description provided in Marvin Miracle's *Maize in Tropical Africa* (1966). The bulk of the chapter is devoted to an analysis of broad trends in maize production, yields, and consumption over the past thirty years, based on secondary statistics. Given the poor quality of statistics in Africa, the conclusions must be treated with caution. Some countries undoubtedly have fairly reliable statistics on food production and consumption, but in many other countries civil war and financial constraints have produced gaps and inconsistencies in the data.

Historical Background

Maize arrived in Africa in the course of the sixteenth century, most likely through Portuguese traders who stopped along both the western and eastern coasts. From the coast, maize slowly moved inland through various routes, particularly through the incursions of slave traders, who valued maize as a storable and easily processed grain (Miracle 1966). Most farmers had little to do with the new crop, however, and for centuries maize was an important food staple in only a few pockets of Africa.

Maize's transition to a major crop occurred in Kenya during World War I, when the colonial government encouraged farmers to plant maize for the war effort. At the same time, a serious disease epidemic in the traditional food crop, millet, led to famine, and stocks of millet seed were consumed rather than saved for planting. By providing farmers with seed of a late-maturing white maize variety, the colonial government sped the transition from a millet- to a maize-based food economy. After the war, the development of export markets encouraged maize production, and by the

1930s maize was established as the dominant food crop in much of Kenya and Tanzania (Taylor 1969; Gerhart 1975).

In southern Africa (outside of South Africa), commercial maize production began when white settlers moved into the area in the late 1890s. Maize became a major crop for commercial farmers, especially because of the increasing demand for grain in the mining areas. Colonial governments found it attractive to promote maize for African wage earners because it was easy to process compared to the traditional staples, millet and sorghum (Miracle 1966). Finally, the booming starch market in England provided an outlet for white maize exports from Zimbabwe (Masters 1994).

Maize gradually became a staple of the African population in eastern and southern Africa, beginning with those who were most exposed to commercial activities. By the 1930s, maize was important in smallholder agriculture as both a subsistence and a cash crop. The expanding infrastructure, especially the railways, further encouraged the expansion of maize. The nearly complete changeover in the diets of millions of Africans from traditional sorghum and millet to maize in less than two generations represents a remarkable revolution in food production and consumption patterns. In Malawi, 80% of the cultivated area is now planted to maize, though the crop was virtually unknown a century ago.

In most of western Africa, maize never reached the level of importance it acquired in eastern and southern Africa, but production grew sufficiently for western Africa to become a net exporter of maize in the early part of the twentieth century. The most dramatic expansion of maize production in western Africa has occurred over the past two decades in the savanna areas, as improved technology, development of rural roads, and urban demand fueled a rapid increase in maize area (Kennedy and Reardon 1994).

Trends in Food Production and Consumption

The role of maize in Africa's food economy must be viewed against overall trends in food production and consumption across the continent. The food production trends have been well documented. The index of per capita food *production*, stagnant in the 1960s, began moving steadily downward in the 1970s. This trend may have bottomed out around 1985, but overall the index is considerably lower in Africa than in South Asia, where per capita food production has risen steadily since the green revolution of the 1960s. The trend in per capita calorie *consumption* in Africa has been less pronounced, showing only a slight decline from the 1960s. Shortfalls in domestic food production have been made up through a growing reliance on imports, especially of rice and wheat, which surged in the 1970s (Figure 2.1). For maize, Africa has shifted from being a net exporter in the 1960s to a net importer in the 1990s. Maize imports have been highest in drought years, especially 1992, when maize imports exceeded rice imports.

All recent projections show a widening gap between food supply and demand in Africa. Africa imports about 10 million tonnes of food each year, almost half of it under various food aid programs. With current trends in production and no change in per capita consumption, this gap is projected to widen substantially by 2020.

Aggregate indicators on the input side also point to the difficult situation confronting African food producers. The amount of arable land per agricultural worker has diminished steadily with population growth, although the amount of available land varies substantially from one country and region to another. Aggregate data on population density, however, can be misleading as an indicator of land scarcity, because large areas in Africa are marginal for agriculture. Figure 2.2 shows the distribution of countries with respect to population per unit of land, standardized for production potential following Binswanger and Pingali (1988). The figure also shows cultivated land as a percentage of estimated potential arable area, based on Cleaver and Donovan (1995). By both measures, several countries—particularly Nigeria, Kenya, and Ethiopia—now surpass major Asian countries in intensity of land use. At the same time, several countries—such as

Figure 2.1 Africa: Net Imports of Wheat, Rice, and Maize

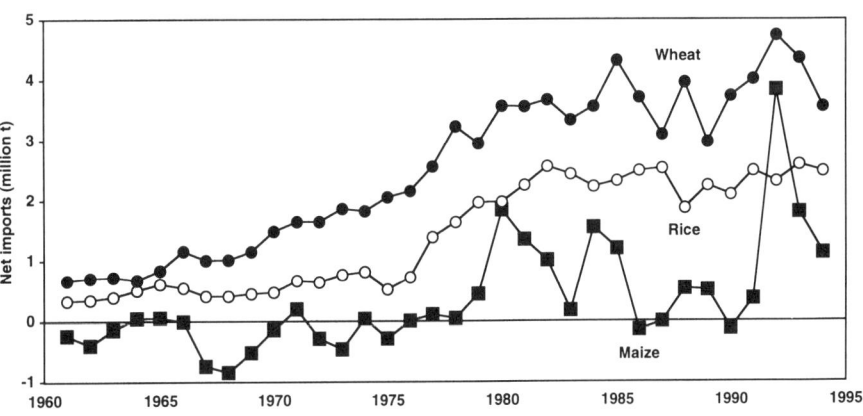

Source: Computed from FAO data tapes.

Figure 2.2   Africa, Selected Countries: Adjusted Population Density and Proportion of Arable Land Utilized, Mid-1980s

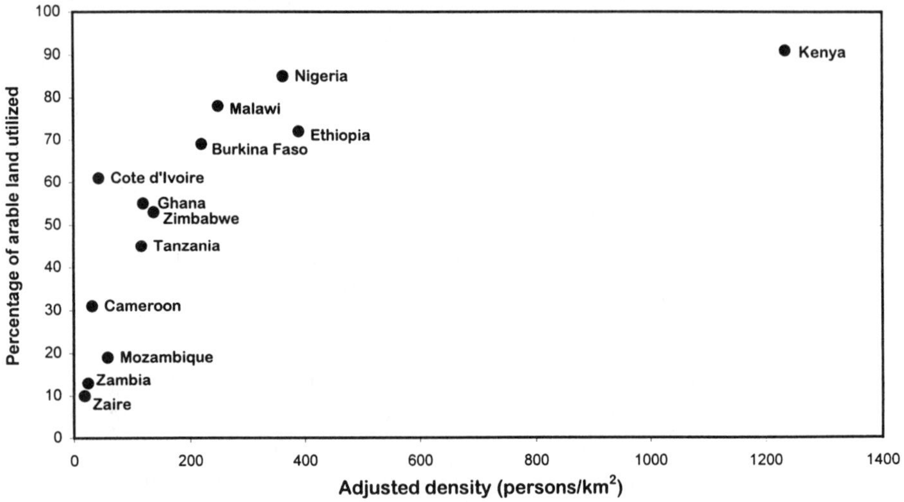

[Scatter plot with x-axis "Adjusted density (persons/km²)" ranging 0 to 1400, and y-axis "Percentage of arable land utilized" ranging 0 to 100. Data points: Kenya (~1300, 90); Nigeria (~350, 85); Malawi (~250, 80); Ethiopia (~400, 72); Burkina Faso (~350, 70); Cote d'Ivoire (~100, 60); Ghana (~150, 55); Zimbabwe (~170, 53); Tanzania (~180, 48); Cameroon (~90, 30); Mozambique (~110, 20); Zambia (~50, 12); Zaire (~40, 8).]

Source: Computed from Binswanger and Pingali (1988) and FAO data tapes.

Zaire, Zambia, and Mozambique—have substantial potential to increase the area under cultivation.

Increasing intensification of land use is characteristic of most developing regions undergoing rapid population growth. In most regions of the developing world, intensification has been accompanied by the use of land-saving technologies, especially improved seed-fertilizer and irrigation technologies. But in Africa intensification has lagged. Although farmers' use of modern varieties increased rapidly in the 1980s, especially for maize (an estimated 40% of the area is now planted to improved varieties; see Chapter 9), fertilizer use per hectare is very low and has stagnated since about 1980. In 1989, fertilizer use per hectare on rain-fed land in India was more than three times that in Africa. Given the breakdown of the bush-fallow system under population pressure and soil degradation resulting from the mining of soil nutrients, the limited use of fertilizer in Africa is a critical policy issue in increasing food production.

A final indicator of the precarious state of African agriculture is the steady decline in agricultural output per worker over the past 30 years, the result of slow growth in output combined with increasing population pressure (Pardey and Roseboom 1991). Whereas most regions of the developing world have enjoyed increases in agricultural output both per worker and per hectare, output per hectare in Africa has remained unchanged and output per worker has fallen.

Distribution of Maize Production in Relation to Ecological Regions

Annual maize production in Africa (excluding South Africa) averages about 26 million tonnes on more than 20 million hectares (Table 2.1). In global terms, Africa is a relatively small producer; by comparison, the U.S. state of Iowa alone harvests almost 50 million tonnes of maize. Seven African countries produce more than 1 million tonnes of maize each year; together, they account for over 70% of the maize produced in Africa. (Five of these seven are included as case studies in this book; Ethiopia and Tanzania are the exceptions.)

A large area of Africa is classified by the Food and Agriculture Organization (FAO) as suitable or very suitable for maize production (Figure 2.3). The maize belt extends from southern Africa through eastern Africa and across the savanna of western Africa. The maize belt is bounded on

Table 2.1 Africa: Major Trends and Distribution of Maize Production

	Western and Central Africa	Eastern Africa	Southern Africa	All of Africa
Maize area, 1993–1995 (million ha)	10.2	5.2	5.2	20.6
Maize yield, 1993–1995 (t/ha)	1.2	1.6	1.1	1.2
Maize production, 1993–1995 (million t)	11.9	8.2	5.7	25.8
Regional share of maize production (%)	46	32	22	100
Percentage of area in region that is[a]				
Lowland	90	12	14	28
Mid-altitude	8	47	85	54
Highland	2	41	1	18
Total	100	100	100	100
Growth in maize area (%/yr) in[b]				
1975–1985	2.3	–0.4	0.4	0.6
1985–1995	3.2	0.7	0.1	1.3
1975–1995	2.6	1.4	0.6	1.6
Growth in maize yield (%/yr) in[b]				
1975–1985	1.7	0.9	–1.1	0.2
1985–1995	1.4	0.5	–4.0	–0.6
1975–1995	1.9	0.9	–1.2	0.5
Growth in maize production (%/yr) in[b]				
1975–1985	4.1	0.5	–0.8	0.8
1985–1995	4.6	1.1	–3.9	0.7
1975–1995	4.5	2.3	–0.5	2.0

Notes: a. Lowland areas are less than 900 meters above sea level (masl); mid-altitude areas, 900–1,800 masl; and highland areas, above 1,800 masl.

b. Growth rates for western and central Africa exclude Nigeria. Growth rates for southern Africa exclude Angola, Mozambique, and South Africa. Growth rates for all of Africa exclude all four countries. Although South Africa is the largest maize producer in Africa, accounting for about one-third of total production, more than 85% of the maize produced in South Africa is grown on large commercial farms. Since we generally focus on small-scale producers, South Africa is not included here.

Figure 2.3 Areas of Sub-Saharan Africa Classified as Suitable for Maize Production

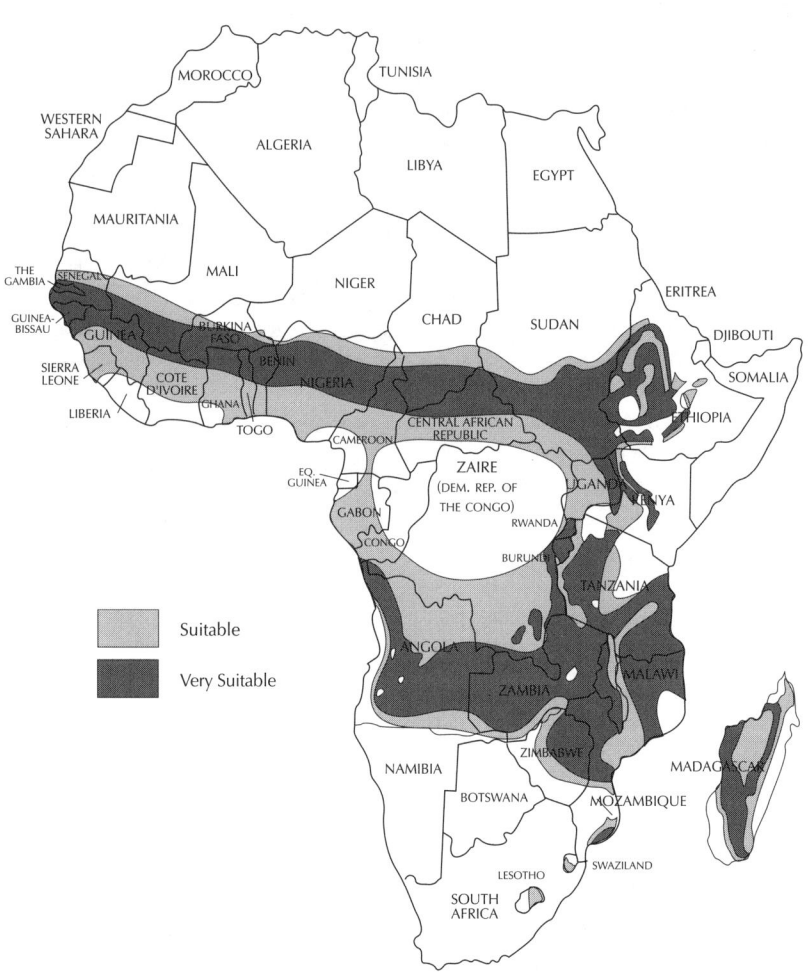

Source: USDA, 1981

one side by the forest zone, where reduced solar radiation limits maize production, and on the other side by semiarid areas, where a short growing season and frequent drought favor sorghum and millet. Throughout this century, however, maize has steadily pushed into the drier areas, displacing the traditional grains. Much of the recent expansion of maize area has occurred in the semiarid areas, especially in eastern and southern Africa.

The International Maize and Wheat Improvement Center (CIMMYT) uses three basic criteria to classify the range of production environments in which maize is grown: altitude, maturity, and grain color and texture.

- Altitude (as a proxy for temperature): Generally, three categories are used for altitude: lowland tropical areas, which are less than 900 meters above sea level (masl); mid-altitude areas, at 900–1,800 masl; and highland areas, above 1,800 masl.
- Maturity: The amount of rainfall and its distribution determine the length of the growing season and the appropriate maturity class for maize (early, intermediate, and late).
- Grain color and texture: Most maize produced for food in Africa is white, although in parts of western Africa yellow maize is grown. Commercial farmers in Zimbabwe produce yellow maize for animal feed. The preferred grain texture varies from relatively soft dent to flint.

The overall distribution of maize by the major altitude environments is shown in Table 2.1. Clearly, there is a major difference between western and central Africa, where 90% of the maize area is in the lowland tropics, and eastern and southern Africa, where 47% of the maize is grown in the mid-altitude areas and 41% is produced in the highlands (mostly in eastern Africa). Maize yields are significantly higher in the mid-altitude and highland environments, in part because of the use of improved technology. Throughout Africa, maize is produced in areas where rainfall is relatively favorable, although some favorable areas are subject to infrequent but very serious droughts.

In eastern and southern Africa, a further important characteristic for distinguishing among maize production systems is the type of farmer producing maize (CIMMYT 1990). The most important category in terms of production is the medium-scale farmer, cultivating 3–10 ha of land using animal power. The second-most important type of farmer, the small-scale farmer, cultivates 1–3 ha manually with a hand hoe or cutlass (machete). Finally, about 5% of the maize area is planted by commercial producers, who usually cultivate 50 ha or more. These distinctions are important for targeting technology development and dissemination. For example, many small- and medium-scale farmers lack adequate labor and draft power to plant on time, so for these farmers the length of the maize growing season is effectively reduced.

The Role of Maize in Production and Consumption

Maize accounts for a little over 20% of domestic food production in Africa, a proportion that has increased over time as maize has replaced other food staples, particularly sorghum and millet. Production of wheat and especially rice has also tended to increase in relative importance over time. The importance of maize as a staple, however, varies widely across regions of Africa. In southern Africa, maize is by far the dominant staple, accounting for over 50% of the calories consumed and as much as 80% of the cultivated area in Malawi. Maize dominates the food economy of southern Africa, enjoying the same position as rice in southeastern and eastern Asia and wheat in western Asia. Per capita consumption of maize in several southern African countries (Malawi, Zimbabwe, Zambia, and Swaziland) averages more than 100 kg per year and is as high as 160 kg in Malawi. Only in Mexico and Guatemala—where maize probably originated—do people consume a similar amount of maize.

In eastern Africa, maize is the most important food staple, and its importance is growing with time. Maize accounts for about 30% of calorie consumption in the region, where Kenya is the leading maize consumer. Maize provides over 40% of calories in Kenya, and per capita consumption is 125 kg per year. In the interior countries of this region—Uganda, Rwanda, and Burundi—maize is a secondary staple in a diversified diet that includes plantains, potatoes, and other cereals.

In western and central Africa, maize is generally much less important in aggregate food consumption, averaging 23 kg per capita and only 13% of calories consumed. Here rice and wheat (together) provide the largest share of calories, followed by sorghum, millet, roots and tubers, and then maize. Maize, however, is the dominant food crop in some areas, especially parts of the coastal savanna and higher areas. Maize has also expanded rapidly as a commercial crop in the Derived and Guinea Savanna belt, where annual rainfall surpasses 750 mm. Much of this maize is destined for the market, especially the rapidly expanding urban population of the region.

The six country studies provide a cross-section of the overall importance of maize in Africa (Figure 2.4). In the four countries in eastern and southern Africa, maize accounts for more than 40% of calories. By contrast, in the two western African countries, maize provides 16% of food calories in Ghana and only 6% in Nigeria.

The demand for maize in the years to come will depend on many factors, including population growth, the relative price of maize, the effect of incomes and urbanization on consumption patterns, and the potential use of maize in livestock feed. At present, only about 5% of maize is used for animal feed; the major exception is South Africa, where half of all maize is fed to animals.

Figure 2.4 Africa, Selected Countries: Contribution of Maize to Total Calorie Consumption

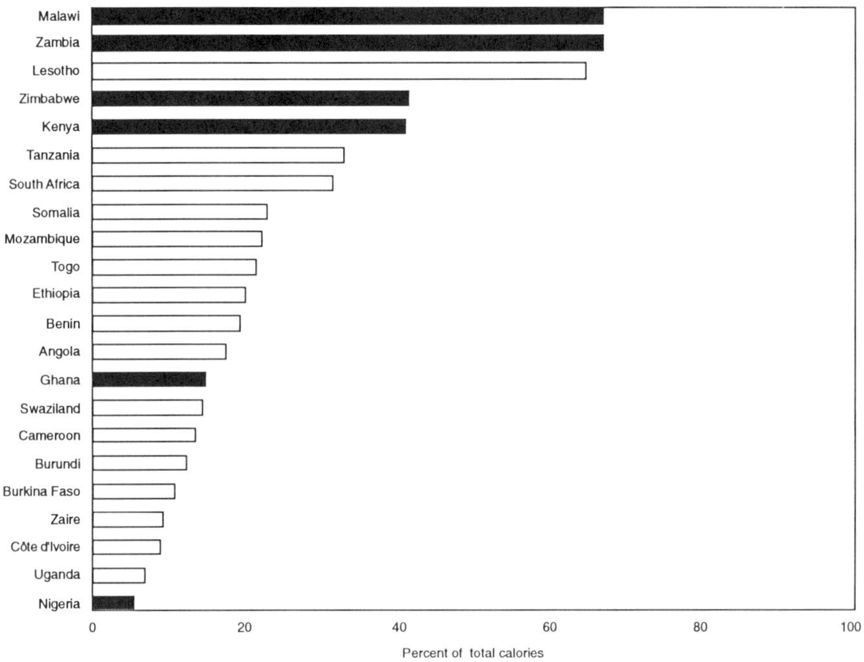

Country	Percent of total calories
Malawi	~68
Zambia	~67
Lesotho	~64
Zimbabwe	~42
Kenya	~41
Tanzania	~33
South Africa	~31
Somalia	~23
Mozambique	~22
Togo	~21
Ethiopia	~20
Benin	~19
Angola	~18
Ghana	~15
Swaziland	~14
Cameroon	~13
Burundi	~12
Burkina Faso	~10
Zaire	~9
Côte d'Ivoire	~9
Uganda	~7
Nigeria	~5

Source: Computed from FAO data tapes.
Note: Countries with shaded bars are included in country case studies in part two of this book.

As incomes rise and urbanization increases, maize appears to maintain its place in consumers' diets. In general, consumers shift from sorghum and millet to rice and wheat, whereas the share of maize in the diet falls only slightly. Thus, given population growth and projected income growth, the demand for maize is likely to grow at 3.0–3.5% annually over the next 10 to 20 years (the higher figure allows for some growth in feed demand) (Rosegrant, Agcaoili, and Perez 1995). In other words, to meet future demand, maize production must increase at a faster rate than occurred during the past two decades.

Trends in Maize Production

Any analysis of trends in maize production in Africa has to be treated cautiously for a number of reasons. First, official statistics can be unreliable.[1] Second, because of the extreme variability of maize yields, estimates of growth rates over time are sensitive to the choice of the time period of analysis. Because southern Africa experienced its worst drought of the century in 1992, the use of 1992 in the end period of analysis results in unrealistically low yields and yield trends. Third, official statistics show a sharp decline in production in countries affected by civil war. The estimates presented here attempt to take account of these various difficulties.

The years 1975–1995 were selected as the period of reference for analyzing production trends. During this period, maize production increased by 2.0% annually. This performance is not significantly better than that for other food staples and is less than growth in production of rice, the only crop whose per capita production increased. What is more important is how this maize production growth came about: More than two-thirds of the increase was achieved by expanding the area planted to maize. Yields of maize in Africa grew at only one-third of the rate achieved in other regions of the developing world. One reason for this difference is that much of Africa is at an earlier stage of land intensification, in which area rather than yield is the dominant source of production growth (Byerlee and Heisey 1996).[2]

The average yield of maize in Africa in the years 1989–1991 was 1.2 t/ha, which is double the maize yield estimated by Miracle (1966) for the 1950s, before improved technology became widely available. Yields were highest in eastern Africa, reflecting a favorable growing environment, as well as use of improved technology.

As noted earlier, yields have grown slowly. When countries that suffered prolonged civil war, such as Angola and Mozambique, are eliminated from the computations, overall yield growth is just under 0.5% annually. The use of improved maize seed and application of fertilizer to maize are not strongly correlated with yields. This absence of a clear relationship between maize yields and the use of modern inputs is an important puzzle that will be examined later in this book. In rain-fed areas of Asia and Latin America, where the use of green revolution–type technology is extensive, yield growth has averaged more than 2.5% annually (and has often been much higher) for two decades or longer.

In some cases, errors in statistics may explain apparent anomalies for specific countries. Some of the largest maize-producing countries in Africa, however, especially Zimbabwe, Malawi, and Kenya, have good systems for reporting agricultural statistics. Another factor may be the rapid shifts in the structure of maize production in some countries during the period of analysis. The proportion of maize produced by smallholders in Zimbabwe grew from less than 25% to exceed 70% in the late 1980s. A similar pattern has been observed in Zambia. These production patterns represent a shift to farmers who generally use fewer inputs and farmers who occupy less fertile land in more marginal rainfall environments. Growth in smallholders' maize yields in Zimbabwe averaged 1.5% annually over the past two decades, a period when the national average maize yield grew at only 0.3% annually.

Another distinguishing characteristic of maize yields in Africa is their high variability. Figure 2.5 shows maize yield plotted against the adjusted coefficient of variation in maize yields for countries producing over 400,000 t of maize annually.[3] The degree to which African countries are grouped by *lower* absolute yields and *higher* yield variation is remarkable. Even among developing countries that have approximately the same maize yields, the variability of yields is nearly always higher in African countries. Climatic factors are responsible for much of this variability, but price variability may also play a role. Yield variability is most pronounced for the smallholder sector in Zimbabwe, where the coefficient of variation of yields exceeds 50%.

Evolution of Institutional and Policy Changes

Government Intervention in Maize Markets

Ever since maize started to become an important crop in Africa, colonial and independent governments have intervened heavily in the maize economy.

Figure 2.5 Developing Countries in Africa and Other Regions: Mean Yields and Yield Variability

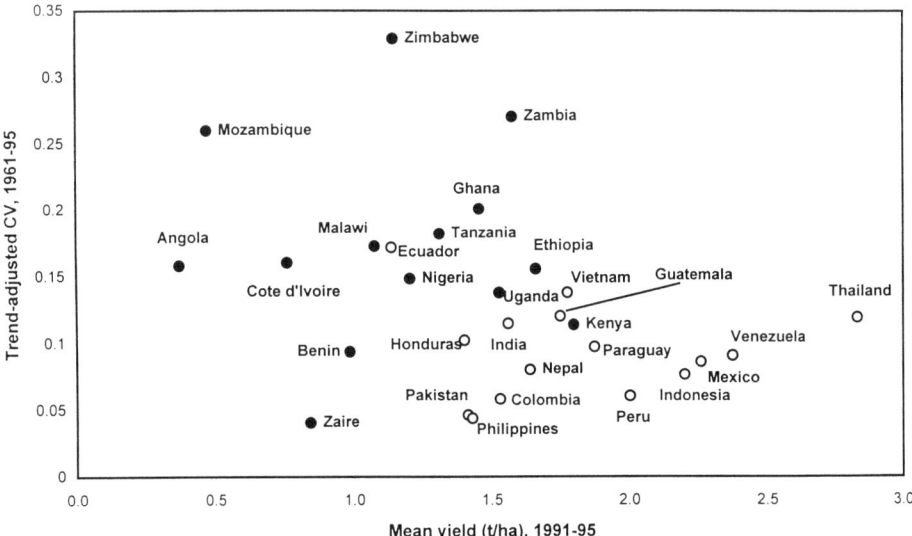

Source: Computed from FAO data tapes.

Colonial governments promoted maize production, initially on the large-scale farms of settlers but later on African farms, during and after World War I. The construction of railways facilitated the movement of maize and its growing commercialization, especially for the expanding mining sector and international trade. As maize grew in importance, governments intervened more heavily to control production, prices, and imports. Maize marketing boards—endowed with the sole authority to purchase maize at prices fixed by the board, as well as to import or export maize—were established during the 1930s in most countries of southern Africa (e.g., Lukanty and Wood 1990), as well as in Tanzania and Kenya in eastern Africa. In western and central Africa, however, maize marketing has remained largely in the private sector and is subject to relatively little government regulation.

Intervention in maize marketing was usually accompanied by government control over input markets, especially following independence when governments and donors redoubled their efforts to modernize the smallholder sector. In most countries, seed was produced by the public sector through a government seed monopoly, and fertilizer was largely imported and distributed by the government. Prices of seed and fertilizer were often controlled and in many cases subsidized to promote adoption of these inputs. In Nigeria, for example, fertilizer subsidies were as high as 85% during the 1980s. Much of this fertilizer was directed to the rapidly expanding maize area.

The Era of Liberalization

With the advent of structural adjustment programs in the 1980s, the policy environment began to shift in many African countries. These programs aimed to remove policy distortions through exchange rate devaluation and liberalization, fiscal austerity, reductions in export taxes, liberalized trade, and reform of agricultural input and product markets. For Africa as a whole, real exchange rates were 20% lower in 1993 relative to 1987 (Donovan 1995). Seventeen countries liberalized food crop marketing. The status of liberalization with respect to maize markets was rather variable. Finally, 16 countries reduced or eliminated fertilizer subsidies and liberalized fertilizer marketing.

For several reasons, the effects of these policy reforms on agricultural growth rates have been less than anticipated. First, policy reform is still incomplete in most countries; in some cases, the process has been reversed. Second, the transition to private marketing has been slow and difficult, especially for markets for key inputs such as fertilizer. Finally, public-sector investments in rural infrastructure and agricultural research and extension have declined in some countries. An important issue is the extent to which policy reforms have affected incentives for food crop production.

We reviewed trends in real maize prices in eight countries and discovered no consistent correlation between policy reform and improved price incentives. Indeed, a country such as Ghana, which implemented reforms over a sustained period, had a sharply negative trend in maize producer and consumer prices. In addition, the withdrawal of government parastatals from food marketing has increased price uncertainty to producers in some countries, since the private sector has not always effectively filled the vacuum.

Conclusions

All indicators suggest that Africa's food production crisis deserves urgent action. The downward trend in per capita food production observed from 1971 to 1985 may have stabilized, but the low level of food consumption, high incidence of rural poverty, and rapidly increasing population all add to the urgency of the problem. Since maize is the most important food crop in Africa, improvements in the performance of maize production will be crucial to solving Africa's food security problems and alleviating poverty.

We predict that maize production will have to grow by about 3.0–3.5% annually to satisfy demand. This rate of growth substantially exceeds the 2.0% growth in production recorded over the past two decades. A rate of 3.5% cannot be achieved without expanding maize area, as well as intensifying production to increase yields. Fortunately, maize is produced largely in relatively favorable areas, some of which possess considerable potential for area expansion. At the same time, a large potential exists to increase yields in most maize-producing environments. Yields in all but a few areas are extremely low by world standards, whereas yield variability is very high in Africa relative to other regions. This means that a maize production strategy requires an increase in yields, as well as efforts to improve yield stability.

The performance of the maize subsector in any given country is the result of a complicated set of interactions over time among agroclimatic factors, historical circumstances, technological change, and the policy and institutional environment. This complexity limits the ability to devise broad policy prescriptions for Africa. The policy environment is now more favorable for agriculture in Africa than it was during the 1970s and 1980s. Price discrimination against the agricultural sector has been reduced in many countries, and liberalization of the markets is providing incentives for more private-sector participation. Many countries, however, continue to struggle with the transition to private-sector input and output marketing, and uncertainty in input supply and producer prices must still be addressed. These issues are analyzed in the following chapters.

Notes

1. For example, maize yields in Tanzania show a sharp discontinuity in 1976, apparently reflecting a revision of the statistical estimation procedure. Nigerian maize statistics also exhibit sharp discontinuities.

2. A closer examination of trends by decade and region within Africa indicates that maize area expanded most rapidly in western and central Africa, especially between 1985 and 1995 (Table 2.1). In that same decade, maize production expanded by 4.6% annually in western and central Africa, which also had the best yield performance. This situation reflects rapid growth in maize in Nigeria, Ghana, and other countries, especially in the savanna areas. In southern Africa, production has fallen over the past two decades, as both area and yields have stagnated or fallen.

3. The adjusted coefficient of variation is computed as $\sigma(1-R^2)^{0.5}/\mu$, where σ is the standard deviation, μ is the mean, and R^2 is the coefficient of determination of the trend regression line adjusted for degrees of freedom.

Part 2
Country Studies

3

Zimbabwe's Emerging Maize Revolution

Carl K. Eicher & Bernard Kupfuma

The green revolution was launched in 1960 in Zimbabwe (five years before it got underway in India), when Zimbabwe released the high-yielding maize hybrid, Southern Rhodesia #52 (SR52), after 28 years of indigenous research. Because this first maize revolution in Zimbabwe was spearheaded by white commercial farmers and not replicated by smallholders, it was ignored during the 1960s by the leaders of Africa's newly independent nations.

At independence in 1980, Zimbabwe's new majority-ruled government introduced a number of programs to help smallholders increase food and cash crop production. Drawing on hybrid maize varieties, expanded access to credit, higher guaranteed government maize prices, and marketing subsidies, smallholders initiated Zimbabwe's second maize-based green revolution. Smallholder maize production doubled between 1980 and 1986. Today, virtually 100% of Zimbabwe's maize area is under hybrids, and short-duration hybrids perform better than other food grains in many low-rainfall areas. Now the biggest challenge for Zimbabwe is to develop cost-effective marketing policies and institutions to sustain its smallholder maize revolution.

The challenge for countries trying to replicate Zimbabwe's success is to focus on fulfilling four interrelated preconditions: political, technological, economic, and institutional.[1] Zimbabwe's maize-based green revolution demands scrutiny by African governments, the academic community, and donor organizations for several reasons. First, as mentioned earlier, Zimbabwe's green revolution pre-dated Asia's and was based on many years of indigenous research. Second, Zimbabwe's green revolution was based on maize grown under rain-fed conditions, whereas Asia's green revolution was based on irrigated wheat and rice. Zimbabwe's maize-based green revolution challenges those who contend that maize is a "fashionable crop" that is "vulnerable in years of low rainfall and competitive for land with safer crops" (Lipton 1989:362). Zimbabwe generated a reliable maize surplus and exported maize in 19 of the 21 years from 1970 to 1991.

Third, Zimbabwe's efficient seed distribution system contradicts assertions that public and private seed companies cannot be relied upon to deliver hybrid maize seed cost-effectively to resource-poor farmers. Fourth, Zimbabwe developed an integrated system of development institutions that facilitated maize production, including politically active farm organizations, technological and institutional innovations, marketing subsidies, and aggressive export marketing programs. Fifth, Zimbabwe's experience demands scrutiny because of the sharp difference in the speed with which hybrid maize was adopted in Zimbabwe compared to neighboring countries. These markedly contrasting rates of adoption raise important questions for researchers, policymakers, and donors who seek to reproduce the Zimbabwean success story.

The purpose of this chapter is to analyze Zimbabwe's maize-based green revolutions and discuss the preconditions for replicating this success in other African nations. We have chosen to define and analyze Zimbabwe's green revolution in terms of a sharp increase in maize production, not in terms of access to food and food distribution. Although these last two issues are critically important, they are beyond the scope of this chapter.

Maize in Zimbabwe's Economy

Zimbabwe is a nation of 11 million people with an average life expectancy of 53 years, near-universal primary education, and an annual per capita gross national product of U.S.$500 (World Bank 1996a). Although the agricultural sector contributes only 13% to the gross domestic product, it is the major employer in the country and accounts for 45% of total exports (Rukuni and Eicher 1994).

Zimbabwe's dualistic agricultural sector has two subsectors: the large-scale commercial farming sector and the smallholder sector. In 1996, there were 4,000 large-scale commercial farms, down from 5,000 at independence in 1980. The smallholder sector encompassed 1 million communal farms, 8,500 small-scale commercial farms, and 50,000 recently created resettlement farms.[2] The distribution of land between large-scale farmers and smallholders by Natural Agro-Ecological Region is shown in Table 3.1 (note that Region I has the highest annual rainfall and Region V the lowest). Table 3.1 also shows that large-scale farmers control almost 64% of the land in Natural Region II, the region of prime agricultural land in Zimbabwe. Most smallholders are located in Natural Regions III, IV, and V, the driest parts of the country.

Maize is the national food staple. Smallholders historically have cultivated the largest area of maize; in the 30 years from 1965 to 1994, the area planted to maize by smallholders accounted for 70% of the national maize

Table 3.1 Zimbabwe: Land Classification by Natural Region, 1995

Subsector	Natural Region[a]					
	I	II	III	IV	V	Total
Smallholder (%)[b]	47.3	36.3	65.6	78.1	68.8	65.2
Large-scale commercial (%)	52.7	63.7	34.4	21.9	31.2	34.8

Source: UZ/RF LTSP (1995).
Notes: a. The highest annual rainfall is in Natural Region I and the lowest is in Natural Region V.
b. Smallholder areas include communal farms, small-scale commercial farms, and resettlement farms.

area. The average maize yield obtained by smallholders, however, has been around one-fourth of the average yield obtained on large farms. Figure 3.1 reveals that the average smallholder maize yield increased during the 1980s but that yield variability increased as well because of recurring drought. As noted previously, the large increase in smallholder maize production in the 1980s was a product of higher prices, a backlog of maize varieties, yield increases, and an increase in market collection points in areas farmed by smallholders.

The First Maize-Based Green Revolution: Large-Scale Commercial Farmers, 1960–1980

The political, institutional, and technical foundation for Zimbabwe's first green revolution was laid early in the twentieth century. The first settlers in Zimbabwe focused on maize production, and maize exports grew at the phenomenal annual rate of 18.8% between 1909 and 1930, mainly to satisfy the demand for white maize in England's starch industry (Masters 1994). The settler farmers subsequently gained control over prime agricultural land through the passage of a series of Land Ordinances that "guaranteed white economic dominance and black poverty during the 90-year colonial period" (Herbst 1990b). The colonial strategy of confiscating land and depressing the wages of farm workers and migrant laborers and the profitability of small-scale farms created a favorable macroeconomic environment and a labor reserve for large-scale farms.

Between 1903 and 1919, the government established agricultural research stations in response to pressure from white commercial farmers. These same farmers formed provincial associations that subsequently evolved into several powerful farm organizations, including the Zimbabwe Tobacco Association, representing the tobacco farmers, and the Commercial Farmers Union, representing the large-scale commercial farmers (Blackie 1987). During World War II, the government secured the cooperation of

Figure 3.1 Zimbabwe: Average Maize Yields on Smallholder and Large-Scale Farms, 1965–1994

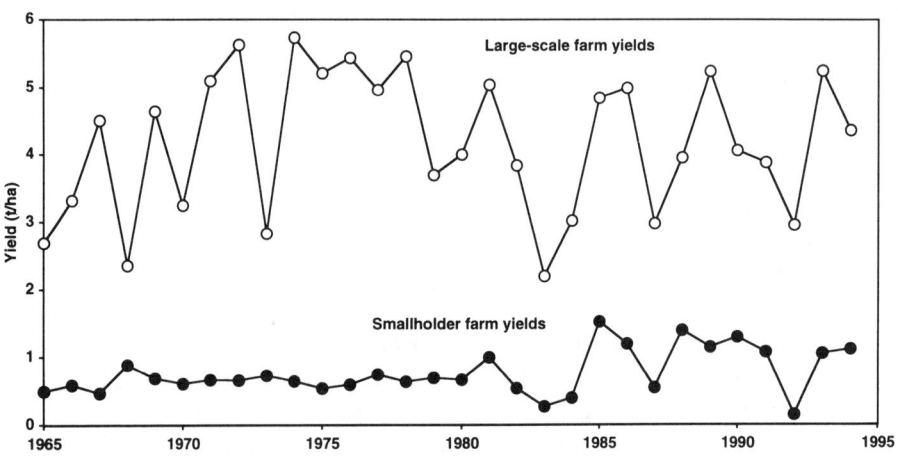

Source: DR&SS, 1995

commercial farmers in increasing food production in exchange for passage of the Licensing Act of 1942. This act made it mandatory for all commercial farmers and ranchers to buy a license from the newly formed Rhodesian National Farmers Union, which was renamed the Commercial Farmers Union after independence in 1980. The Licensing Act has been described as a "stroke of organizational brilliance," because it assured the union of a sound financial base (dues from farmers and ranchers) that allowed white farmers to finance research and lobbying efforts (Herbst 1990b). The union currently has a salaried staff of 120 and occupies a 10-story building in Harare (Bratton 1994). The union is led by a farmer-president, who leaves farming to work full time at the union headquarters for a two-year term of office.

From 1920 to 1950, the technical and institutional preconditions for the first green revolution were developed through large public and modest private investments in the four prime movers of agricultural development:

- New technology produced by long-term public and private investments in agricultural research
- Human capital and managerial skills produced by investments in schools, training programs, and on-the-job experience
- Capital investments in infrastructure, such as dams, irrigation, telecommunications, and roads

- Investments in farmer support institutions, such as marketing, credit, fertilizer, and seed distribution systems[3]

Research on hybrid maize was initiated by H. C. Arnold at the Harare research station in 1932. After 17 years of research, a hybrid variety, Southern Rhodesia 1 (SR1), was released to commercial farmers in 1949 (Mashingaidze 1994). Over 22% of the maize area in commercial farming areas was planted to this new variety in the first year after it was released (Weinmann 1975), but yields of SR1 proved unstable, so maize breeders continued to develop and release improved hybrids. In 1960, they hit the jackpot with the release of SR52, the high-yielding, long-duration hybrid (150 days to maturity) that launched Zimbabwe's first green revolution. Farmers who used SR52 increased their maize yields by 46% over yields of Southern Cross, the most common improved open-pollinated variety (OPV) available locally (Weinmann 1975; Tattersfield 1982).

Without question, SR52 is the most famous food crop variety in southern Africa. It was quickly adopted by commercial farmers in Zimbabwe who lived on fertile land with high rainfall. Within eight years of SR52's release, it covered two-thirds of the maize area planted by commercial farmers (Rattray 1969).

Two political developments of the 1950s and 1960s contributed to the rapid adoption of hybrid maize in Zimbabwe and neighboring Zambia. The first was Great Britain's decision in 1953 to establish a regional political federation consisting of Northern Rhodesia (now Zambia), Southern Rhodesia (now Zimbabwe), and Nyasaland (now Malawi). The federation lasted only 10 years but facilitated the formation of a regional research network and an exchange of hybrid maize varieties in the three countries. The second political event that influenced the adoption of hybrid maize was Rhodesia's illegal assumption of independence from Great Britain in 1965, which provoked international sanctions against Rhodesia's exports, including tobacco, the leading agricultural export. As a result, tobacco prices fell, and commercial farmers scrambled to diversify away from tobacco to maize, cotton, wheat, soybeans, and coffee (Blackie 1987). The sharp fall in tobacco prices placed considerable pressure on Zimbabwe's researchers to develop a short-duration maize hybrid that could replace tobacco on sandy soils in low-rainfall areas. After a few years of research, a new series (R200, R201, and R215) of three-way maize hybrids was released by government breeders in the early 1970s.[4] Fortuitously, these new short-duration hybrids met the needs of both commercial farmers who had once grown tobacco and of a large number of smallholders. This is one of many examples of how technological spillovers from commercial farming have helped smallholders.

Zimbabwe's commercial farmers developed efficient farmer support organizations to facilitate the rapid uptake of hybrid maize. The seed industry is a case in point. In 1940, a small group of commercial farmers established the Zimbabwe Seed Maize Association to produce certified maize seed under the supervision of the Ministry of Agriculture (Tattersfield and Havazvidi 1994). In 1949, the association distributed the first hybrid maize seed to commercial farmers. Today, around 160 farmers produce hybrid maize seed, which is sold by the Seed Co-op Company of Zimbabwe domestically, as well as in a dozen countries throughout Africa. Zimbabwe's seed distribution system currently provides maize seed to small-, medium-, and large-scale farmers. The Seed Co-op Company is the crown jewel of the seed systems in Africa.

The excellent physical infrastructure in large-scale commercial farming areas also encouraged commercial farmers' rapid adoption of hybrid maize. These areas have a good network of paved roads and are close to the railway lines. All farms are electrified and are well served by a good telecommunication network. Input and output markets, crop intake depots, and banks are easily accessible. Such a system does not exist in the smallholder sector.

To summarize, Zimbabwe's first green revolution, which extended from 1960 to 1980, was spearheaded by politically active commercial farmers who lobbied successfully for a strong national agricultural research system, institutional innovations, public investments in roads and other infrastructure, guaranteed farm prices, and aggressive export marketing schemes.

The Second Maize-Based Green Revolution: Smallholders, 1980–1986

Although we have pinpointed 1980 as the starting date for the smallholder green revolution, smallholders began to adopt hybrid maize seed in the early 1970s after the short-duration (and more drought-tolerant) three-way hybrids were released.[5] But throughout the 1970s, smallholder maize production was constrained by the civil war.

At independence in 1980, roughly 5,000 commercial farms controlled half the arable land in Zimbabwe, and 700,000 smallholders (communal farmers) occupied the other half (Blackie 1982). The new government declared its political support for a smallholder road to development and agreed to honor the terms of the independence agreement, which stipulated that commercial farm land would be sold on a "willing buyer—willing seller" basis for the next decade. The political decision to maintain a strong commercial farming community helped ensure a reliable food surplus and "bought" time to restructure basic agricultural institutions (e.g., credit, research, and extension) to serve the majority of farmers—black smallholders.

Smallholders with an average of 2–4 ha of arable land spearheaded the second maize revolution by rapidly adopting hybrid maize and fertilizer. In the six years from 1980 to 1986, smallholder maize production doubled.[6] This unexpected success is attributed to a combination of factors, including peace in the countryside, which enabled many smallholders to bring land abandoned during the civil war back into cultivation; a backlog of short-duration hybrid maize varieties available from the research system; a sharp increase in guaranteed producer prices; the removal of racial and institutional barriers to credit, which gave smallholders access to seed and fertilizer; and the expansion of subsidized government marketing services (e.g., grain buying points) in rural areas (Rohrbach 1989).[7] Zimbabwe's second green revolution generated broad-based benefits for tens of thousands of rural families. This achievement garnered the Africa Leadership Prize for President Mugabe in 1988 for showing that smallholder farming was profitable and efficient.

Without question, Zimbabwe's smallholders benefited from technological and institutional spillovers from large-scale maize farmers. For example, the national infrastructure of roads, research, and farmer support institutions—developed and nurtured by commercial farmers from the 1930s until independence—helped jump-start the smallholder-led maize revolution of the 1980s (Blackie 1990). At independence, the new government inherited one of the most productive public agricultural research systems in Africa, which included high-quality programs for maize, tobacco, cotton, and livestock research. The estimated annual rate of return to public investment in hybrid maize research was 43% from 1932 to 1990 (Kupfuma 1994).

Who benefited from Zimbabwe's second green revolution? Unlike Asia, where irrigation is synonymous with the early stage of the green revolution, in Zimbabwe crop production is critically dependent on rainfall.[8] Smallholders in the higher-rainfall areas, located in the heart of the maize belt in the three Mashonaland Provinces, were responsible for 70% of the total smallholder maize sales to the Grain Marketing Board in the 1980s (Stack 1994).[9] But large welfare gains undoubtedly accrued to farm households that planted high-yielding R200, R201, and R215 hybrids in lower-rainfall areas. These hybrids typically yielded more than local maize and sorghum, and they enabled farmers to release land and labor to expand oilseed production, livestock production, and rural nonfarm activities.

Fueled by higher producer prices, improved seed, credit, and fertilizer and marketing subsidies, Zimbabwe was awash with maize by the mid-1980s. This explains why an analysis of Zimbabwe's second green revolution must go beyond production issues and examine how the government managed its national food economy during several vastly different food policy scenarios, ranging from overflowing grain silos in the mid-1980s to the catastrophic drought of 1992. In 1985, good rainfall and excellent growing conditions produced a record crop of 3 million tonnes of maize, an amount equivalent to three years of domestic consumption. But the added cost of financing the government's maize reserve (2 million tonnes in 1985) forced a reappraisal of producer pricing policy and the level of maize reserves. The government subsequently announced a policy decision to curb maize production for the 1985–1986 crop year. Commercial farmers were encouraged to reduce the area under maize and diversify into oilseeds, game ranching, and horticultural crops for export. Both commercial farmers and smallholders slowly reduced the area under maize in the late 1980s, but in the 1990s the area under smallholder maize increased.

Zimbabwe's maize experience since independence is instructive not only for its successes: The country also experienced some setbacks that should be carefully studied by other countries in Africa. Soon after independence, the government directed its national agricultural research service to develop relevant technology for smallholders, especially those living in low-rainfall areas where maize, sorghum, and millet were the food staples. This shift in research mandate was hampered, however, by the loss of experienced research officers and the government's failure to provide adequate funds for financing field trials in the smallholder farming regions. The slow erosion over the past decade of Zimbabwe's public research system, one of the country's national treasures, is now being addressed in a number of studies (Rukuni 1996).

Zimbabwe's credit experience is also of great relevance to farmers and governments in southern Africa. A rapid increase in government credit for smallholders in the early 1980s was a major contributor to the smallholder-led maize revolution. The main government credit agency, the Agricultural Finance Corporation, has its origins in the Land Bank of 1911, which had a mandate to serve white commercial farmers. At independence, the government decided to expand credit to smallholders, especially those producing maize and cotton. Credit was perceived as a magic wand, an entitlement, and the Finance Corporation increased the number of loans to smallholders from 18,000 in 1980–1981 to 77,000 in 1985–1986. By 1990–1991, however, the number of loans had to be reduced sharply because of several problems. First, managerial and loan supervision problems arose in "scaling-up" the credit program from 18,000 to 77,000 loans. Second, recurrent droughts increased risk and the rate of default. Third, the delinquency rate was high, partially because of the speed in extending the loans and inadequate supervision. In January 1990, 80% of the smallholders borrowing from the Finance Corporation were in arrears (Chimedza 1994). The Finance Corporation responded to these problems by becoming more selective and reducing the number of loans to smallholders to 30,000 in 1990 and 23,000 in 1994, which represented only 2.3% of the 1 million smallholders in the country.

Fifteen years after independence, commercial farmers still receive a large share of government credit. The total value of Finance Corporation loans to 1,133 commercial farmers in 1990–1991 was more than seven times the total value of loans to 30,190 smallholders (Chimedza 1994). This situation raises a fundamental political question for a government committed to leveling the playing field for smallholders: Why should 1,133 commercial farmers receive more government credit than 30,190 smallholders?

The story of Zimbabwe's smallholder maize revolution would not be complete without a discussion of the use of fertilizer by communal farmers. At independence in 1980, smallholders accounted for less than 10% of national fertilizer use. Improved access to credit facilities and marketing outlets increased smallholders' share of fertilizer sales to 34% in 1985–1986, but fertilizer use declined thereafter and leveled off to around 25% in the 1990s (Conroy 1990). The drop in smallholders' use of fertilizer corresponds to the decline in loans for smallholders.

Most fertilizer used by smallholders is applied to maize in the high-rainfall regions (Conroy 1990). A 1990–1991 survey revealed that 85% of smallholders in low-rainfall areas purchased inputs, but only about 4% had access to Finance Corporation loans. Moreover, smallholders in low-rainfall areas used one-eighth as much fertilizer as smallholders in high-rainfall areas (Table 3.2). Fertilizer use patterns partially explain the productivity differences between areas of different agroecological potential. Table 3.2 shows that the difference in maize yield between high- and low-rainfall areas seems to correspond to the difference in fertilizer application rates in the two areas.

Most smallholders who use fertilizer apply less than the recommended levels because these levels are inappropriate, particularly for maize production in semiarid zones (Conroy 1990). Deriving appropriate fertilizer recommendations will require more location-specific research by the public research organizations. Unless farmers are provided with solid information on the payoff to fertilizer, they will apply it at low rates. The challenge to

Table 3.2 Zimbabwe: Smallholders' Use of Inputs, Loans, Fertilizer Application Rates, and Average Maize Yields in Low- and High-Rainfall Areas, 1990–1991

	Low-Rainfall Areas	High-Rainfall Areas
Farms buying inputs (%)	85.0	98.0
Farms with Agricultural Finance Corporation loans (%)	4.0	36.0
Total fertilizer used on maize (kg nutrients/ha)	16.0	119.0
Average maize yields (t/ha)	1.0	3.6

Source: Government of Zimbabwe (1995).

researchers in the public and private sectors is not only to develop appropriate recommendations on fertilizer use but to stabilize yields, given the low and erratic rainfall regimes in smallholder farming areas. Zimbabwe's public research system, however, is under severe financial stress. For all of this research to be accomplished, more resources, particularly from the public purse, have to be committed to agricultural research for smallholders.

Several researchers have expressed concern that the 1980–1986 revolution in smallholder maize production may prove to be a one-time historical event if steps are not taken to revitalize research and development (R&D) and farmer support services (Jayne et al. 1993). Zimbabwe's post-independence development experience illustrates how difficult it is for a new government to restructure agricultural credit, research, and extension institutions that primarily serve commercial farmers and shift the priority of these institutions to serve hundreds of thousands of smallholders, especially those in marginal areas. In other parts of Africa over the past 35 years, donors have helped many governments design and implement farmer support projects (credit, extension, seed, and fertilizer) to serve pockets of smallholders. But most ministries of agriculture have had great difficulty acquiring the managerial capacity and financial resources to replicate "successful" smallholder projects on a regional or national level. Zimbabwe's experience with smallholder credit provides additional insight into the complex managerial problems involved in scaling-up farmer support institutions to serve the rural majority, the smallholders. The scaling-up problem has also plagued the Sasakawa-Global 2000 (SG 2000) food production project in Ghana[10] and similar projects in other countries.

To summarize, Zimbabwe's second maize revolution occurred as smallholders doubled maize production between 1980 and 1986. Zimbabwe's achievement, however, cannot be attributed to a single factor, such as higher prices, improved technology, or greater access to credit. Zimbabwe's second green revolution, like its first, was achieved by crafting a system of policies and improved technology and institutions to serve smallholders (Blackie 1994a).

Lessons and Insights

Zimbabwe's maize revolution sheds light on the strategic importance of an active government role in the early stage of agricultural development. The state was the organizer and risk taker in developing Zimbabwe's agricultural research system, all-weather road network, and integrated extension service. Nongovernmental organizations did not develop Zimbabwe's impressive roads, nor did private seed companies conduct research for 28 years to produce the hybrid maize varieties that fueled Zimbabwe's first and second green revolutions. But commercial farmers also played an important role by developing a politically powerful farm organization that made the case for sustained government investments in the prime movers of agricultural development, including the development of productive public research and farmer support organizations. Also, commercial farmers created Zimbabwe's maize seed distribution system. Since independence, the private sector has slowly taken on a greater role in maize breeding, seed distribution, and the marketing of new high-valued export crops.

The critical lesson that emerges from Zimbabwe's qualified maize success story is the need to avoid dogmatism on what should be done by the state or the private sector in getting agriculture moving. The issue for policymakers is to focus on the changing roles of the public and private sectors in technology development and diffusion of innovations over time and to nurture public-private partnerships. Without question, Zimbabwe's first maize revolution, led by commercial farmers, was nurtured by large public investments in research, roads, and subsidized loans and by favorable input and product prices. But at independence, Zimbabwe's smallholder sector lacked the basic infrastructure for development. The private sector rarely invests in development of the physical infrastructure because of the difficulties in internalizing the benefits from such investments; thus the improvement of rural infrastructure is the responsibility of the public sector. The use of public funds to improve the infrastructure for smallholders in the future should be seen as an investment rather than a subsidy. The costs to the state of establishing input and output marketing networks, telecommunications, roads, and other infrastructural services in smallholder areas may appear to be high, but there are many economic, social, and political benefits to a dynamic smallholder sector.

Zimbabwe's green revolution experience also sheds light on the vexing issue of the fiscal sustainability of government support programs, especially fertilizer distribution and grain marketing and storage programs, and the cost and managerial problems associated with carrying over large grain reserves. In the early 1980s, the new government extended the services of the Grain Marketing Board to smallholders in resource-poor areas and increased credit, marketing, and consumer food grain subsidies. By the mid-1980s, however, it was clear that the marketing subsidies and the interest charges on carrying over a national maize reserve of 2 million tonnes (equivalent to two years of domestic consumption) could not be sustained, and farmers were urged to reduce maize production. Grain marketing and consumer subsidies were scaled back in the early 1990s.

Preconditions for Replication

Zimbabwe's green revolutions were achieved by fulfilling four preconditions: political support for achieving national maize self-sufficiency under normal weather conditions, technological innovations, institutional innovations, and a favorable macroeconomic environment for agriculture. These interlinked innovations, institutions, and policies created a system of national development institutions and policies with a capacity to develop new technology, adapt it to local ecosystems, diffuse it to farmers, and manage a national food economy in times of abundance and scarcity. What can other countries in Africa do to replicate Zimbabwe's smallholder-led green revolution? The most important lesson is to focus on the basic preconditions for increasing food production rather than to stress one "missing" factor such as fertilizer, credit, or extension. Of all the preconditions, the most important is political leadership for agriculture, the scarcest ingredient in African political circles.

Political Leadership

Political support for a sustainable green revolution can be generated from indigenous farm organizations, agribusinesses, professional agriculturalists, and political leaders (usually from rural backgrounds). Political leadership for agriculture has three critical roles to play in getting agriculture moving. First, special attention should be given to creating a broad-based agrarian structure that is capable of putting people to work. Abundant empirical evidence from the Third World shows that a smallholder agrarian structure can achieve both efficiency and equity objectives. The smallholder model can help ensure that the benefits of increased agricultural productivity are broadly distributed throughout rural society through employment and agricultural growth linkages. Second, political leadership is required to take the hard decisions on mobilizing and reinvesting some of the agricultural surplus (taxes) back into essential public investments in the agricultural sector—such as rural infrastructure, rural electrification, and agricultural research—to achieve a higher rate of growth of the agricultural sector in the future. Third, political leadership is crucial in facilitating the participation of the rural majority in making the case for its economic interests in the political arena.

Without question, several political factors were of central importance in the first and second green revolutions. Zimbabwe's commercial farmers developed a strong national farm organization (the Commercial Farmers Union) and affiliated commodity associations that have made the case for commercial farmers' economic interests in the political arena. For example, commercial farmers successfully lobbied the government to establish the Maize Control Board in 1931 (subsequently renamed the Grain Marketing Board) (Muir 1994). In 1945, the Zimbabwe National Farmers Union was formed to represent the interests of small-scale, black commercial farmers. At independence, the new government made a political commitment to pursue a smallholder road to development and helped smallholders set up their own farm organization, the National Farmers Association of Zimbabwe.[11] The National Farmers Union and the National Farmers Association merged in 1991, forming a new organization, the Zimbabwe Farmers Union, to represent the interest of both smallholders and small-scale commercial farmers. At the time of the merger, about 10% of the smallholders in the country were paid-up members of the new farm organization (Bratton 1994).[12] Zimbabwe's experience is a sobering reminder of how difficult it is to mobilize hundreds of thousands of dispersed smallholders into politically active farm organizations that can counter the "urban bias" of urban consumers, trade unions, the military, and the civil service (Bates 1993). The challenge now is to open up the political system and encourage smallholders and rural people to make the case for their economic interests in the political arena.

The Macroeconomic Environment

The economic precondition for a sustainable green revolution includes the interrelated issues of farm profitability and fiscal sustainability. The foundation of a sustainable green revolution is profitable farming on a recurring basis, including the creation of a set of expectations among farmers that it will be profitable for them to invest their resources (especially labor) in farm and land improvements and the adoption of new technology. Zimbabwe's experience reveals that hybrid maize is profitable to many smallholders at low input levels and in low-rainfall areas.

An important unresolved issue is the fiscal sustainability of Zimbabwe's green revolution. At independence in 1980, the new government unified the two extension services and expanded credit for smallholders. The government Grain Marketing Board expanded the number of seasonal grain buying points from 5 in 1980 to 148 in 1985 so smallholders would have ready access to the market. These buying points were costly to operate, however, so the number was scaled back to 42 in 1989 and to 9 in 1991. The Grain Marketing Board's maize operations required government subsidies of U.S.$30–$90 million annually in the mid-1980s (Masters 1994). The lack of fiscal sustainability of Zimbabwe's second green revolution program explains why major economic reforms were launched in 1991 to reduce the subsidies on the country's maize program.

Technological Innovation

A green revolution in maize is not likely to take root in Africa simply as a result of importing a "miracle" maize variety from another continent. Maize is less robust than wheat in terms of international technology transfer. African agroecological conditions and production systems are very diverse. Research managers in Africa should assume that maize will require substantial location-specific research by biological scientists in different target agroecologies. Research by social scientists is also needed on such issues as the storage quality of maize at the farm level and consumer taste and color preferences.

A continual stream of new technology from home *and* abroad is a precondition for a green revolution. The starting point in technology development is to blend improved farm practices with technology developed by public and private research systems. A country's ability to borrow scientific knowledge and technology from abroad and blend it successfully with indigenous technology requires the same kind of scientific capacity that is needed to invent new technology.

Zimbabwe's public agricultural research system developed a formidable capacity to borrow and develop new maize technology. Because inbred and hybrid maize lines from the U.S. Corn Belt were found to be ill-suited to conditions in Zimbabwe, researchers imported maize varieties from Central America and South Africa to use in their breeding program. Stable government funding for research provided the continuity of investigation that was essential for the development of SR52 and the R200, R201, and R215 series.

Zimbabwe's experience illustrates how much time is required to develop new technology for a green revolution. As a rule of thumb, breeders estimate that it takes an average of 7 to 10 years to develop and farmer-test new crop varieties. The experience of the United States is instructive on this point. The theory of hybridization dates back to 1905, public research expenditures on hybrid maize began in 1910, and production of hybrid maize spread rapidly in the 1930s.

In Zimbabwe, small teams of highly motivated and well-paid local scientists[13] devoted their entire careers to research on one or two commodities. Zimbabwe's favorable scientific atmosphere is illustrated by the fact that four senior maize breeders were in charge of hybrid maize research over a period of 56 years, from 1932 to 1988 (Eicher 1990). Each senior breeder had at least five years of overlap with his predecessor (Olver 1988). An important lesson that flows from Zimbabwe's experience is that the technological leadership of a nation can quickly evaporate when a government fails to maintain continuity of research funding and scientific leadership. Government expenditures on research and extension were roughly the same in the early 1980s, but expenditure on extension was increased relative to research during the 1980s. By 1991–1992, the annual government extension expenditure was more than double that of research (Zimbabwe 1991).

The number of scientists in the Department of Research and Specialist Services (DR&SS) increased from 92 in 1980 to 170 in 1990 and declined to 127 in 1992. The real DR&SS budget (adjusted for inflation), however, declined from 1981 to 1983, increased steadily to 1987, and then declined again.[14] This decline in real expenditures on public agricultural research has contributed to a quiet crisis in agricultural research in Zimbabwe. The decline occurred at a time when the DR&SS research mandate had expanded significantly to cater to the diverse needs of smallholders. Also, the DR&SS lost most of its experienced white research officers soon after independence. By 1984, 75% of the white officers had left the department, and only 18 researchers with more than five years of experience remained. Although the number of newly recruited black researchers increased during the 1980s, they had the added responsibility of expanding the scope of research programs to fulfill the new DR&SS mandate.

Zimbabwe's experience also sheds light on the issue of the appropriate size of the agricultural research enterprise. Unlike many other African countries, in Zimbabwe the government decided against significantly increasing the number of agricultural scientists following independence. Table 3.3 shows that the DR&SS had 92 research scientists in 1980 and 127 in 1992.[15]

Zimbabwe's public agricultural research system is presently under great financial stress. The system started feeling the financial squeeze after 1987, when its real budget began to fall even though the size of the DR&SS in terms of the number of scientists remained unchanged. Although the decision not to expand the staff establishment out of proportion to available resources is to be applauded, clearly more resources—financial and human—are required to meet the research needs of smallholders, especially those in low-rainfall areas.

Currently, Africa has many weak public agricultural research systems. Many countries should reduce the size of their public research systems (number of scientists, research programs), increase the scientists' salaries, and place greater emphasis on borrowing technology from neighboring countries and regional and international research networks (Eicher 1989; Maredia and Eicher 1995).

Table 3.3 Zimbabwe: Number of Scientists in the Department of Research and Specialist Services (DR&SS), 1980–1992

Year	National	Expatriate	Total
1980	92	0	92
1985	135	17	152
1990	152	18	170
1991[a]	140	18	158
1992	117	10	127

Source: Roseboom et al. (1995).

Notes: The DR&SS has about 60% of the agricultural scientists in Zimbabwe. These figures do not include agricultural scientists from other institutions.

a. The number of scientists has fallen since 1991 because of the freeze on filling vacated posts as part of the structural adjustment program's goal to reduce the size of the civil service.

Institutional Innovations

The third precondition for a green revolution involves the art of developing an efficient system of farmer support institutions (public and/or private) to diffuse improved technology to farmers and to market the increased agricultural output. As noted earlier, an important institutional lesson that emerges from Zimbabwe is how the roles of the public and private sectors change over time and how the private sector comes to substitute for the public sector in the development process.

The evolution of agricultural research and seed distribution illustrates these points. Zimbabwe's public research system performed admirably during its formative period from 1920 to 1960, when it achieved important breakthroughs in maize and cotton research. In 1973, during the civil war, commercial farmers set up and financed a private research station (Rattray-Arnold) in anticipation that the public research system would shift its emphasis to smallholders after independence. In 1982, commercial farmers launched another private venture, the Agricultural Research Trust, on a 260 ha research farm outside Harare for the purpose of testing new technology for commercial crops. Today, the public and private research systems coexist, demonstrating how private research has substituted for and supplemented public research over time.

Since Africa's rural landscape is littered with inefficient government seed organizations, it is interesting to examine Zimbabwe's experience. Because hybrid maize seed must be replaced every year, it is important for an efficient seed service to be on line to diffuse new hybrids as they are released by researchers. In Zimbabwe, a group of commercial farmers organized a seed cooperative in 1940, nine years before the first hybrid maize variety was released. The Seed Co-op Company was given exclusive access to all varieties developed by government researchers in exchange for storing adequate seed for replanting in case of a national crop failure. Pannar, a private seed company based in South Africa, entered the Zimbabwean market in 1979, followed by Pioneer Hi-Bred International in 1985. Zimbabwe's Seed Co-op signed an agreement with the U.S. DeKalb Genetics Corporation in the early 1990s to gain access to DeKalb's global germ-plasm base and international marketing skills.

Zimbabwe's experience demonstrates that private seed companies can be relied upon to deliver hybrid seed to farmers before planting season and at prices resource-poor farmers can afford. These factors help explain the rapid adoption of hybrid maize in Zimbabwe. The growing importance of private investment in research and seed distribution in Zimbabwe illustrates the need for social scientists to analyze the evolving roles of the public and private sectors over time. Social scientists are also needed to examine what incentives and property rights can encourage the private sector to play an increasingly important role in R&D and farmer support services for the more than 1 million smallholders in the country (Eicher and Rukuni 1996).

Conclusions and Implications

After 1960, most of Africa's newly independent nations imposed heavy taxes on farmers and treated agriculture as a national parking lot for the poor. Every nation, however, needs to develop the capacity to deal simultaneously with short-term food emergencies and long-term agricultural growth. Because of increasing population pressure and land constraints, future food production in many countries will have to come from raising crop yields rather than expanding crop area. Although it is currently fashionable to draw lessons for Africa from Asia's green revolution, much can be gleaned from Africa's experience in generating its own green revolutions.

Zimbabwe's first maize-based green revolution from 1960 to 1980 was spearheaded by several thousand white commercial farmers who developed a powerful national farm organization that made the case in the political arena for a strong national public research system, efficient public and private farmer support institutions, a favorable economic environment for commercial farmers, and an export-oriented farm policy to ensure overseas markets for the country's maize surpluses. But Zimbabwe's first green revolution failed to capture world attention because it was not replicated by smallholders. International press coverage focused largely on trade sanctions and the civil war from 1965 to 1979.

At independence, Zimbabwe's new majority-ruled government helped level the playing field for smallholders by extending access to credit and marketing, integrating the two racially divided extension services, and directing the national public research system to reorder priorities in favor of smallholders. Zimbabwe's second maize-based green revolution was led by smallholders, who doubled maize production from 1980 to 1986. Without question, some of the institutions that facilitated the first green revolution helped jump-start the second one. The rapid expansion of smallholder maize production in the 1980s provides solid evidence that smallholders will respond to new production opportunities if markets are available and four interrelated preconditions are met: political leadership for a smallholder road to development, appropriate technology, efficient public and private farmer support institutions, and a favorable macroeconomic environment for agriculture.

The usefulness of Zimbabwe's smallholder green revolution model for other African countries increases in direct proportion to our appreciation of its limitations. Zimbabwe's mixed record in reforming its farmer support institutions should be studied by countries seeking to replicate its second, smallholder-led green revolution. The Zimbabwean experience demonstrates how difficult it is to restructure and scale-up farmer support organizations to serve hundreds of thousands of dispersed smallholders. For example, 15 years after independence, only 2.3% (23,000) of the 1 million smallholders currently receive government credit. Another reason the smallholder-led green revolution in Zimbabwe represents a qualified success is its lack of fiscal sustainability. Zimbabwe's biggest challenge now is to develop cost-effective marketing policies and institutions to sustain its green revolution.

The economic history of Zimbabwe reinforces the basic point that each nation must address a number of specific problems arising from its past history, as well as develop a capacity to manage a national food economy in times of abundance and scarcity. These country-specific problems illustrate why donors should stop flooding Africa with generalized policy prescriptions and standard institutional models, such as the T&V extension model. The contrasting experiences of Zimbabwe's smallholder-led green revolution and Malawi's delayed green revolution (Chapter 5) reveal that there is no blueprint for jump-starting green revolutions in Africa. Country-specific models will be shaped by agroecological conditions, agrarian structure, the small events of local history, the ability to mobilize political support and investments in research and farmer support organizations, and national and international market opportunities. Case studies of agricultural intensification under rapid rates of population growth in Africa are needed to help deepen our understanding of the sequencing of institutional arrangements and of developing technological and institutional innovations and fiscally sustainable economic policies.

Notes

1. This chapter draws heavily on Eicher (1995) and Kupfuma (1994).

2. On average, commercial farms are about 2,000 ha in size. Smallholder farmers cultivate between 3 and 5 ha, small-scale commercial farms about 200 ha, and resettlement farmers around 12 ha of arable land.

3. By 1950, however, there was relatively little private investment in dams, irrigation, and so forth. Commercial farmers made large private investments in these activities in the 1950s and 1960s (P. A. Donovan, personal communication, 5 April 1994).

4. These varieties were developed from inbred lines from South Africa (Rattray 1988).

5. Rohrbach (1989) has reported that 20 to 30% of communal maize area in Zimbabwe was planted to hybrids before 1980.

6. About one-third of the increase in smallholder maize production from 1980 to 1986 came from bringing idle land back into cultivation after the termination of the civil war in 1979. The remaining two-thirds of the increase came from higher yields (Rohrbach 1989).

7. The years 1980 and 1986 were chosen for comparative purposes because they had fairly normal maize growing seasons. In 1980, a year of slightly below normal rainfall (700 mm), smallholders planted 931,000 ha of maize, harvested

738,000 t, and delivered 89,000 t to the Grain Marketing Board. In 1986, a year with normal rainfall and growing conditions, smallholders planted 1.1 million ha of maize, produced 1,338,000 t, and sold 682,000 t to the Marketing Board (Jayne et al. 1993).

8. About 5% of the cultivated land in Africa is under irrigation, compared with about 35% in India and 60% in Indonesia.

9. Three-fourths of the increase in maize production from 1980 to 1985 occurred in 18 of the 150 smallholder areas in the country (Jayne and Rukuni 1993).

10. Yudelman et al. (1991) reported that the loan repayment rate fell sharply when the SG 2000 food production program scaled-up its credit program in Ghana.

11. The National Farmers Association of Zimbabwe pre-dates independence. Before 1980 it was a nonpolitical master farmer organization in Fort Victoria (now Masvingo Province).

12. At the time the two farm organizations merged in 1991, the National Farmers Association of Zimbabwe had approximately 65,000 paid-up members, and the Zimbabwe National Farmers Union had around 4,500 paid-up members.

13. The director and chief research officer of Zimbabwe's public agricultural research system were paid as much as the highest ranking civil servant (secretary of agriculture) in the Ministry of Agriculture from 1930 to 1964 (Kupfuma 1994).

14. The real budget for the DR&SS increased by an annual average of 7.8% from 1964 to 1974 and declined by one-fourth from 1980 to 1990 (Kupfuma 1994).

15. The percentage of expatriate scientists in the DR&SS averaged around 10% during the 1980s.

4

Zambia's Stop-and-Go Maize Revolution

Julie A. Howard & Catherine Mungoma

For more than 80 years, Zambia has provided a unique laboratory for examining the impact of policies, and the organizations that influence and implement them, on the development and dissemination of maize technology. Beginning in the 1920s, increasing commercial demand for maize to feed urban mine workers hastened the replacement of traditional flinty[1] varieties with higher-yielding, open-pollinated white dent varieties (OPVs) imported from South Africa and the United States. Zambian maize production increased nearly fourfold from the early 1960s to the late 1980s because of a combination of surplus land,[2] new varieties better suited to smallholders' conditions, favorable input prices, and the physical availability of input and product marketing outlets (Figure 4.1). Government expenditures in support of maize were not sustainable, however, consuming 17% of the government budget by the late 1980s. Since then, the implementation of structural adjustment programs, which liberalized marketing and financial services formerly dominated by the government, has contributed to a decline in maize area and production.

This chapter explores how the development, spread, and, more recently, dis-adoption of maize technology has been shaped by Zambia's policy and organizational environment. The interaction of technology development with policies and organizations is detailed in Table 4.1, p. 47. The chapter also traces the political motivations that drove the establishment of the policy and organizational framework and impeded the efficient operation of the parastatal-managed marketing system. Options facing Zambian policymakers for increasing food production in the future under evolving political and economic conditions are discussed in light of these past experiences.

Socioeconomic and Agroecological Characteristics

Maize was originally introduced to southern Africa by Portuguese traders in the sixteenth century. This maize was flinty, low-yielding, and varied

Figure 4.1 Zambia: Maize Area and Production, 1962–1995

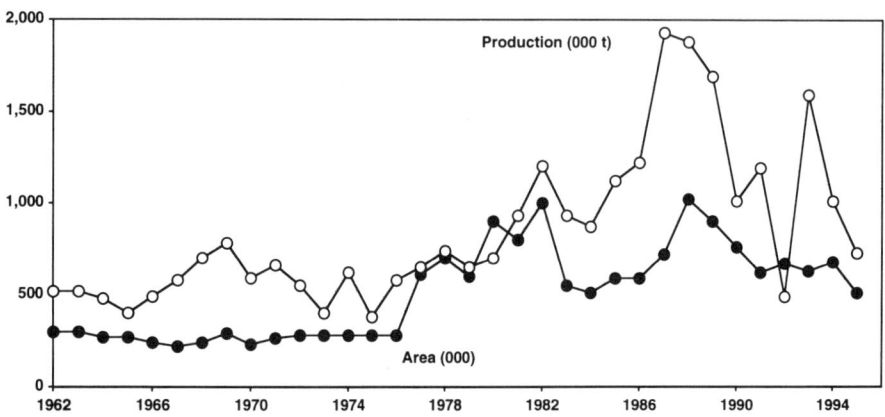

Sources: Data for 1962–1990 from USDA; data for 1991–1994 from CSO/MAFF (1995); and estimate for 1995 from CSO/MAFF.

in color. Small farmers traditionally grew maize as one of a mixture of crops that also included sorghum, millet, pumpkins, and groundnuts. Flinty maize was an important staple in Zambia by the end of the 1700s, but maize did not become dominant in most systems until the arrival of European colonizers in the 1900s (van der Bijl 1987; Blackie 1994a). Today, maize is the leading food grain in Zambia, accounting for 70% of cereals area and a similar share of total calories (FAO 1994b).

Colonization also introduced large-scale farming systems that evolved alongside small-scale traditional systems. Three-quarters of Zambia's farmers today are smallholders who plant less than 5 ha, use mainly hand hoes and few external inputs, and consume most of their produce. More than 60% of Zambia's total cropped area is managed by smallholders. The remainder is farmed by medium (5–20 ha) and large commercial farmers (mainly European settler farmers) who use improved seed, fertilizer, and animal draft power or tractors and who sell most of their production.

Zambia is located in the savanna ecological zone, and maize is grown in all three of its major agroecological regions. Rainfall varies from under 700 mm annually in the southern Zambezi Valley near the Zimbabwean border in Region I to over 1,400 mm annually in parts of Northern Province in Region III. Region II's annual rainfall (800–1,000 mm), good soils, and proximity to Lusaka and the Copper Belt markets make it the most favorable region for producing maize.

Table 4.1 Zambia: Maize Technology and the Policy/Organizational Environment

	1500–1700	1900	1910	1920	1930	1940	1950	1960	1970	1980	1990
Technology	• Flinty, variable color OPVs introduced by Portuguese traders. • Maize becomes important in African farming systems.		• Testing of imported white dent OPVs at Mazabuka Experimental Gardens.				• Federation established; many agricultural research activities transferred to Zimbabwe.	• SR52 rapidly adopted by large farmers, yield doubles. • Federation dissolved, 1963. • Zambia receives SR52 parent lines.	• First Zambian hybrid released. • First Zambian joins maize program. • Breeder discontinuity leads to SR52 contamination.	• Donors increase support to maize program. • 10 hybrids and 2 improved OPV varieties released by Research Branch, 1984–1992. • Golden Valley Agricultural Research Trust created.	
Policy/ Organizational Environment		• Arrival of European colonizers. • Railway line built connecting Zimbabwe with Zambian copper belt. • British South Africa (BSA) secures prime agricultural land for European settlers.		• Copper industry develops. • Maize becomes an important wage good.	• Maize Control Act restricts grain movement, establishes monopoly state buying stations in commercial areas, establishes dual pricing structure for European and African maize.		• Most commercial maize processed into refined maize meal by large-scale mills because of controls on sales, milling.	• Independence, 1964. • Zambia Seed Producers Association (ZSPA) produces hybrid maize seed, 1968.	• Price subsidies on fertilizer. • Outlets for credit, seed, fertilizer, maize marketing expanded. Zamseed established, 1978.	• Lima Program trains farmers in hybrid seed, fertilizer use. • Fertilizer use quadruples, 1960s–1980s. • Maize subsidies total 19% of Zambian budget. • Fertilizer price subsidies end.	• Great drought, 1991–1992. • Marketing, processing fully liberalized. • Input distribution partially liberalized, multinational seed company enterprise. • Maize area, seed, fertilizer use drop.

Technological Change, 1910–1992

The key maize technological advancements between 1910 and 1992 were the importation and spread of white dent OPVs that outperformed traditional flints; access to Zimbabwean SR52, a high-yielding white dent hybrid; and the development of Zambian shorter-season hybrids more suitable for smallholder conditions.

Importation of White Dent Varieties

In 1913, the colonial agricultural department in Zambia established facilities at the Mazabuka Experimental Gardens to test imported maize varieties from the United States, South Africa, and Zimbabwe. Higher-yielding white dent cultivars such as Hickory King were initially used by the large-scale European settler farmers who cultivated land adjacent to the rail line,[3] and they began to replace the traditional flinty cultivars among both large farmers and smallholders beginning in the 1920s (Weinmann 1972). Extension programs promoted Hickory King among smallholders during the 1950s. Acceptability of dents by farmers and consumers was important, because it opened possibilities for scientists to carry out further work with the pool of available germ plasm from South Africa, the United States, and Europe—which was dominated by populations with denty rather than flinty grain quality characteristics.[4]

Access to Zimbabwean SR52

After the formation of the Federation of Rhodesia and Nyasaland in 1954, many agricultural research activities were transferred from Zambia to Zimbabwe's Department of Research and Specialist Services. Improved access to Zimbabwe's superior research facilities and improved varieties, especially the new dent hybrids, clearly benefited Zambian farmers but stunted the development of Zambia's own maize research program. Zimbabwe's hybrid research program began in 1932, and SR1, the first Zimbabwean hybrid, was released in 1949 (Eicher 1995). The spectacularly successful SR52 followed in 1960. The long-season, high-yielding dent had immediate and powerful impacts on Zambia's commercial farm sector: Maize yields on large farms doubled from 1.3 to 2.7 tonnes/ha between 1949–1953 and 1959–1963 (Makings 1966).

Smallholders, who could not plant on time, continued to use the earlier maturing "local" OPVs, which were mixes of traditional flinty varieties and improved denty OPVs such as Hickory King and Southern Cross. Local varieties required lower levels of management than the early imported hybrids, but they yielded less, averaging 1.0–1.5 tonnes/ha. Smallholders also began to try SR52, often mixing the hybrid with local seed.

During the early 1980s, the extension service's Lima Program intensively promoted SR52 and fertilizer packages among smallholders.

Development of Smallholder-Appropriate Zambian Hybrids

SR52 was almost universally adopted by large-scale farmers in Zambia, but only about 30% of smallholder area was ever planted to SR52 or its advanced generations (Howard 1994; Kumar 1994). A serious problem facing smallholders interested in planting SR52 was the long growing period that required early planting. SR52 was especially risky for smallholders in drier, drought-prone regions. Even in areas with better rainfall, smallholders tended to plant commercial maize late because (1) it is extremely difficult to prepare fields with hand hoes before the first rains have softened the surface; (2) fields hoed early in the season may have to be weeded a second time later on, a problem given seasonal labor constraints; and (3) they usually wait to plant commercial maize until after subsistence crops have been planted, although doing so results in a 1–2% yield loss for each day of delay. In summary, smallholders required maize seed that was shorter-season and better adapted to conditions in the different agroecological regions.

After the dissolution of the federation in 1963 and independence in 1964, Zambia began the task of rebuilding its national maize research team. Expatriate breeders on short-term contracts headed the maize team through the late 1970s. Although expatriate researchers made important contributions,[5] they had short tenures, each adviser brought a different view of maize breeding priorities, and there were breederless periods between advisers. As a result, Zambia's strategic breeding objectives seesawed between yield improvement and qualitative factors, such as sterility and protein quality, and this lack of continuity led to the improper maintenance of breeder's seed and the eventual contamination of both parent lines of Zambian SR52 (Howard 1994; Rusike 1995).[6]

In the late 1970s, donors and international organizations—including the Swedish International Development Authority (SIDA), the United States Agency for International Development (USAID), the Food and Agriculture Organization of the United Nations (FAO), and the International Maize and Wheat Improvement Center (CIMMYT)—significantly increased their support for the maize program. More emphasis was given to addressing the needs of smallholders through coherent, sustained research programs and to attracting and training Zambian researchers. Although smallholders had already begun to adopt SR52, questions remained about the appropriateness of hybrid varieties for smallholders, given their uncertain access to seed and fertilizer. The maize program was subdivided into OPV and hybrid sections, funded by different donors. Work on hybrids was led by SIDA-funded breeder Dusan Ristanovic, who had begun work in Zambia in 1977 as the senior maize breeder at the Yugoslav Maize Research Institute winter nursery farm near Mazabuka. There he noticed that the Zambian parent lines of SR52 had become contaminated and began reconstituting them. The resulting hybrid from cleaned parent lines, MM752, was released in 1984 (Howard 1994).

An additional nine dent hybrids and two OPVs were developed and released in Zambia between 1984 and 1992.[7] Yields of the new hybrids were on average 60% higher than yields of local materials and 20% higher than those of SR52. The characteristics of the new hybrids also addressed several issues of the OPV versus hybrid debate. First, the new varieties were disease resistant and up to seven weeks earlier than SR52. Farmers could plant late and still get good yields. Second, in all but the most adverse environments, hybrid maize out-yielded local (and improved) OPVs, even without fertilizer (Gibson, personal communication; *Productive Farming,* October 1995). Third, unlike the single cross SR52, several new hybrids were double or three-way crosses, which meant their yields were more stable if farmers planted second-generation seed. Finally, the yield in seed production from three-way crosses was as much as three times higher than that from single crosses, thus reducing the unit cost of maize seed production.

The new hybrids were extremely popular among smallholders. Their potential yields were 5–8 tonnes, compared to 1–2 tonnes for local varieties and only slightly more for the improved OPVs. From the mid-1980s to the early 1990s, smallholder hybrid maize area doubled from 30 to 60% of total maize area.[8] Between 1975–1979 and 1985–1989, maize production increased by 137%, and the smallholder share rose from 60 to 80% (Howard 1994; GRZ 1991).

The Policy and Organizational Environment, 1910–1992

The availability of new maize seed and fertilizer technology was essential but, by itself, not sufficient to make increases in large commercial farmer and smallholder productivity possible. New technology cannot raise agricultural productivity if farmers fail to adopt the technology because of inadequate marketing and pricing policies (Bonnen 1990). In Zambia, four factors were important in the creation of a policy and organizational environment that facilitated widespread adoption of maize dent hybrids by large farmers and smallholders through the 1980s: commercial demand for maize and the resulting political pressure for improving the policy and organizational environment; the subsequent changes in marketing and price policies; development of the seed industry; and increased access to seed, fertilizer, and credit for smallholders.

*Influence of Commercial Demand and
Political Interest on Policy and Organizational Change*

The growing demand for maize in Zambia's urban centers was the primary catalyst for the replacement of traditional flinty OPVs with imported, higher-yielding denty white OPVs and later hybrids. With the rapid development of copper mining beginning in the 1920s, maize became an important wage good for urban workers. The large commercial demand for maize in urban centers lent political muscle to industrialists and farmers attempting to create an enabling environment of policies and organizations to promote and control the development, production, marketing, and processing of maize and maize inputs.

After independence in 1964, maize development programs played an important role in solidifying political support for Kenneth Kaunda's governing party, the United National Independence Party (UNIP), whose supporters were urban dwellers and remote farmers who had been excluded from economic development during the colonial period. The government's general aim was to increase maize production to supply the urban mining areas with cheap maize meal. A second objective was to reduce reliance on the European large-scale farmers by increasing the participation of African farmers in commercial agriculture and raising rural incomes. A third objective was to improve regional equity by increasing market participation by farmers in remote, less agriculturally advanced provinces (Howard 1994). The expanded consumer and producer subsidies were perceived as part of a "social contract" between President Kaunda's UNIP government and Zambian citizens. The politicization of the maize program made it extremely difficult for Kaunda to implement needed reforms in the 1980s.

Marketing and Price Policies

Maize price controls and subsidies to maintain market infrastructure began with the colonial government during the 1930s. The colonial government wanted to supply maize to the mining centers while protecting the large white commercial farmers' share of the maize market. The newly independent Zambian government continued to control and subsidize maize marketing, first through the National Agricultural Marketing Board (Namboard) and later through the Zambian Cooperative Federation (ZCF) and its member societies. The independent government still wanted to deliver maize to the cities but also sought to integrate smallholders into the commercial maize market. To do so, the government subsidized the establishment of marketing services to smallholders throughout the country and implemented pan-territorial, pan-seasonal pricing policies. Over 60% of maize production was marketed through official channels until the early

1990s. Maize bought by official marketing organizations was resold to parastatal milling companies in urban areas, where it was processed into maize flour and other products, which were then sold at controlled prices to urban consumers.

Price controls and the overvalued exchange rate kept maize prices far below border levels, discouraging production in areas close to the line of rail with established market access. Guaranteed prices and markets for maize accelerated production in more remote areas, where farmers had few, if any, commercial crop alternatives; as production shifted to more remote areas, the costs of providing marketing services rose dramatically. The policies discouraged local trade, nonmaize agricultural production, and on-farm storage. For many producers, it made more sense to sell their maize grain and buy back processed (and subsidized) meal instead of storing their own production (Jones 1994).

Development of the Seed Industry

The nucleus of Zambia's seed industry was a group of large commercial farmers who possessed the management skills needed to produce technically demanding hybrid seed. These seed producers formed the Zambian Seed Producers' Association (ZSPA) in 1963. After the breakup of the federation, the Zambian government was given 8 lb of SR52 parent seed by the Zimbabwean government. Because of the wary relationship between the new government and commercial farmers after independence, ZSPA members were not allowed access to the SR52 parents for seed production until 1968. During the early 1970s, the government denied permission for ZSPA to launch a private seed marketing company (B. Landless, personal communication).

Instead, in the early 1980s SIDA began to fund maize research and helped establish a parastatal seed company, Zamseed. The ZSPA initially resisted the formation of Zamseed but was eventually persuaded to become a major shareholder (B. Landless, personal communication).[9] SIDA's simultaneous funding of hybrid maize research and Zamseed created strong linkages between the two organizations. To ensure the new company's viability, Zamseed and its technical advisers pushed for the development of hybrids. New hybrid seed would have to be purchased every year, whereas OPV seed could be saved on the farm and planted the following season. Maize represented 70 to 90% of the total volume of Zamseed's sales and 60% of revenue by the late 1980s. Maize seed sales grew from 2,000 t to 15,000 t between 1970 and 1989 (Figure 4.2).

Improving Smallholder Access to Credit and Fertilizer

The expansion of the cooperative depot system also made it possible for smallholders throughout the country to obtain subsidized credit through

Figure 4.2 Zambia: Sales of Fertilizer and Improved Maize Seed, 1962–1995

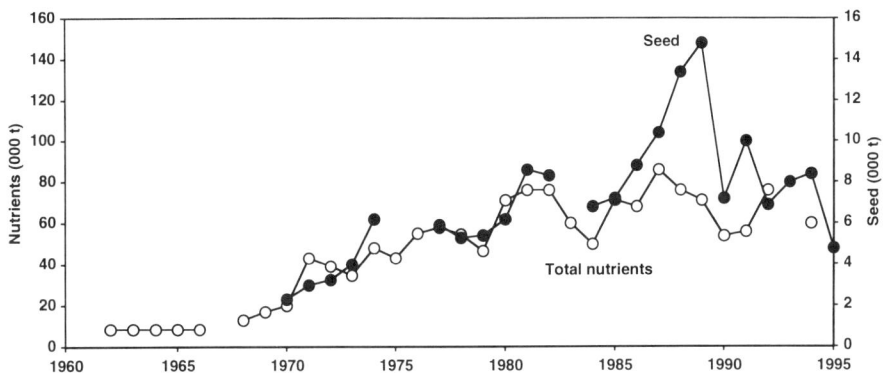

Sources: Data on seed sales for 1971–1981 from Ministry of Agriculture, Food, and Fisheries (MAFF), cited in Rusike (1995); for 1982–1994 from Zamseed; for 1995 from CSO/MAFF (1995) and Muliokela (personal communication). Estimated total fertilizer nutrient applied to maize is 90% of total consumed in Zambia (Williams and Allgood 1990). Data on sales for 1962–1979 and 1989–1992 from FAO (1971, 1979, and 1993); for 1980–1988 from GRZ (1989); for 1994 from CSO/MAFF (1995).

government programs and greatly expanded smallholder fertilizer use. Almost all of the credit was restricted to seasonal maize inputs. By contrast, large commercial farmers could secure short-, medium-, and long-term credit for a variety of enterprises through private commercial banks. About one-quarter of small and medium farm households received loans each year until the early 1990s (GRZ 1991). Fertilizer prices were also directly subsidized, cutting prices by 30 to 60% of landed cost during the 1970s and early 1980s (Jansen 1988).

The combination of improved access to seed and fertilizer through local depots, availability of credit, and direct price subsidies had a striking impact on smallholders. Fertilizer use quadrupled between the 1960s and the late 1980s (Figure 4.2), and fertilizer consumption in the remote areas increased from 15 to 39 percent of the total (Sipula 1993). A survey of major maize growing regions carried out in 1992 found that 88% of small-scale adopters of Zambian hybrids had used fertilizer in at least one season, and fertilizer application rates for maize were the second-highest in Africa in the late 1980s (Howard 1994; CIMMYT 1990). In addition, 64% of small-scale farmers sold maize, 42% had received credit for maize, and 47% had been visited by an extension agent. The dependence of these farmers on local, as opposed to regional, depots is an indication of how widespread and localized service provision became throughout Zambia's maize growing areas. Eighty-two percent of the farmers using Zambian hybrids obtained their fertilizer at local depots, 86% sold their maize there, and 80% purchased maize seed locally (Howard 1994).

The Process of Reform in the 1990s

By the mid-1980s, it was obvious that the government-controlled system of maize marketing and consumer subsidies was unsustainable. Maize-related subsidies consumed 17% of the total government budget by 1988, and the ailing copper sector was no longer able to provide a financial cushion. Although the average rate of return (or payoff) to the package of maize investments[10] from 1978–1991 was negative and therefore uneconomic, the maize programs had important effects on equity. The marketing subsidies and price controls redistributed maize production from large-scale to small-scale farmers and from areas adjacent to the line of rail to more remote and drier regions (Howard 1994).

Skyrocketing marketing board costs and consumer price subsidies were the primary factors contributing to the financial unsustainability, and the resulting problems snowballed throughout the system. Marketing agencies paid producers late, so producers could not repay loans on time. Liquidity problems for the credit and input agencies led to late procurement and delivery of inputs (Howard 1994). Maize research and extension also weakened throughout the 1980s. Real government and donor funding for research and extension fell by more than 70% between 1985 and 1990. Maize research grew increasingly dependent upon donor funding: Government contributions made up 75% of the maize research budget in 1979 but only 17% in 1990.[11]

External Pressures for Reform

The International Monetary Fund (IMF), World Bank, and donor agencies began to promote agricultural market reforms in the mid-1980s, conditioning future loans on the implementation of reforms that promoted both liberalization (the removal of regulatory restrictions on the private sector) and privatization (the withdrawal of the state from direct marketing functions) (Chapter 14). Price controls were removed on all commodities with the exception of maize meal, and some restrictions on foreign exchange, imports, and exports were lifted. Because maize represented a "social contract" between the UNIP government and Zambian citizens, however, it was difficult for Kaunda to implement maize market reforms except in a piecemeal, stop-and-go fashion. Maize meal price increases sparked urban riots in 1986 and 1990, and discussions between the Zambian government and the IMF broke off.

The pace of structural adjustment gained momentum with the election of Frederick Chiluba to the presidency in late 1991. Chiluba, a former Copper Belt labor leader, made maize system reform a key part of his political platform during the 1991 presidential campaign. Shortly after taking office, Chiluba accelerated and expanded the reform process begun by the Kaunda government—removing import and export restrictions, completely liberalizing the foreign exchange market, and reducing the size of the civil service. Implementation of the most challenging item on the reform agenda, maize sector reform, was temporarily delayed by the severe 1991–1992 drought.

Maize Marketing Reforms

Following the drought, the reform process was plagued with problems resulting from inconsistent policy formulation and implementation by the government and from the collision between maize market reforms and reform impacts from other sectors, especially finance. Beginning in 1992–1993, the government largely ended the parastatal system, through which farmers countrywide had been assured of guaranteed markets and prices for their maize at local cooperative depots, and began to encourage greater private-sector activity. To allay concerns that a weak private-sector response would create a maize market vacuum, however, for the next two seasons the government continued to intervene in maize marketing to a lesser extent (e.g., appointing buying agents, setting floor prices for grain, suggesting into-mill prices), thereby creating confusion and distrust among private traders (GRZ 1995; Scott 1995).

The 1994–1995 season was the first in which the government refrained from announcing any guaranteed maize prices and the private sector played a dominant role in input and commodity marketing. Nearly full regional and seasonal differentiation of maize and other crop prices occurred, based on market conditions and marketing costs. Real maize prices, which had fluctuated since 1992–1993, began to recover in 1994–1995. Private-sector confidence rose with the consistent messages from the government about nonintervention and the introduction of a marketing credit revolving fund. The government also began to lease many of its storage warehouses to private traders and transporters (GRZ 1995).

By 1996, the development of networks of large and small localized traders was proceeding rapidly in the traditional line-of-rail provinces, where some large commercial farmers established businesses to perform crop marketing, processing, and input supply functions for area farmers. Similar businesses began to emerge in Eastern Province, which has a relatively high population density and a long-established Asian/Zambian business community. The development of trading networks has been much slower in remote Northern and Western Provinces, which have lower population densities, weaker infrastructure, and little tradition of trading.

Fertilizer, Seed, and Credit Market Reforms

Reforms implemented in the late 1980s and early 1990s ended price subsidies and parastatal domination of fertilizer and seed production, importation, and distribution. The parastatal Nitrogen Chemicals of Zambia (NCZ) was slated for privatization, and foreign-based fertilizer companies such as Omnia, Kynoch, Sasol, and Norsk Hydro began to import fertilizer in 1992. The entry of the new companies greatly expanded the number of fertilizer products available in the country. Under the NCZ's monopoly, only six compounds and two nitrogen sources were available. Omnia alone introduced over 30 different compounds between 1993 and 1996, many of which have higher concentrations of nutrients, reducing transport, handling, storage, financing, and application costs (Rusike 1996a, 1996b).

Similarly, the number of maize hybrids and open-pollinated varieties available to Zambian farmers doubled between 1992 and 1996, following the entry of Cargill, Pannar, Pioneer Hi-Bred International, and Carnia to the seed market. These companies are increasingly important players in the seed market. In 1996, Zamseed controlled 65% of market share, Cargill 20%, Pannar 3%, and Carnia 3% (Rusike 1996a).

The private seed companies have a strong international research base, and many of the hybrid breeding responsibilities that had been carried out by the national maize research team are now effectively devolved to multinational firms. The major maize breeding work of these firms is done outside Zambia. For example, Pannar and Cargill have released flinty hybrids in response to farmer requirements for materials that will stand up better to on-farm storage conditions. Seed and fertilizer companies are also carrying out their own trials of fertilizers and new varieties but only in sites close to the line of rail. Public research organizations are also evolving: Commercial farmers, donors, and the government recently established the Golden Valley Agricultural Research Trust, which will carry out varietal testing and agronomic trials for private firms, nongovernmental organizations, and farmer associations (Rusike 1996a, 1996b; S. Muliokela, personal communication).

Distribution of inputs and provision of credit to smallholders have emerged as the greatest challenges facing the government during the reform process. Under the parastatal system, the smallholder credit system was coordinated with the maize marketing structure. Groups of smallholders received credit in the form of maize inputs and signed agreements authorizing the local cooperative depot to deduct the repayment from the sale of their harvest. As parastatals, Zamseed and the NCZ were compelled by the government to distribute seed and fertilizer through the cooperative unions, even to unions that were going bankrupt and could not repay (Rusike 1996a, 1996b).

The main sources of subsidized smallholder credit—the Lima Bank, the Credit Union and Savings Association (CUSA), and the Zambian Cooperative Federation—began to collapse in the early 1990s. In 1994, the Ministry of Agriculture, Food, and Fisheries (MAFF) established the Agricultural Credit Management Program to develop private-sector delivery of credit to smallholders. Two private organizations were appointed to manage credit and input distribution. In 1995, repayments were very poor, averaging only 27%—similar to the recovery rates by the ZCF, CUSA, and the Lima Bank. A key problem is that farmers do not feel a social obligation to repay loans they perceive to come from the government. Debt collection problems are also affecting private fertilizer companies, such as Sasol and Omnia, which are evolving new channels for fertilizer distribution in response. Omnia is experimenting with distributing fertilizers directly to farmers along the line of rail using sheds the company leases from the government. The company exchanges maize and soybeans for fertilizer and seed delivered to the nearest shed and provides grain bags (Rusike 1996b).

The Supply Response to Structural Reforms

By 1996, structural reforms to the agricultural sector had elicited two distinct responses. On one hand, areas served by established transportation and trading networks largely benefited from the reforms. The liberalization of foreign exchange markets removed the bias against agricultural exports (produced mainly by large-scale farmers and agribusinesses), which more than doubled in value between 1992 and 1996 and became more diversified. A variety of private contracting arrangements also emerged to facilitate input provision, commodity production, and trade—mainly near the line of rail and in Eastern Province. Some of these programs were oriented to maize, for example, the Omnia fertilizer-for-maize barter arrangement, but most focused on export crops. Outgrower schemes for tobacco (Burley Tobacco Growers Association), cotton (Lonrho), and a private exporter of paprika prefinanced crops or provided inputs, offered advisory services, and collected and marketed the crop (GRZ 1995).

On the other hand, in the more remote and drier regions (Northern, Western, and Southern Provinces), there has been a dramatic reduction in maize area and use of inputs and a reversion to subsistence crops such as sorghum, millet, groundnuts, and mixed beans. By 1995, total area cultivated in Zambia had declined by more than 15% from the 1985–1990 average, mainly because of the reduction of maize area (GRZ 1989, 1995). Input use also dropped as a result of the contraction in credit (Figure 4.2). Fertilizer use, which peaked in 1986–1987 at nearly 85,000 t of nutrient, declined to 60,000 t in 1994–1995 (CSO/MAFF 1995). Maize seed sales declined from 15,000 t in 1989–1990 to 2,500 t in 1994–1995 (CIMMYT 1993; CSO/MAFF 1995; Rusike 1996a). Studies conducted by the Adaptive Research Planning Teams in Western, Northern, and Eastern Provinces confirmed that private dealers faced with poor transport conditions and high market costs were unwilling to deliver inputs to remote areas, and key inputs were readily available only in district centers (Bangwe 1995).

Conclusions and Policy Implications

In 1996, the World Bank's Task Force on Poverty in Sub-Saharan Africa concluded that economic growth alone is not sufficient to reduce poverty unless the pattern of growth leads to increased production and employment opportunities for the poor (World Bank 1996b). Zambia's experience with maize sector development over the past 80 years shows the power of the policy and organizational environment to influence the pattern of growth by determining the agenda for technology development and establishing who will benefit from technology spread.

During the colonial era, growing commercial demand for maize to feed urban mine workers led to the replacement of traditional flinty OPVs with higher-yielding denty OPVs and hybrids and to the implementation of services and controls to protect large farmers' share of the maize market. At independence, the new majority government expanded market and credit services to meet twin goals of assuring cheap maize for urban residents and increasing equity by drawing more remote smallholders into the commercial maize market.

The initial results of the implementation of structural reforms suggest that Zambia is regressing to a bimodal pattern of agricultural development, with commercial farming along the line of rail and in other areas with established trading infrastructure and subsistence farming in more remote areas. A key question in 1996 is how technology, policies, and organizations can contribute to the creation of an economic pie that is not only growing but is shared by a large part of the population. This will involve the creation of an enabling environment for the development of high-value nonmaize commercial crops in remote areas, linked to the dissemination of food grain technology. There are five important areas: (1) strengthening smallholder organizations; (2) increasing support for public research and extension; (3) improving mechanisms for contract enforcement; (4) encouraging linkages between food grain and export crop channels; and (5) carrying out economic analyses to determine benefits, costs, and the distribution of gains from investments in export and improved food grain production in remote areas.

Strengthening Smallholder Organizations

The experience of large-scale farmers in both the colonial and postindependence periods underscores the importance of having a seat at the policymaking "table." Large-scale farmer organizations played an important role in encouraging the development of dent OPV and hybrid technology and influenced the formulation of policies to protect their markets and set prices at acceptable levels. Beginning in the late 1980s, the Commercial Farmers Bureau (now the Zambian National Farmers Union [ZNFU]) racially integrated its leadership and actively sought the participation of smallholders. By 1996, however, ZNFU members represented only 5% of the total farming population. One issue is whether it is possible to represent such diverse interests under the same organizational roof.

Smallholder associations in remote areas can also play an important coordinating role in input and product marketing—for example, aggregating seed and fertilizer orders to reduce transportation costs and organizing joint marketing or forward contracting arrangements with distant buyers. They can also educate their members to use credit responsibly and can facilitate group lending programs with penalties for nonrepayment enforced by the association.

Support for Public Research and Extension

In the past, public research and extension were vital to the development and dissemination of maize and fertilizer packages appropriate for large farmer and smallholder use. Today, maize research and extension are rapidly evolving toward the system common in developed economies, where breeding and extension activities are shared by the public and private sectors. It is already apparent that private seed and fertilizer companies will concentrate on areas where they can realize a profit: hybrid maize and export crops but not OPV maize or other food grains. They will confine distribution to areas close to the line of rail and other major transportation arteries. This leaves to the public sector a wide swath of important research and extension areas that are critical for increasing smallholder food grain productivity—for example, OPV breeding, coordination with seed producers, agronomy, and soil fertility and nonmaize food grain systems.

Improved Contract Enforcement

Policies during the 1980s introduced many smallholders to the benefits of hybrid maize and fertilizer use, but the system also encouraged a culture of lax repayment of state-subsidized loans for these inputs. Today, private-sector financing and delivery of inputs for maize and other commodities are seriously threatened by repayment problems. Both businesses and farmers need a working and accessible judicial system to enforce contracts—for example, for repayment of input credit on the businessman's side and for delivery of quality inputs and payment for commodities on the farmer's side. Arrangements such as Omnia's barter of maize for fertilizer increase transactions costs for both farmers and firms. Farmers assume the risks of storing fertilizer over several months, for example, and Omnia, a company specializing in fertilizer, has to become a grain trader.

Linking Export Crops with Food Grains

In areas where maize and other bulky food grains are not commercially viable on their own, production of higher-value export crops or joint export/food grain production may be economic. Cotton companies in Mali and Mozambique have twinned export crops with food grains to meet the food requirements of smallholder producers of export crops. The companies facilitated the adoption of food grain technology by extending the production, input, and product marketing services already established for cotton to food grains. Transportation, storage, and handling costs are shared by the two enterprises. The risks of providing credit for food grain inputs are diminished because farmers are tied to the single-channel market for the export crop, cotton.

Economic Analysis of Investment Alternatives

Following independence, Zambia's government tried to improve incomes for urban residents and those in remote areas who had been excluded from colonial development programs. Encouraging production of commercial maize countrywide proved to be an extremely uneconomic way of meeting this development objective. In the future, MAFF and Ministry of Finance economists, working with the private sector, can play a critical role in better matching development objectives with concrete programs. The economists can carry out economic analyses to identify potential investment opportunities in remote areas and can construct collaborative public-private action plans to alleviate policy and organizational constraints to investments. Analysis is needed to determine the economics of production, processing, and marketing of alternative commodities in more remote areas; key policy, institutional, and organizational constraints to investment; and benefits, costs, and the distribution of gains from potential investments by the public sector to reduce the constraints—for example, through improvement of infrastructure.

Notes

1. The texture of maize grain ranges from hard (flint) to soft (dent). "Dent" maize has a characteristic depression in the top of the kernel, which comes from

the proportion of hard or vitreous endosperm in the kernel to the soft or floury endosperm. The "dent" is formed because the soft endosperm collapses inward as the kernel dries (Blackie 1994a).

2. Less than 25% of Zambia's arable land is cultivated.

3. Agricultural areas adjacent to the railway line bisecting the country have been favored recipients of state investment since the colonial era. The British South Africa Company (BSA) built the railway line extending north from Zimbabwe in the early 1900s to serve the mining areas in northern Zambia and Zaire. The BSA negotiated with tribal chiefs to obtain land close to the railway line for European settlement. State agricultural resources were concentrated in these fertile "line-of-rail" areas in Southern and Central Provinces, because maize produced there could easily be transported to urban markets (Jansen 1988:6–7).

4. Most South American maize varieties are flinty, but germ plasm is not well adapted to mid-altitude conditions in Africa.

5. J. B. Abington inaugurated Zambia's maize germ plasm collection and in the early 1970s developed four open-pollinated varieties and the first Zambian hybrid, ZH1. None of these varieties was very widely adopted. ZH1's growing season was still fairly long, and yields were inferior to SR52; additionally, no seed distribution organization was in place to make improved varieties available to smallholders (Chibasa, personal communication).

6. Zambia was given the parent lines of SR52 at the dissolution of the federation. By the late 1970s, contamination of these lines resulted in a yield loss of about 15% in the Zambian SR52 compared to the original Zimbabwean version (Ristanovic, Gibson, and Rao 1986).

7. MM501, MM502, MM504, MM601, MM603/604, and MM752 hybrids and MMV400 and MMV600 OPVs were released in 1984. Subsequent hybrid releases were MM612 (1988), MM441 (1992), and MM62 (1992).

8. On the remainder of maize area, smallholders continued to plant local OPVs for their own storage and consumption.

9. Zamseed was organized in 1981, with the GRZ, the Zambia Seed Producers' Association, the Zambia Cooperative Federation, Svalöf (Sweden's largest seed company), and the Swede Fund as the major shareholders.

10. These included investments in research, extension, seed, and marketing. The average proportions of total cost represented by each for the period 1978–1991 were additional production costs associated with the new technology (33.7%), maize-related costs of research (2.3%), extension (3.9%), the seed industry (1.3%), and marketing (58.9%).

11. Calculated from Howard (1994).

5
Maize Technology and Productivity in Malawi

Melinda Smale & Paul W. Heisey

More than any other people in the world, Malawians depend on maize as a staple food: Two-thirds of the food calories consumed daily in Malawi come from maize. Ninety percent of the cropped area is planted to maize, which is produced almost entirely by smallholders cultivating under 5 ha.

Increasing population pressure and declining farm size leave little doubt that the additional maize production required to feed Malawi's people in the future must be obtained almost entirely through higher maize yields. Since maize production has not kept pace with population growth, per capita consumption has declined steadily over the past two decades. The rural wage, denominated in terms of the maize price—a good indicator of rural poverty—has also fallen. Higher maize yields are therefore essential to reverse the downward trend in both income and consumption (Heisey and Smale 1995).

Until recently, most of the maize area was planted to local varieties, but Malawi's distinctive combination of physical and cultural endowments suggests that high-yielding maize varieties have great potential for increasing productivity. Semiflint hybrids released in the early 1990s have been well accepted by smallholders. These hybrids are attractive to both commercial and subsistence producers, because they satisfy the grain texture preferences of farmers who produce maize for their own consumption[1] and generally yield more than unimproved varieties, even under low-input, low-management conditions. Declining soil fertility and inadequate institutional support for seed research and input distribution, however, constrain the prospects for attaining the substantial yield increases Malawi will need to feed its people year after year.

To provide a basis for comparison with other maize-producing nations of Africa described in this book, this chapter will outline some of the factors affecting both the development of improved maize seed in Malawi and maize productivity through the early 1990s. The next section, an overview of key historical and cultural issues affecting maize productivity in

Malawi, explains critical features of the institutional structure for distributing improved seed and fertilizer through 1993, when the government and institutions changed. Diffusion patterns for seed and fertilizer, and evidence on the performance of recent research releases, are summarized in the second section of this chapter. Prospects for the future are discussed in the final section.

Historical, Cultural, and Institutional Factors

Land Use and Population

Most of Malawi is well suited to rain-fed maize cultivation. Historically, the relative fertility of the soil has contributed to migrations of African populations from surrounding regions. In the twentieth century, changes in cultivation intensity have accelerated the depletion of much of the country's natural fertility (Heisey and Smale 1995), and declining soil fertility is now a major constraint on maize production.

From 1961 through 1991, maize output increased by 1.8% per annum, or a total of 55%. Three-quarters of the estimated increase in total production resulted from an expansion in maize area. Yields overall appear to have increased at a small but statistically significant rate of 0.4% per annum (Figure 5.1). Per capita maize production declined over the past three decades, because production lagged behind the annual population growth rate of over 3.0% (Figure 5.2).

Figure 5.1 Malawi: Maize Yields, 1961–1995

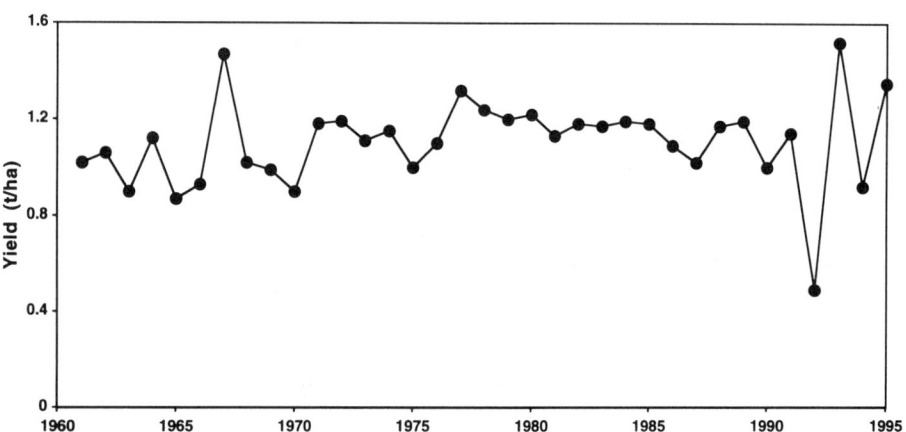

Source: Calculated from FAO (various years).

Figure 5.2 Malawi: Maize Production and Utilization Per Capita, 1961–1995

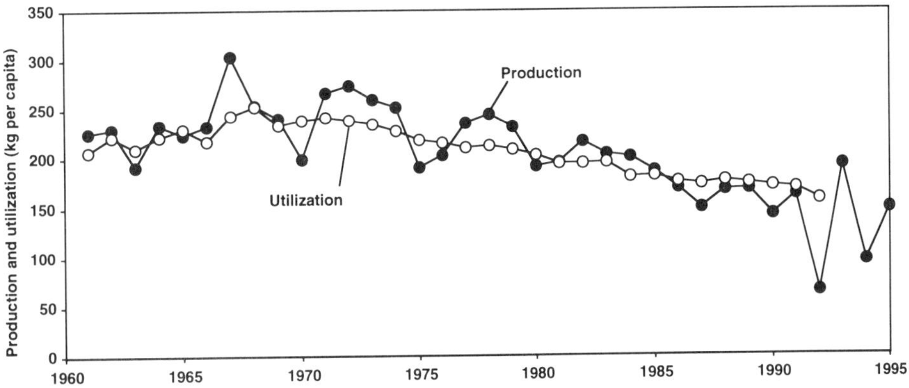

Note: Utilization per capita after accounting for net trade and stock changes.
Source: Calculated from FAO (various years).

Malawi is often characterized as a "land-scarce" country. Although additional land has been planted to maize over the years, much of this new area, at least in the more densely populated Southern and Central Regions,[2] was probably shifted from other crops rather than taken out of long-term fallow[3] (Heisey and Smale 1995). In recent decades, the ratio of the agricultural population to cultivated land area has continued to increase, meaning that the size of farmers' holdings has declined and the distribution of holding sizes has been increasingly concentrated in the smaller size categories. In 1990, more than 90% of farm households cultivated less than 3 ha.

One major source of the downward pressure on the size of smallholders' farms is generally believed to be the diversion of land for producing export crops on estates.[4] Local land shortages occurred after 1900, partly because as much as 15% of the land in the Southern Region was occupied by European planters (Pachai 1973). During the 1950s, the colonial government acquired substantial amounts of European-held freehold land for redistribution to Africans. After independence in 1964, the government encouraged Africans to establish estates for producing export crops. The number of estates proliferated in the 1980s, although their average size diminished.

Along with rising population densities and the changing structure of land use, the sheer dominance of continuously cropped maize in the farming system contributes to declining soil fertility. The area planted by smallholders to maize and maize mixtures rose from 65 to 75% of cultivated area in the early 1960s to 75 to 85% of cultivated area 30 years later

(Heisey and Smale 1995). Survey data for three of the five major maize-producing Agricultural Development Divisions suggest that less than 10% of smallholder maize area is planted after fallow, the median length of fallow is only three years, and a mere 0.5% of the area planted to maize had been in long-term fallow.

The growing population and unfavorable transport situation, both aggravated by the war in Mozambique, have also contributed to the declining availability of maize per capita. From 1970 to 1990, per capita calorie consumption from maize fell by 0.7% annually (Figure 5.2). If increasing incomes had led to more diversified diets, decreasing consumption of maize might be considered a favorable development. The evidence does not support this interpretation, however. Over much of the 1980s, per capita income in Malawi actually declined (World Bank, various years). Rural wages, denominated in terms of the maize price, rose from the 1950s to the late 1960s, only to decline since then.

Since maize production has not kept pace with population growth, the staple food appears to have become more costly. As maize becomes more dominant in the cropping system, alternative crops that could improve smallholder incomes are relatively less likely to be planted. Maize yield increases may be necessary not only to reverse the decline in per capita consumption but also to stabilize income. As cropping patterns have become less diverse, the risk of poor nutrition in the population, which already experiences serious child malnutrition problems, has likely increased (Ferguson, Millard, and Khaila 1990).

The Cultural Significance of Maize in Malawi

For Malawi's smallholders, "maize is life" (*chimanga ndi moyo*). The ideal of producing sufficient maize for the household to use in the stiff porridge (*nsima*) that is the staple dish "informs everyone's actions and rationales for their actions before, during, and after the maize harvest."[5] Each "hungry season," when their maize stocks have been depleted, many farm households face undernutrition as maize prices rise prohibitively and supplies in local markets fluctuate. Dietary preferences and the risks associated with relying on official markets imply that a major objective of farm household decisionmaking is to produce enough maize to satisfy annual subsistence needs.

Maize is believed to have replaced sorghum and millet as the dominant staple in Malawi sometime after 1900. Although evidence is inconclusive, most of the dominant maize land races loosely referred to as "local maize" (*chimanga cha makolo*)[6] are probably descended from maize brought by the Portuguese to the eastern coast of Africa in the sixteenth century to supply garrisons and caravans (Miracle 1966). Literally, *chimanga cha makolo* means "maize of the ancestors."

Farmers have long emphasized flint grain texture in selecting their maize seed. Flint maize types have a higher proportion of hard starch granules in the kernel than dent maize types. When processed using traditional methods, flint maize has a higher flour-to-grain extraction ratio in the production of refined white flour, because the germ separates more easily from the bran when the grain is pounded in the mortar.

Traditionally, rural women produce the refined white flour (*ufa woyera*) they prefer for preparing *nsima* through a long process in which harvested maize is shelled by hand, dehulled by pounding it with a mortar and pestle, winnowed, fermented by soaking (which gives it a unique flavor), washed, dried in the sun, and pounded again once or twice. The introduction of low-cost hammer mills has been widespread, but rural women continue to produce flour by the taxing conventional method, most substituting the grinding mill for the mortar in the final (second or third) stage of pounding (Smale et al. 1991).[7] The family's annual maize harvest is typically stored in the husk, without chemical treatment, in a raised cylindrical structure called the *nkhokwe*. Farmers also state that the husk cover and harder grain of the dominant local maize varieties protect them longer from the weevils that compete for the family's stored grain supply (Ellis 1959; Sibale 1988).

The vast majority of farm households produce local flint maize for home consumption. The preference of farm households for consuming flint maize (like *chimanga cha makolo*) reflects their observation that the yield of improved dent maize is significantly reduced by on-farm processing and storage losses.

Despite their suitability for home food processing and their ability to withstand storage insects, local varieties have some important disadvantages. The plants are tall, take a long time to mature, and yield poorly even when chemical fertilizer is applied. Yields of local maize in farmers' fields (including fertilized and unfertilized maize) average about 1 t/ha. The typical family of five must plant more than 1 ha of these varieties to meet annual subsistence requirements. Today, more than two-thirds of all smallholders cultivate less than 1 ha of land (House and Zimalirana 1992).

The poor yields of local varieties compared to improved materials may result in part from the limited range of genetic content (compared to that found in a center of origin for a crop species) and possible deterioration in landraces over time. The "yield gap" between the maximum maize yields obtained on experiment stations and national average yields is about 9 t/ha (Edmeades 1990). Part of this gap reflects measurement and statistical error, but various estimates indicate that between 25% and 45% of the "true" gap can be attributed to genetic differences (Edmeades 1990; Kydd 1989). In the trials cited by Kydd (1989), hybrid maize yields were approximately 4 t/ha higher than local maize yields when both kinds of maize were grown under favorable management and fertility conditions.[8]

Maize Seed Research[9]

The colonial agricultural department began testing maize lines imported from the United States, South Africa, and Zimbabwe at the beginning of the twentieth century (Rusike 1995). In the 1940s, S. Hoyle began collecting maize landraces and inbreeding local materials to produce pure lines (Kydd 1989). After the Great Famine of 1949, the administration initiated research into higher-yielding food crops, including hybrid maize. Chitedze Agricultural Research Station was established, a chief agricultural research officer was appointed to coordinate national research, and R. T. Ellis was assigned to begin research on maize synthetics and hybrids. Ellis and other colonial administrators recognized the importance of grain texture to smallholders, and many of these first synthetics and hybrids were semiflint types (Ellis 1959; Rusike 1995).

Following independence, the post of plant breeder was filled intermittently by a series of expatriates and Malawians on short-term assignments (Ellis had resigned in 1959). Research focused on testing lines rather than new materials, and breeding lines deteriorated because of poor seed maintenance practices related to staff vacancies and insufficient supplies and funds. From 1967 to 1977, the hybrid maize breeding program was discontinued. During that period, the breeding program switched emphasis toward the development of improved open-pollinated varieties (OPVs) with flint grain texture for smallholders. To satisfy the limited demand of estate owners and larger smallholders for commercial maize seed, seed of high-yielding dent hybrids such as SR52 (the highest-yielding maize hybrid in the region at that time) was imported from Zimbabwe or South Africa.

By 1977, the hybrid maize breeding program had been officially restored, and several dent hybrids, as well as semiflint composites, were subsequently released. The three senior breeders, however, were sent for training overseas during the 1980s, and technicians maintained breeding lines. In the late 1980s, the World Bank and other donors exerted pressure on the maize research program to emphasize flint grain texture in the hybrid breeding program. The national maize research team initiated its flint hybrid breeding program in 1987. To obtain the semiflint hybrids MH17 and MH18, the team top-crossed Malawian single-cross hybrids with a flint variety obtained through the International Maize and Wheat Improvement Center in Mexico. By using a top-cross (a nonconventional hybrid) and building on its earlier work with dent hybrid lines and flint OPV materials, the team was able to release the new semiflint hybrids within the relatively short period of three years.

There are several ways in which the release of MH17 and MH18 represents a plant breeding success. First, after years of technical service, national scientists were eventually given training and awarded a leadership role. Second, the release of MH17 and MH18 represents a fruitful collaborative effort between national scientists and scientists working for an international research institution. Third, the breeding effort demonstrates the positive effects on research of stable, although relatively modest, donor funding. Fourth, MH17 and MH18 represent a deliberate scientific effort to incorporate the concerns of small farmers into plant breeding strategies. The internal rate of return to maize research in Malawi is likely to be fairly high, even with an adoption ceiling of only 25% (Smale and Heisey 1994).

*The Institutional Structure for
Distributing Improved Seed and Fertilizer*

In the past, hybrid maize seed has been diffused in fixed proportions with a recommended level of fertilizer as a technical package; the package came in one size suitable for production on one acre (0.4 ha). The package was distributed through credit clubs or formal associations of farmers, and credit was typically recovered when maize was delivered to the nearest outlet of the Agricultural Development and Marketing Corporation (ADMARC).[10] This system was effective in delivering a technology to a minority of relatively advantaged smallholders and for drawing maize surpluses into the official marketing and storage system for later redistribution to food-deficit households.

Credit clubs were generally fairly small. Club members shared common socioeconomic characteristics: They were among the most advantaged of villagers. Default by any group member resulted in loss of eligibility, so the group's continued access to credit depended upon the compliance of each member. Club members were bound by stringent repayment procedures and the threat of punitive measures to return the loan in-kind at the end of the season. In recent years, an estimated 65 to 70% of all fertilizer and 70 to 80% of all hybrid seed used by smallholders was supplied to them annually through these clubs (HIID/EPD 1994). Extension advice on how to use hybrid seed and fertilizer was also largely restricted to members of credit clubs;[11] this advice emphasized a uniform set of recommendations for seeding rate, pure stand cultivation, and fertilizer application.

In 1993, following several years of erratic rainfall and in the midst of political turmoil over the vote for multiparty elections, the smallholder credit system was plagued by widespread defaulting for the first time in its history. The credit system collapsed, and sales of seed and fertilizer contracted severely. The effect of the disrupted credit program on seed and fertilizer sales continues to be evident. The collapse of the credit system, however, was more a result of important recent political changes within Malawi than of problems inherent in the credit system's design.

There are also technical problems associated with delivering seed and fertilizer as a package. Until recently, neither the composition of the package nor its size could be varied to suit the different objectives and needs of smallholders. Fertilizer costs are high[12] and constituted the major portion of the cost of adopting the seed-fertilizer package. Yet, results from farmer-managed demonstrations have shown that the maize hybrids currently grown in Malawi yield more than local varieties, even with low to zero levels of nitrogen. Partial budget analysis also suggests that for many farmers, the economically optimal level of fertilizer may be far below the level that has been promoted through the package; for others, the adoption of hybrid seed alone may be economically optimal (see the next section). In other words, more widespread adoption of seed could occur with a more flexible combination of technologies. Adoption of seed alone could improve the social welfare of smallholders, even though it might not generate the dramatic yield differences that occur when hybrid seed is adopted with high levels of management and fertilizer.

No easy solution exists to the input distribution problem in Malawi. The diffusion of hybrid maize seed has been constrained not only by offering seed, fertilizer, and extension advice through a credit package but also by certain features of seed supply and marketing. Production of seed relative to demand appears to have been limited over the years (Cromwell and Zambezi 1993). Until 1978, when the National Seed Company of Malawi was formed, ADMARC and the Ministry of Agriculture were responsible for producing seed. The National Seed Company was initially majority owned by ADMARC and minority owned by the Commonwealth Development Corporation, an institution of the British government. The transfer of seed production and other functions to the National Seed Company resulted in a number of related institutional improvements. In its initial years of operation (through 1980), however, the National Seed Company's sales of hybrid maize seed were sufficient to plant less than 5% of the smallholder maize area (Heisey and Smale 1995; Cromwell and Zambezi 1993). Since 1990, Cargill, a large, privately held multinational company, has been the majority owner of the National Seed Company; ADMARC and the Commonwealth Development Corporation retain minority shares.

Turning over seed multiplication to a private company has not solved seed supply problems. The National Seed Company's production costs appear to exceed ADMARC's sale price to farmers (Cromwell and Zambezi 1993). The company's largest client has been ADMARC, which absorbs the cost-price differential through a general government subvention. A competitor, Lever Brothers, began selling directly to farmers in 1992 and is now also permitted by law to sell seed to ADMARC. The National Seed Company also has begun to market seed directly to farmers in rural areas, although charging prices that are competitive with the ADMARC price may cause short-term losses because of cost-price differentials. As long as ADMARC remains the major client for the seed companies, seed sales will depend upon how well ADMARC functions. Detaching seed sales from the ADMARC-smallholder credit system assumes that the physical and legal infrastructure for private marketing of seed is already in place. Although private traders now purchase grain from farmers and sell it to ADMARC in some areas, there is little evidence as yet of significant seed sales in rural areas by private traders.

Privatization is also too simplistic a solution for fertilizer distribution in Malawi. Large economies of scale in fertilizer production mean that Malawi, like most nations in Africa, will remain an importer and a price taker in a world market with considerable price variability. The high costs of transporting a bulky commodity from the African coast to landlocked Malawi and then to a large population of smallholders who farm less than 1 ha imply that distributors will require substantial capital and flexible financial arrangements. Institutional mechanisms other than the private market may be necessary to ensure fertilizer delivery to farmers. In a recent study of fertilizer policy in Malawi, the Harvard Institute for International Development (HIID/EPD 1994) concluded that although the marketing of fertilizers at subsidized prices through ADMARC makes private-sector marketing unattractive, fundamental problems with access to domestic capital and foreign exchange will continue to inhibit private-sector participation, even if price subsidies are removed.

Price and Policy Environment

Historical factors contributed to the slow formation of policies and institutions designed to support maize research and development. Before independence, most incentives for smallholder production were directed to the promotion of export crops, the most successful of which was tobacco. Maize was a subsistence crop produced or exchanged by Africans to feed themselves. The need to invest in maize improvement did not assume importance for policymakers until the Great Famine of 1949.

Since independence, estate owners have generally grown maize as a secondary crop in rotation with tobacco. Maize produced on estates is used to feed laborers or is sold after harvest to official markets, to be stored by the official marketing agency or industrially processed. Estate owners therefore have had little interest in promoting the breeding of a maize variety that satisfies the grain texture preferences (related to processing and storage qualities) of smallholders, who consume the maize they produce.

Over most of this period, the effects on smallholders of input and output pricing policies for maize could be generally considered secondary compared to policies regarding the pricing and licensing of export crops. Several features of the maize policy environment merit special note. Although African traders have been exempted from restrictions on trading produce since 1957, Asian traders were subject to restrictions from the

1970s, and ADMARC increasingly dominated agricultural marketing and enjoyed monopsonist/monopolist status for outputs and inputs. The consumer, purchase, and input prices of ADMARC have generally been pan-territorial and pan-seasonal, which effectively subsidizes producers and consumers in more remote areas.

In theory, ADMARC's maize producer price was to be set in the wide band between export and import parity. In practice, the producer price loosely followed the export parity standard, primarily because ADMARC had to export at a loss in years when relatively high prices helped to draw out a maize surplus. Input pricing was marked by a subsidy of about 25% on nitrogen fertilizer, although debate spurred by aid donors over phasing out the subsidy began in the 1980s. The most important policy factor affecting seed-fertilizer technology in this period was the decision to move to urea as a lower-cost source of nitrogen (primarily owing to lower transport costs).

Diffusion of Improved Maize Technology

An estimated 0.3% of aggregate maize area in Malawi was planted to hybrids in 1970–1971.[13] By 1980, no more than 7% of the maize area was planted to composites (probably first-year seed) in any of the major maize-producing Agricultural Development Divisions. The limited anecdotal and survey evidence suggests that farmers did not perceive large yield advantages with many of the improved OPVs; only small amounts of improved OPV seed were available for diffusion; and other problems, such as lodging, were associated with some of the taller improved OPVs.

The diffusion of improved maize seed was sluggish in the mid-1980s. The lowest adoption rates occurred during the 1986–1987 season, the year following a financial crisis that prevented ADMARC from purchasing the hybrid maize farmers offered for sale. Two years after their release in 1990, the semiflint hybrids MH17 and MH18[14] covered an area that surpassed the total area planted to maize hybrids during the mid-1980s (Heisey and Smale 1995). Another downward fluctuation occurred in 1993–1994, when the credit system collapsed. In recent years, however, hybrid maize has accounted for about half of the total maize output (Figure 5.3), which underscores its importance in national food security.

Because MH17 and MH18 are both high yielding and have semiflint grain texture, they are attractive to farmers for either home consumption or sale, and they suit the objectives of a wider range of smallholders. In a 1991–1992 survey in which 150 farmers compared the performance of dent and semiflint hybrids with local maize in their own fields, mortars, and granaries, MH17 and MH18 ranked as well as the previously released dent hybrids in terms of yield and nearly as well as local maize in terms of processing and storing characteristics (Table 5.1).

Figure 5.3 Malawi: Diffusion of Hybrid Maize, 1981–1996

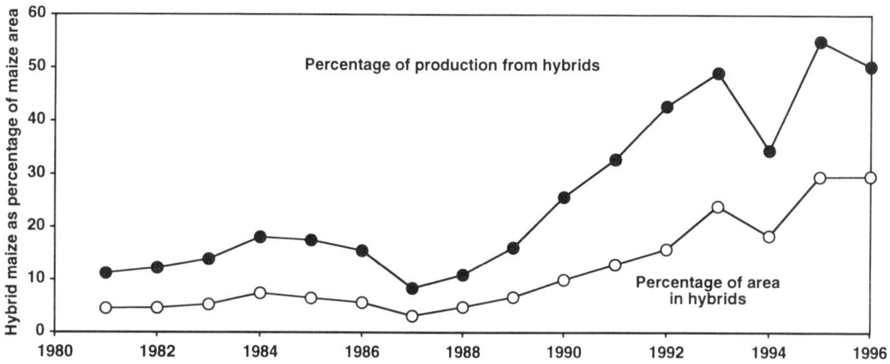

Sources: GOM/NSO (1984) and Ministry of Agriculture (1984–1996).

Table 5.1 Malawi: Farmers' Ranking of Yield, Processing, and Storage Characteristics in Maize Hybrids and Local Maize, 1991–1992 (percentage)

Ranking	Semiflint Hybrid MH18	Dent Hybrid MH12	Fertilized Local	Unfertilized Local	All Local Maize
Yield					
First	62	44	0	7	—
Second	27	44	31	14	—
Third	11	12	69	58	—
Fourth	0	0	0	21	—
Flour-to-grain extraction					
First	83	8	—	—	92
Second	17	4	—	—	8
Third	0	88	—	—	0
Insect resistance in storage					
First	82	17	—	—	100
Second	18	66	—	—	0
Third	0	17	—	—	0

Source: CIMMYT/Ministry of Agriculture Farmer Evaluation Survey, 1991–1992.

In the same survey, farmers reported obtaining higher yields from hybrids (both semiflint and dent) than from local maize in their own fields, under their own management conditions and input choice, in a drought year (Table 5.1; Smale et al. 1993). Objective evidence of the yield advantages of Malawi's hybrids is provided by four years (1989–1990 to 1992–1993) of farmer-managed, researcher-supervised demonstrations in the Central Region. During the drought year, as well as in other years, unfertilized hybrid maize yielded more than unfertilized local maize (Table 5.2). Coupled with the greater fertilizer responsiveness of hybrid maize, this finding suggests that at *any* level of fertilizer application, the expected yield of hybrid maize is greater than that of local maize (Jones and Heisey 1994). The results of other trials, whether conducted on farmers' fields or on experiment stations, appear to confirm this conclusion (Heisey and Smale 1995). The yield advantage of unfertilized hybrid maize over unfertilized local maize is likely to hold for at least three of the five ecological zones in Malawi (70% of the maize area).

The data also show that for any year and for any treatment for which spatial yield distributions could be compared, whenever hybrid maize and local maize yields are compared for equal levels of fertilizer, the cumulative probability of obtaining a yield lower than a given level is less for hybrids than for local maize. The stability of results over the years leads us to conjecture (with caution) that for many farmers in the main maize growing areas, the yield risk with Malawi's maize hybrids is less than with unimproved local varieties at any fertilizer level, including zero.

Results of partial budgeting analysis (Jones and Heisey 1994) further suggest that with subsidized prices, in all normal years hybrid maize at the recommended fertilizer levels is the economically optimal choice for all types of farmers. With unsubsidized prices, hybrid maize is always more profitable than local maize for maize-deficit households, although recommended fertilizer levels are not always economically optimal. For maize-surplus households facing unsubsidized prices, no particular technology alternative appears economically optimal in all seasons. Even when the

Table 5.2 Malawi: Mean Yields (t/ha) of Maize Hybrids and Local Maize, With and Without Fertilizer, in "Normal" and Drought Seasons, 1989–1993

	Unimproved Local Maize			Hybrid Maize		
"Normal" Season Fertilizer rates (N-P)	1989–1990	1990–1991	1992–1993	1989–1990	1990–1991	1992–1993
0-0	1.0	1.1	1.1	1.6	1.6	1.6
40-10	1.8	1.8	1.8	n.a.	n.a.	2.5
95-37	n.a.	n.a.	n.a.	3.8	4.0	3.8

Drought Season Fertilizer rates (N-P)	Unimproved Local Maize, 1991–1992	Hybrid Maize, 1991–1992
0-0	0.4	0.8
40-10	0.9	n.a.
95-37	n.a.	1.9

Source: Ministry of Agriculture/United Nations Development Program/Food and Agriculture Organization of the United Nations, Fertilizer Demonstration Program. Data were used from 21 sites in 1989–1990, 89 in 1990–1991, 101 in 1991–1992, and 136 in 1992–1993.

Note: n.a = data not available.

assumption of a high degree of risk aversion was imposed using a more general theoretical approach, in nearly all cases the economically superior treatments did not change from the risk-neutral case outlined here (Heisey 1994).

Several conclusions follow from these results. First, maize-deficit households, which may be less likely to have experience with seed-fertilizer technology, have greater economic incentives to adopt the technology than maize-surplus households. Second, in cases where hybrid and local maize can be compared at the same level of fertilization (no fertilizer or 40–10 N-P_2O_5 in 1992–1993), hybrid maize is more profitable under nearly all circumstances. Even in cases where it is not, using hybrid maize covers the cost of the seed. When local maize and hybrid maize are consumption substitutes (Smale et al. 1993), it no longer makes economic sense to fertilize local maize. The large aggregate increases in fertilizer use on local maize in the late 1980s were associated with the use of dent hybrids, which were not consumption substitutes with local maize. Third, the principal effect of subsidies is on the use of fertilizer rather than of seed. Subsidies also appear to have the effect of encouraging maize-surplus households to use fertilizer at the recommended levels rather than encouraging adoption by maize-deficit households, which may not use fertilizer for the reasons summarized earlier.

Under most assumptions, therefore, the yield advantages from using hybrid seed and fertilizer translate into economic advantages. Input-output price ratios and the type of growing season appear to be more decisive factors than farmer aversion to risk in determining which technology choices are economically optimal.[15]

The central finding that represents the most radical departure from the general preconceptions found in the literature about the use of high-yielding varieties is that even with low to zero levels of nitrogen and under modest management levels, and often in a year of moisture stress, the maize hybrids currently grown in Malawi yield more and in most cases are more profitable than the local maize currently grown. These advantages are associated with high-yielding varieties that are often thought, incorrectly, to "require" fertilizer.[16]

Seed and fertilizer are a divisible technology, meaning a farmer can use very small amounts of them (a handful of seed, a cup of fertilizer) and the cost per unit of land does not change with the extent of area planted. This means no economies of scale are associated with the technology itself, and it is neutral with respect to farm size. What introduces economies of scale into the adoption of seed-fertilizer technology in Malawi (what makes the technology more likely to be adopted by large-scale rather than small-scale farmers) are its packaging and its association with extension advice and the credit system. This distortion is reflected over time in the skewness of the adoption pattern by farm size (Table 5.3). In recent years, there is some evidence that through cash purchases, adoption of hybrid maize seed and/or fertilizer has increased, even in the smallest farm-size categories and among farm households headed by women (Smale et al. 1991; Peters and Herrera 1989). A significant amount of fertilizer (up to 45%) was also purchased with cash in the past (see Williams and Allgood 1990), before the closing of ADMARC's more remote market outlets and the expansion of the credit system in the early 1990s.

Both aggregate data and farmers' reports suggest that smallholders have tended to adopt fertilizer on local maize before adopting improved maize seed (see Heisey and Smale 1995; Table 5.3). This pattern may reflect the strong influence of smallholders' grain texture preferences on seed adoption before the release of the semiflint hybrids or the limited availability of improved seed, relative to fertilizer, in previous decades. Farmers' perceptions of declining soil fertility may also explain the pattern. Farmers who have little experience growing improved maize may not perceive differences in the yield response of improved and local maize grown with low levels of fertilizer.

Differences in cultural practices for maize in Malawi appear to be associated with variety, fertilizer use, or both. When small-scale farmers intensify their maize production through the use of high-yielding hybrid seed or inorganic fertilizer, they tend to increase their management levels through timelier planting and weeding, higher plant densities, or planting after a rotation crop.

Table 5.3 Malawi: Adoption of Maize Technology by Farm Size, 1985 and 1990

	Farm Size (ha)					
	<0.5	0.5–1.0	1.1–1.5	1.6–2.0	2.1–3.0	>3.0
Percentage of households headed by women						
1985	42	34	24	18	10	8
1990	51	29	19	9	15	9
Percentage of maize area planted to improved seed						
1985	2	2	6	8	15	25
1990	4	10	19	20	27	26
Percentage of farmers using inorganic fertilizer						
1985	9	16	25	40	44	54
1990	33	48	70	72	86	76
Average dose of fertilizer applied to hybrid maize (kg nutrients/ha)						
1990	46	64	62	81	83	61

Sources: Data for 1985 taken from Kydd (1989) and Sahn and Arulpragasam (1991). Data for 1990 were calculated from CIMMYT/ Ministry of Agriculture survey data, 1990–1991.

Implications for the Future

The one essential ingredient for improved maize productivity in Malawi is a real and renewed political commitment to the development of smallholder agriculture. Over the past few decades, the development strategy has favored estates at the expense of smallholders. In 1993, Malawians voted—for the first time in their history—in favor of a multiparty political system, and major institutional and economic changes have followed this decision. These recent political changes mark an appropriate juncture for rigorously analyzing the implications of policies on smallholder productivity.

Institutional

Several institutional changes are required to support technical innovations in Malawi over time. Perhaps the most urgent need is a national commitment to maize research. The maize program has been beset since its inception with both low absolute levels and unreliability of public funding. This problem expressed itself in staffing discontinuities that resulted in scientific stagnation, the loss of momentum in research, and the loss of breeding material. At a bare minimum, continuity in staffing and funding is necessary to maintain existing breeding material. Maintaining a flow of improved materials, or attaining better solutions for crucial crop management problems such as declining soil fertility, cannot be accomplished without this continuity. Lack of national commitment, in turn, exposes the maize research system to the whims of donor financing and donor-driven priorities.

The evolution of complementary roles for seed production and distribution systems is also critical to assure that technology reaches farmers. The very marketing arrangements that increased the National Seed Company's short-term profits (such as sales to smallholders indirectly through ADMARC and the credit club system) also increased its vulnerability to policy changes. Private seed companies need to develop their own markets, and a more decentralized structure may be needed to successfully distribute seed among Malawi's many smallholders. Furthermore, without long years of publicly funded maize research, neither the National Seed Company nor Cargill would have had maize hybrids suitable for dissemination. Interaction and cooperation between public and private institutions continue to be important in ensuring that existing hybrids are maintained and new ones developed. In the future, the pricing and structural implications of multinational dominance of the national seed system will need to be analyzed.

Technical

The yield advantages from using hybrid seed and fertilizer in Malawi have been confirmed by farm surveys and aggregate data. Each year from 1981 through 1995, the reported aggregate hybrid maize yield ranged from 2.4 to over 4 times the yield of unfertilized local maize. If the use of seed-fertilizer technology has increased, as we have documented here, then why have average maize yields not increased?

First, from 1981 through 1991, yields of unfertilized local maize fell by 2.0% annually; yields of hybrid maize fell by 1.1% annually. This pattern suggests that the increasing use of improved seed and fertilizer has only served to counteract underlying factors that negatively affect yield, such as declining soil fertility. Weather variability in recent years has also dampened the effects expected from greater use of a higher-yielding technology.

What will it take to sustain technical change in Malawi's maize production? Given the role of maize as a wage good and food staple in the economy, one simple but useful definition for sustained technical change is a rate of increase in maize output that matches population growth. Even under fairly optimistic assumptions for adoption ceilings and yield increases, projections suggest that to maintain the desired level of maize calories consumed by Malawi's growing population, further technological advances in maize production, as well as imports, are likely to become necessary.

Improved seed is not enough. Declining soil fertility and low management levels among smallholders need to be addressed in the longer term. A better understanding is needed of the dynamics of soil organic matter and the importance of other nutrients. With continuing deterioration in soil fertility and the high cost of imported fertilizer, solving the soil fertility problem may now be the highest maize research priority.

Notes

1. Most farmers produce maize for both home consumption and sale.
2. These are two of the three administrative regions of the country.
3. Fallow lasting more than 20 years.
4. The agricultural sector consists of the estate and smallholder sectors. Smallholders retain their land-use rights under customary tenure, which implies that land is held collectively and use rights are allocated through traditional authorities. Estates are defined legally in terms of their leasehold or freehold tenure status rather than farm size. Maize produced by estates accounts for less than 5% of total maize area, or 10% of total production (Mkandawire, Jaffee, and Bertoli 1990).
5. From villagers' statements reported in Peters and Herrera (1989:47–48).
6. "Local" maize refers to maize that has not been scientifically bred, although some maize farmers call "local" is produced from the seed of varieties released by the national research system. In general, in Malawi "local" refers to unimproved, open-pollinated maize varieties for which the seed is retained and managed by farmers.
7. Alternatively, either dent or flint maize can be processed on the farm as a coarser, less prestigious, whole-grain flour called *mgaiwa*. When women pound grain into *mgaiwa*, there is no difference in the flour-to-grain extraction rate for

dent and flint maize. The relative nutritional strengths of *mgaiwa* and *ufa woyera* have been debated since the colonial period (see Kydd 1989). Women report that with some dent maize types, simple modifications in pounding methods may reduce processing losses (Smale et al. 1991).

8. Three other facts of maize physiology may or may not contribute to recent varietal deterioration among Malawi's landraces. First, unlike self-pollinating crops, maize has a high propensity to cross-pollinate. Traits expressed by one genotype may be difficult to maintain because of the high risk of contamination with pollen of other genotypes. Second, the dominance of maize in local farming systems means it is difficult to isolate fields to reduce outcrossing. Third, survey data suggest that a substantial proportion of the seed planted as "local maize" was actually obtained from other farmers in labor exchange or from the maize food grain market (Smale et al. 1991). Since most maize grain sold at official market outlets is now hybrid maize, "local maize" includes second-generation hybrid seed.

9. For details and supporting evidence, see Kydd (1989) and Smale and Heisey (1994).

10. A parastatal marketing board, ADMARC delivers inputs; collects, transports, and stores smallholder crops; and sells maize grain to rural consumers.

11. An estimated 40% of extension agents' time was used in credit-related activities, according to the HIID/EPD study.

12. Fertilizer costs are high despite subsidies and the gradual shift from low- to high-analysis fertilizer.

13. Calculated from seed sales data reported by Quinten and Sterkenburg (1975) and FAO area figures (FAO, various years).

14. Except for some early work on semiflint hybrids under the colonial administration, all of the maize hybrids released by the national maize program from independence until 1990 were dent hybrids. Nearly all regional research with maize hybrids (Kenya, Zambia, Zimbabwe, and South Africa) has also produced dent hybrids.

15. It is important to remember, however, that "partial" budget analysis, as opposed to "whole-farm" analysis, assumes that one technique is substituted for another on all of a farmer's maize area. In fact, farmers in Malawi generally prefer to use a combination of seed types and fertilizer levels and have multiple reasons for doing so (Smale, Just, and Leathers 1994).

16. This conventional wisdom of the nonscientific literature is not borne out by research, which shows that improved maize with high yield potential yields more than maize with low yield potential across normal maize environments, with or without fertilizer.

6
Increasing Maize Production in Kenya: Technology, Institutions, and Policy

Rashid M. Hassan & Daniel D. Karanja

Kenya's economy is predominantly agrarian, and one crop—maize—dominates all national food security considerations. Almost every farm produces some maize. Maize supplies 40 to 45% of the calories and 35 to 40% of the protein consumed by the average Kenyan. The crop, which is grown under a wide range of ecological conditions—from the wet highlands to the semiarid zones and the humid coastal lowlands—accounts for more than 20% of all agricultural production and 25% of agricultural employment. Smallholders produce about 70% of the nation's maize, although large-scale commercial farms contribute a significant proportion of the marketed maize.

Maize was first grown in east Africa on the islands of Zanzibar and Pemba in the sixteenth century (Miracle 1966). Arab traders introduced maize from the islands to Kenya's coastal areas; later, through the movement of European settlers, maize spread further into the mainland, where it remained a minor food crop until the turn of the twentieth century. In 1903, maize occupied 20% of Kenya's food crop area; by 1960, this area had risen to 44% (Meinertzhagen 1957; Kenya 1966). In the intervening decades, maize was used to offset food shortages resulting from disease epidemics, drought, and locust invasions that decimated the traditional food crops, sorghum and millet. Aside from being more resistant to pests and diseases, maize grew in popularity because it was easier to store and process than traditional food crops. As export markets for maize continued to expand, the colonial government imported maize seed, provided higher prices for maize, and subsidized transport costs (Njoroge et al. 1992; Taylor 1969; Gerhart 1975).

Responding to the continued need to increase maize production, the government initiated a comprehensive maize research program in 1955. This program developed and released more than 20 high-yielding maize hybrids and varieties in the ensuing four decades. Farmers' adoption of these and related maize technologies in the mid-1960s sparked substantial growth in maize production through the 1970s, but growth slowed in the

1980s and 1990s, and Kenya frequently resorted to imports to meet local demand for maize. Given the limited area of arable land in Kenya and the rising demand for food, fueled by one of the fastest-growing populations in the world, the task of increasing maize production in Kenya is urgent and formidable.

This chapter analyzes the role and impact of research, institutions, and policy in increasing maize production in Kenya. In each instance, we discuss the factors that led from early advances in maize production to eventual stasis or decline. The findings presented here may help identify the policy and institutional reforms that are needed to generate more appropriate maize technologies, improve their transfer, and raise national maize production.

Evolution and Impacts of Maize Research in Kenya

As we have pointed out, systematic maize research in Kenya dates back to 1955, when the government, responding to the demands of large-scale maize farmers, initiated a maize improvement program at Kitale in western Kenya. The government hired a full-time breeder, M. N. Harrison, to develop late-maturing hybrids for the wet highlands.[1] Recognizing the growing importance of maize in people's diets, the government later expanded the research mandate to encompass germ plasm suitable for the semiarid, mid-altitude, and coastal regions as well.

By crossing imported germ plasm and locally adapted varieties, maize researchers developed and released the first Kenyan hybrid, H611, in 1964, after nine years of research. H611 yielded 40% better than Kitale Synthetic II, one of its parents (Harrison 1970). Over the next 15 years, 13 hybrids and open-pollinated varieties (OPVs) for diverse agroclimatic conditions were released. Specific agronomic recommendations were generated for these materials so farmers could exploit their genetic potential (Allan 1971). Table 6.1 lists hybrids and varieties released by the government's maize research program between 1961 and 1995.

Large-scale farmers in the high-potential areas rapidly adopted the new hybrids (Gerhart 1975). About half of the large-scale farmers adopted improved seed and inorganic fertilizer, which was a major factor in the growth in maize yields between 1963 and 1974 (Table 6.2, Figure 6.1).

A second phase of maize technology diffusion, from 1975 to 1984, was sparked by a surge in smallholders' adoption of improved seed, particularly in the high-potential areas (Table 6.2, p. 84), where adoption rates eventually equaled those of large-scale farmers. Yield gains were smaller in this period, however, partly because many smallholders adopted improved seed and not fertilizer and partly because of the unfavorable policy environment and severe drought in 1979–1980 and 1983–1984.

Between 1985 and 1991, improved seed was adopted by smallholders in low-potential areas, but fertilizer use remained low. The growth in

Table 6.1 Kenya: Maize Varieties Released, 1961–1995

Variety	Year Released	Maturity	Elevation	Yield (t/ha)[a]	Percentage of Maize Area (1992–1993)[b]
Kitale Synthetic II	1961	Late	High	3.4	—
Katumani Synthetic II	1963	Early	Medium	2.0	—
H611	1964	Late	High	4.5	—
H621	1964	Late	High	4.1	—
H631	1964	Late	High	4.5	—
H622	1965	Medium	High	5.2	—
H632	1965	Medium	High	4.5	—
H612C	1966	Late	High	5.9	—
Katumani Composite A	1966	Early	Medium	2.3	—
Katumani Composite B	1968	Early	Medium	2.8	5.3
H511	1968	Medium	Medium	3.6	7.2
H512	1970	Medium	Medium	4.1	3.7
H611C	1971	Late	High	5.9	—
H613C	1972	Late	High	6.0	—
Coast Composite	1974	Medium	Low	3.3	1.0
H614C	1976	Late	High	6.3	—
H625	1981	Late	High	6.8	22.9
H612D	1986	Late	High	6.4	—
H613D	1986	Late	High	6.0	2.1
H614D	1986	Late	High	6.6	41.8
H626	1989	Late	High	6.8	12.8
Dryland Composite I	1989	Early	Medium	2.9	0.2
Pwani Hybrid I	1989	Early	Low	3.8	0.6

Source: Adapted from Karanja (1990).
Notes: a. Yield in research trials.
b. Estimated from 1992–1993 survey data gathered by the Kenya Maize Data Base Project.

Figure 6.1 Kenya: Maize Area and Yield, 1961–1995

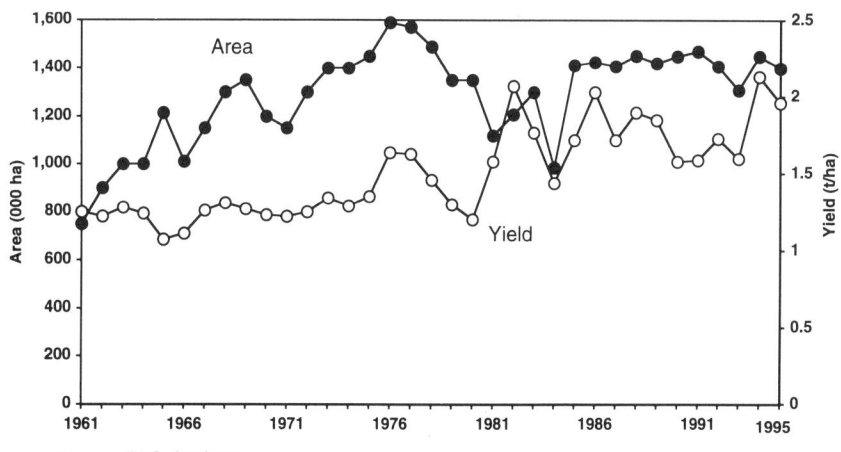

Source: FAO database.

Table 6.2 Kenya: Growth in Maize Area, Yield, and Production and Trends in Adoption of Improved Maize Seed and Fertilizer, 1963–1991

	1963–1974	1975–1984	1985–1991	Total 1963–1991
Growth in				
Area (%/yr)	2.8	−4.3	0.2[a]	0.7
Yield (%/yr)	0.8	1.5	0.3[a]	1.6
Production (%/yr)	3.6	−2.8	0.5[a]	2.3
Number of new varieties released	13	2	6	21
Percentage of farmers adopting improved seed				
Large-scale farmers, high-potential zones	48	72	94	94
Small-scale farmers, high-potential zones	16	58	95	95
Small-scale farmers, low-potential zones	5	17	57	57
Percentage of farmers adopting fertilizer[b]				
Large-scale farmers, high-potential zones	42	60	83	83
Small-scale farmers, high-potential zones	11	35	63	63
Small-scale farmers, low-potential zones	2	5	11	11

Sources: Data on maize area, yield, and production from Government of Kenya, *Statistical Abstracts,* various years, Nairobi, Government Printer. Data on adoption of improved maize from 1992–1993 survey data gathered by the Kenya Maize Data Base Project.
Notes: a. 1985–1995.
b. Cumulative percentage.

maize yields slowed in the late 1980s for several reasons, including reduced competitiveness of maize production, poor weather, and social upheaval in major maize-producing areas (Nyoro 1992). Also, drastic cuts in funding for maize research (Figure 6.2) slowed the development of new varieties and hybrids. The yield advantages of recent releases were less impressive than those of the first improved materials, in part because most varieties and hybrids released in the 1980s were simply variants of the breeding material used in the 1960s (Karanja 1996). Despite renewed interest in and donor support for maize research in the 1990s, inadequate operational funds continue to be a major constraint on KARI's maize research program (Karanja 1996).

Results of a 1992–1993 national survey of 1,400 maize farmers revealed that improved maize seed and fertilizer were adopted more widely in high-potential zones than in zones of lower potential and by more large-scale farmers than smallholders.[2] The benefits of hybrid seed and fertilizer largely accrued to farmers in a favorable area—the Kenyan highlands, where yield response to fertilizer was highest. In contrast, farmers in the lowlands commonly grew the less fertilizer-responsive OPVs (both improved and unimproved). Farmers' major reasons for not using improved seed were that an appropriate variety was lacking, seed was expensive, or they were unaware of improved seed. The reasons limiting farmers' adoption of fertilizer included its expense and unavailability (Hassan, Karanja, and Mulamula, forthcoming).

Increasing Maize Production in Kenya 85

Figure 6.2 Kenya: Public Maize Research Expenditure, 1955–1988 (deflated; 1971=100)

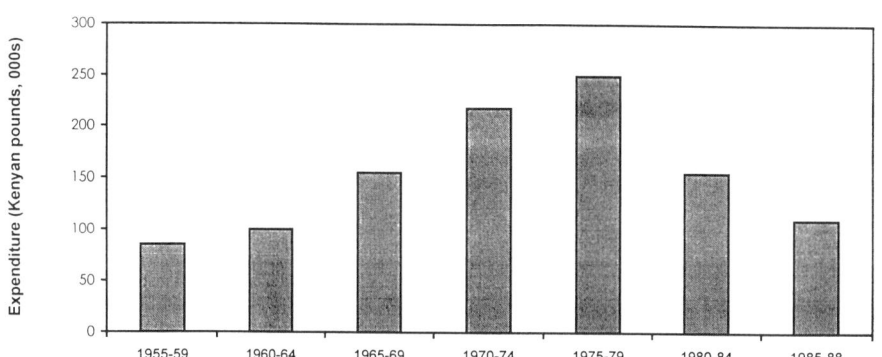

Source: Karanja (1990).

Patterns of maize technology adoption varied by access to farmer support services, access to input markets and credit, and infrastructure. Most farmers reported some contact with extension after the inception of the Training and Visit (T&V) system; even so, extension contact was still higher for large-scale farmers than smallholders, male-headed households than female-headed households, and high-potential regions than lower-potential regions.

Lack of credit was a critical constraint on the adoption of maize technologies. Fewer than 10% of maize farmers received credit in 1992. Large-scale farmers had better access to credit and input markets than did smallholders. Almost all villages in the lowland tropics and semiarid areas lacked village-based input supply mechanisms, whereas more than half of the villages in the high-potential zones enjoyed better access to markets (Table 6.3). Similarly, farmers in the high-potential zones gained access to credit more easily than their counterparts in low-potential zones (Table 6.4).[3] Access to credit also varied by gender, with male-headed households receiving more credit than female-headed households. For some farmers, the availability of cash or credit through the planting of cash crops—particularly tea and coffee—played an important role in the adoption of maize technology, especially fertilizer.

Higher farm-level maize yields were closely associated with greater use of improved seed and fertilizer. In most regions, however, the average farm yields in 1992 were about half of KARI's experimental yields and 25 to 50% lower than yields recorded for researcher-managed trials in farmers' fields.[4] The gap between potential and actual yields was wider for

Table 6.3 Kenya: Adoption of Improved Maize Seed and Fertilizer, Extension and Market Access, and Proportion of Area Planted to Maize by Zone, 1990–1993

	High-Potential Zone	Medium-Potential Zone	Low-Potential Zone
Area planted to maize (000 ha)[a]	731	155	151
Percentage of total maize area[a]	70	15	14
Percentage of farmers using improved seed[b]	93	49	53
Percentage of farmers applying basal fertilizer[b]	68	29	8
Percentage of farmers with[b]			
Extension agent within village	47	17	7
Input supplier within village	74	42	28
All-weather road to market	34	25	14

Notes: a. 1990–1991 season (Otichillo and Sinange 1991).
b. Kenya Maize Data Base Project, 1992–1993.

Table 6.4 Kenya: Percentage of Farmers Adopting Improved Maize Seed and Fertilizer and Having Access to Extension and Credit, by Zone and Farm Size, 1992–1993

	High-Potential Zone		Medium-Potential Zone	Low-Potential Zone	
Percentage of Farmers	Small Farms	Large Farms	Small Farms[a]	Small Farms	Large Farms
Used hybrid seed	92	94	49	13	0
Used OPV seed	0	0	1	40	50
Used credit to purchase seed	5	62	1	1	0
Used basal fertilizer	66	98	29	8	0
Used top-dress fertilizer	17	70	8	2	0
Used credit to purchase fertilizer	4	62	1	2	0
Never received extension advice	34	18	44	57	46
Were reached by extension before T&V[b]	28	41	22	26	28
Were reached by extension after T&V[b]	72	59	78	74	72

Notes: Large farms have more than 8 ha of land; small farms have less than 2 ha.
a. In the medium-potential zone, the "large" farm-size category was dropped because of insufficient data (only three farmers had more than 8 ha of land).
b. Percentage of farmers who received extension advice.
Source: Kenya Maize Data Base Project, 1992–1993.

smallholders and farmers in low-potential regions. All farmers surveyed applied lower than optimal levels of fertilizers, and their use of other sources of plant nutrients was limited. Moreover, maize farmers tended to

Table 6.5 Kenya: Gap Between Maize Yields Obtained in Research Trials and by Farmers and Disparity Between Farmers' Nutrient Use and Recommended Nutrient Levels, by Zone and Farm Size, 1992–1993

	High-Potential Zone		Medium-Potential Zone	Low-Potential Zone	
	Small Farms	Large Farms	Small Farms	Small Farms	Large Farms
Average farm size (ha)	1.9	241.0	2.1	2.4	21.9
Farmers' average yield (t/ha)[a]	2.4	4.7	1.3	1.1	1.7
Potential yield (t/ha)[b]	4.2	6.2	2.8	2.2	2.7
Ratio of farmers' yield to potential yield	0.57	0.76	0.48	0.49	0.62
Nitrogen applied (kg/ha)	14	41	9	3	0
Percentage of recommended level	22	62	16	5	0
Phosphorus applied (kg/ha)	22	57	21	7	0
Percentage of recommended level	40	104	43	15	0
Phosphorus-to-nitrogen ratio	1.5	1.4	2.4	2.4	—

Notes: a. Data from Kenya Maize Data Base Project, 1992–1993.
b. Fertilizer Use Recommendations Project (KARI 1990).

apply a higher ratio of phosphorus to nitrogen than KARI recommended (Table 6.5).[5]

The survey results highlight the considerable potential for increasing maize production through increased adoption of better varieties and fertilizer, particularly the application of correct amounts of fertilizer and the use of more nitrogenous fertilizer. To tap this potential, farmers will need more location-specific varietal and agronomic recommendations and, more important, profitable choices of technologies. Current recommendations are too broad to meet the challenge of changing cropping systems and soil fertility levels. KARI will need to intensify its research on crop management and at the same time inject new energy into its breeding program to ensure future improvements in yield potential.[6] None of this can be done, however, without increased research funding. Current funding, which in real terms equals the levels of the 1960s, must be more than doubled and sustained. As for KARI's human resources, more than 60 scientists work on maize, which is more than sufficient to face the challenge.

The Role of Seed and Extension Services

Seed and extension services were instrumental for rapidly diffusing the new maize varieties and hybrids in the mid-1960s. The relatively developed rail and road network was also important. An expansion of roads and seed distribution networks in the late 1960s and early 1970s further improved farmers' access to inputs and increased smallholders' participation in maize production.

In 1963, one year before the first hybrid was released, the government made an agreement with a local private seed company, the Kenya Seed Company, to multiply and distribute the new hybrid seed.[7] The Seed Company initially used the facilities of an established marketing organization, the Kenya Farmers Association, but as demand for the seed increased, the Seed Company expanded its operation by contracting more seed growers and stockists. From 104 seed stockists in 1966, the Seed Company distribution network grew to 3,000 stockists by the late 1980s (Ndambuki, Kiplagat, and Rubui 1992). Figure 6.3 shows the sharp rise in maize seed sales from the early 1960s to the mid-1980s.

Growth in local seed sales has stagnated since the mid-1980s, however, and export opportunities have remained limited. In the past few seasons, farmers have complained of the high cost and repeated shortages of quality seed, and these complaints have added to the pressure on the government to liberalize the seed industry. Poor seed quality control, poor contractual arrangements with farmers, and pricing regulations were among the factors thought to cause the nation's seed problems (World Bank 1995). After a major review of the industry examined ways to make it competitive, several reforms were enacted, including legislation to establish plant breeder's rights and deregulation of seed pricing. In addition, private seed companies submitted prospective varieties for testing by KARI.

Figure 6.3 Kenya: Sales of Improved Maize Seed, Early 1960s to Mid-1980s

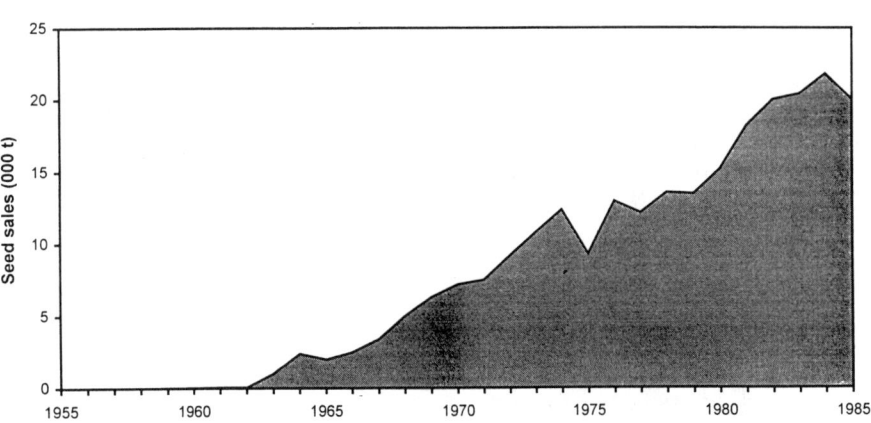

Source: Kenya Seed Company, unpublished data.

At the same time seed of the new hybrids was being produced, the public agricultural extension service waged an aggressive hybrid seed and fertilizer campaign to stimulate farmers' interest in hybrids. Farmers participated in more than 5,000 field demonstrations that showed the remarkable superiority of the new maize technology over farmers' own practice. The Agricultural Information Centre, established in 1966 and affiliated with the extension program, supplied information on hybrids through radio programs and leaflets on recommended production practices for hybrids that were enclosed in packets of hybrid seed (Karanja 1996).

As public resources declined in the 1980s, the extension program was unable to cope with the increased number of farmers seeking assistance. An inadequate operating budget, poor links with research, and poor motivation hampered the program's efficiency. To compensate for these deficiencies, the new T&V system of extension was instituted in 1981 with financial support from the World Bank. The goals of the T&V were to improve extension's capacity to deliver technology to farmers; improve links among farmers, extension, and research; and expand the number of farmers covered by the extension system.[8] A recent evaluation estimated high rates of return for the T&V program (Bindlish and Evenson 1993), but the program's high recurrent costs raise concern about its financial sustainability once donors withdraw their support. Moreover, the program's effectiveness is closely tied to KARI's ability to generate better technologies.

The Role of Marketing, Price, and Credit Policy

The provision of a guaranteed maize market, producer prices, and credit played a key role in stimulating maize production. Maize marketing and pricing were, for a long time, controlled by the government. An agricultural ordinance enacted in 1942 marked the beginning of the government's involvement in maize marketing and the provision of credit to maize farmers, as well as the creation of a maize marketing board—the Maize Control Unit—which purchased all maize at guaranteed prices.[9]

After Kenya gained political independence in 1963, the newly formed Agricultural Finance Corporation offered low-interest credit to farmers, particularly large-scale farmers who possessed sufficient collateral. A 1972 survey indicated that only about 2% of the official credit was issued to smallholders (Heyer, Maitha, and Senga 1976). In recent years, the Finance Corporation has reduced the number of loans to smallholders because of inadequate funds, high transaction costs, and frequent defaults. The Finance Corporation's low interest rate and longer-term loans (compared with loans from commercial lenders), along with its tolerance of late or no repayment of loans, have constituted an implicit subsidy to large-scale maize farmers, who could in fact secure loans from commercial sources (World Bank 1995).

In 1979, the government formed the National Cereals and Produce Board by merging two major parastatal marketing boards, the Maize and Produce Board and the Wheat Board. When maize production became more unstable in the 1980s, the government strengthened the National Cereals and Produce Board's monopoly maize trading status by instituting administrative controls over marketing and pricing activities (Gordon and Spooner 1992).

In the 1980s, the National Cereals and Produce Board established an extensive marketing network of depots and storage silos to improve market access for maize farmers and establish a national strategic reserve capacity. The marketing network, however, proved financially unsustainable. The board consistently generated financial losses, became a burden to the exchequer, caused serious delays in payments to farmers, and undermined its role as the custodian of maize marketing. Faced with mounting losses, the government was pressured to reform maize marketing. A Cereal Sector Reform Program, supported by funds from the European Community, was initiated in 1988 to foster competitive marketing and pricing, remove administrative controls, increase private-sector participation, and rationalize the board's role and marketing network (Gordon and Spooner 1992). The government was slow to adopt these reforms, citing national food security reasons. A worsening macroeconomy and donor conditionalities, however, led to the relaxation of restrictions on maize movement in 1990 and again in 1992. The reduction of subsidies on major farm inputs in 1992 increased maize production costs, and production declined because grain prices were still controlled. In 1994, the government liberalized both internal and external maize trade. Since then, however, on numerous occasions the government has imposed various restrictions on maize trade.

To summarize, Kenya's national maize production campaign was an unqualified success from 1965 to 1980, primarily because of a coherent national maize program including research, seed, and extension programs and supportive marketing, pricing, and credit policies (Karanja 1996). The development and adoption of numerous high-yielding maize varieties that were adapted to a range of agroclimatic conditions led to a doubling of yields, a near tripling of area planted to maize, and a fivefold increase in maize production in a 15-year period (Karanja 1990). This feat improved rural and urban welfare, as well as national food security, by helping to keep food prices relatively low.

Unfortunately, the scientific and institutional cooperation that created the maize success story of the 1960s and 1970s collapsed in the 1980s, as severe economic pressure weakened public financial support for research, extension, and credit. The result has been a general decline in overall maize production, as the number of new technical innovations has dropped and the rate at which farmers adopt current technologies has fallen.

Lessons and Implications

Several key lessons can be drawn from Kenya's maize experience. First, the basic ingredient for a rapid increase in maize productivity was an affordable, profitable technology package that was convincingly superior to farmers' own technology. This package consisted of high-yielding maize seed adapted to different agroclimatic regions and farmer-tested agronomic recommendations.

Second, Kenya's success with maize from 1965 to 1980 resulted from a combination of factors. These included a high-quality public research program that developed high-yielding maize seed and agronomic recommendations for farmers, an extension program that demonstrated production practices to farmers, an effective demand for the new technologies from farmers, an efficient seed company to multiply and distribute seed, an input distribution network to allow timely delivery of inputs to farmers, credit to enable farmers to purchase inputs, and supportive marketing and pricing policies (Eberhart and Sprague 1973; Karanja 1996).

Third, the government chose to play an active, strategic role in the maize story. The government provided political leadership and created an institutional framework and policy environment that mobilized both public and private organizations to support and sustain maize productivity growth to achieve national food policy objectives.

Fourth, Kenya's experience shows that increases in maize production need to be sustained to meet the growing demand for food. In the 1980s and 1990s, Kenya's adverse macroeconomic conditions and erratic weather contributed to the stagnation and decline in maize production.[10] Rapid population growth increased demand on public expenditures for health, education, and other necessities, causing a serious decline in public investment in farmer support institutions such as research, extension, and credit. Also, maize technology development was slowed, less credit was available to farmers through the Finance Corporation, and payments to farmers from the National Cereals and Produce Board were frequently delayed. The cumulative result has been an overall decline in national maize production.

Revitalizing Kenya's Maize Industry

The rise and decline of the maize industry is of critical concern to the government and people of Kenya, most of whom regard maize as their major staple food. The various ills that continue to prey on the industry could provoke a major political and economic crisis unless a bold, coordinated national effort is mounted to combat them.

There are no easy short-run solutions to the problem. The current piecemeal and isolated efforts to restructure the maize industry are commendable but unlikely to provide the dramatic improvement that is needed. The government should take the initiative, just as it did in the 1950s and 1960s, and develop a joint public- and private-sector partnership to increase maize yields and national maize production. The first step is to review the status of the maize sector and lay out a national sectoral plan of action to revitalize the industry. Numerous recent studies can provide the information needed for this initiative, including KARI's maize research priorities (Mills, Hassan, and Mwangi 1995). To be useful, the action plan must develop comprehensive short-, medium-, and long-term technical, institutional, and policy strategies for increasing maize production while at the same time contributing to an increase in overall economic growth. Special attention should be given to policy and institutional reforms that improve the capacity to provide better technologies and ensure long-term commitment and support for the maize industry from both the private and public sectors.

Increased financial support for KARI is required to stimulate the generation of better technologies and region-specific agronomic recommendations. Efforts to increase the productivity of maize in the high-potential region should receive the greatest emphasis, for this region comprises about 70% of the maize area and the payoff to input use is the highest. Considering the growing population pressure on land and the shift of agricultural production into the lower-potential regions, however, a gradual increase in investment in agricultural services will be needed in these areas.

Improved farmer support services should be directed toward smallholders, who account for the largest proportion of maize production and whose maize yields have the greatest potential for improvement. Table 6.5 indicates that the average 1992 maize yield of smallholders in the high-potential zone was about half the yield obtained by large-scale farmers in the same zone. Incremental yield increases within this category of farmers alone would greatly raise national maize production.

The high level of adoption of hybrid seed by smallholders in the high-potential zone negates the thesis that hybrid maize technology is not suitable for African smallholders' circumstances. The challenge ahead, however, is to develop hybrids suitable to smallholders' circumstances in other zones. Farmers in the medium-potential zones would especially benefit from hybrids capable of yielding better than the local cultivars grown under the conditions of poor soil fertility and *Striga,* a parasitic weed.

But raising production will not be sufficient. Researchers must pursue production strategies that improve and conserve the natural resource base. This requires a comprehensive search for alternative sources of plant nutrients and the development of techniques for conserving natural resources. Current low input levels and the lack of nutrient recycling strategies threaten to make a very serious soil fertility problem even worse. Policy and institutional measures to remedy the situation are long overdue. Kenya's future maize policy agenda may have to shift from the current "self-sufficiency" goal to a broader national "food security" goal, which aims at generating a reliable food surplus from domestic sources during bountiful years and judiciously managing its strategic reserve and food imports during drought years. Moreover, future maize production must rely on market-based food security policies rather than government-controlled market inducements. This will require liberalized maize and input markets to offer appropriate and efficient marketing and pricing signals.

Notes

Senior authorship of this chapter is shared.

1. Harrison was supported through a cooperative program of the Ministry of Agriculture, the British Overseas Development Administration, the Rockefeller Foundation, and the Major Cereals in Africa Project through the East African Agriculture and Forestry Organization (Eberhart and Sprague 1973).

2. The KARI/International Maize and Wheat Improvement Center Maize Data Base Project surveyed farmers in 75 villages and 30 maize growing districts in 1992–1993. The project used agroclimatic characteristics to differentiate among six basic maize production zones: the highland tropics, the moist and dry transitional zones, the moist mid-altitude zone, the semiarid zone, and the lowland tropics. These zones were later classified as high-potential zones (the highland tropics and moist transitional zone), medium-potential zones (the dry transitional and moist mid-altitude zones), and low-potential zones (the semiarid zone and lowland tropics).

3. This discrepancy results mainly from a historical bias that favored concentration of public investments in institutions and infrastructure in the high-potential regions among large-scale farmers.

4. The Fertilizer Use Recommendations Project, sponsored by KARI and the Deutsche Gesellschaft für Technische Zusammenarbeit (GTZ), conducted the researcher-managed trials at 70 sites in 30 districts between 1985 and 1989 (KARI 1990).

5. Most Kenyan soils are deficient in phosphorus. The nitrogen and phosphorus nutrient imbalance observed from the survey results can be traced back to the commonly used mono- and diammonium phosphate fertilizers (Kenya 1995).

6. Data and other information from the Maize Data Base and Fertilizer Use Recommendations Projects will be useful in tailoring current recommendations to local conditions.

7. The government owns 51% of Kenya Seed Company shares through a parastatal, the Agricultural Development Corporation.

8. The T&V program started in two districts as a pilot project and within a year had been adopted in almost all major agricultural districts.

9. The first government loans were made available to smallholders in 1948.

10. The annual rate of economic growth declined from 4.8% in the period 1970–1983 to 2.6% for the years 1980–1993 (World Bank 1995:164).

7

Maize Technology Development in Ghana During Economic Decline and Recovery

Robert Tripp & Kofi Marfo

Despite Ghana's increasingly bleak economic prospects during the 1970s and its ailing agricultural sector, the research and extension system nevertheless proved remarkably successful in developing maize technologies that were attractive to farmers. The extent of this success became apparent when the Ghanaian economy began to revive in the mid-1980s and maize production rebounded at a surprisingly rapid rate. In this chapter, we review the circumstances that enabled maize research to progress during these periods of economic decline and recovery and examine factors that are likely to influence the future course and effectiveness of national maize research. We begin with short accounts of the role of maize in Ghana and the history of national maize research.

Maize in the Ghanaian Economy

Maize is the most important cereal crop in Ghana, where it competes favorably with cassava as the cheapest source of calories in the south and with sorghum and millet in the north (Alderman and Higgins 1992). Maize nevertheless contributes less than 20% of calories to the average diet, whereas roots and tubers supply more than 50% of the calories for rural households and almost 40% for urban households (Alderman and Higgins 1992). Even where maize is a principal staple, as in the southern Central and Volta Regions and parts of the Northern Region, it rarely contributes more than 35% of the calories of the average household.

Maize is a heavily traded commodity. Currently, more than one-half of the maize that is produced enters the market, compared with an estimated one-third of maize production in 1957 (Miracle 1966; GGDP 1991; Alderman 1991). Ghanaians consume maize in a variety of forms, favoring thick porridges and gruels made from fermented maize dough. Because these foods require considerable time and skill to prepare, much maize is purchased as prepared food from vendors rather than as grain. Maize preparations are particularly important as urban "fast food," but they are purchased widely in rural areas as well; in fact, a high proportion of maize producers buy maize, mostly as prepared products. A monthly survey in 1987–1988 showed that between 62% and 86% of maize-producing households purchased some maize products in the market (Alderman 1992a).

For the period 1990–1994, annual maize area averaged 590,000 ha. Annual production averaged 823,000 t.[1] The average maize area declined in the second half of the 1970s, in rough parallel with Ghana's worsening economic crisis, and started to improve with the subsequent recovery in the mid-1980s (Table 7.1). Maize yield also began to rise at this time, in large part because of the widespread adoption of improved technology.

Maize is grown throughout Ghana (excluding the far northern Sudan Savanna) in a wide range of environments. Except in northern Ghana, rainfall is generally bimodal, ranging from 800 mm to 1,500 mm annually. The following rough typology includes the majority of maize farming systems in Ghana.

- Coastal Savanna. Maize is usually intercropped with cassava. Most maize is planted in the first season, beginning in March, and drought risk is significant.
- Forest Zone. Maize is grown on scattered plots, usually as an intercrop with cassava and often with other crops as well, such as cocoyam and plantain. Most maize is planted during the major rains, but some maize is also planted during the minor rains.
- Transition Zone. This zone lies between the Forest Zone and the Guinea Savanna. Fertilizer use and continuous cropping of maize are more prevalent. Some maize is intercropped, but most is grown as a monocrop. Maize is planted in both the major and minor seasons.

Table 7.1   Ghana: Maize Area, Yield, and Production, 1950–1994

Year(s)	Area (000 ha)	Yield (t/ha)	Production (000 t)
1950	171	0.99	169
1963	203	0.90	183
1970–1974	421	1.07	452
1975–1979	291	1.05	304
1980–1984	441	0.76	333
1985–1989	504	1.14	573
1990–1994	590	1.40	823

Sources: Data for 1950 from DOA (1959); for 1963 from the Agricultural Census Office, cited in Killick (1966); and for 1970–1994 from the Policy Planning, Monitoring and Evaluation Department, MOA.

- Guinea Savanna. Maize is increasing in importance in the savanna of northern Ghana. Sorghum and millet are the traditional cereals in this zone, but maize is also grown, often in association with the small grains or with groundnuts or cowpeas. Many fields are permanently cultivated, and frequently maize is fertilized.

The relative importance of maize in different areas of the country has changed over time. Maize was traditionally an important crop for home consumption in parts of the Forest and Coastal Savanna Zones. Commercial agriculture was at first concentrated in the southern Forest Zone, where oil palm and, later, cocoa predominated. By the 1940s, cocoa diseases, along with environmental changes from the deforestation brought on by cocoa cultivation itself, reduced cocoa production in some formerly forested areas, and commercial production of maize and other food crops assumed predominance. Considerable investment in the 1960s in improving road links with the Transition Zone and northern Ghana helped transform large areas of the Transition Zone into important suppliers of food crops and attracted immigrant farmers from many parts of the country. Presently, maize production in the Transition Zone and the Guinea Savanna accounts for more than half of the maize planted in Ghana. The development and promotion of maize technology have emphasized these two zones because of their comparative advantage for maize production.

Maize Research and Technology Development

Several institutions have contributed to the development and dissemination of maize technology in Ghana. Before independence in 1957, most agricultural research was the responsibility of the Specialist Branch of the Department of Agriculture. Administrative and organizational changes after independence led to the formation of the Crops Research Institute (CRI) in 1964. The CRI had its principal research station near Kumasi and maintained several small outstations. The Nyankpala Agricultural Experiment Station, near Tamale in the Guinea Savanna, was part of CRI until 1994, when it became an autonomous institution—the Savannah Agricultural Research Institute.

A maize improvement program was launched in 1956. Although several improved maize varieties were introduced in the 1960s, most failed to gain popularity. A number of government and donor projects promoted better maize growing practices in different areas of the country, but it was not until 1979 that Ghana initiated a concerted nationwide effort in maize technology by launching the Ghana Grains Development Project (GGDP).[2] The GGDP aimed to develop and demonstrate practical recommendations for Ghana's major maize growing areas as quickly as possible (Edmeades et al. 1991). To achieve that aim, the project relied upon an extensive program of on-farm research, managed by research staff and extension agents based in towns and villages in all major maize growing regions.

One of the first tasks of the GGDP was to test the improved maize materials the CRI had acquired. Testing was done through an extensive series of on-farm trials, which led to the release of two new varieties in 1984; both varieties have now been replaced by versions resistant to maize streak virus. More recently, an early maturing variety and a quality protein variety were released.

Research on improved varieties was largely conducted on farmers' fields—first as rather large, researcher-managed trials to select among promising materials and later as farmer-managed trials in which farmers tested one or two varieties against their own local variety under their own management.[3] Farmers generally valued the new maize varieties for their superior yield, fertilizer responsiveness, early maturity, and resistance to lodging. These judgments were tempered, however, by uneven marketing experience and problems with storage. The apparent source of the marketing problem was that the new varieties were judged to be somewhat inferior, or at least to require extra time, for use in common maize preparations. Although laboratory tests failed to demonstrate food quality differences between local and improved varieties (Plahar et al. 1987), traders nevertheless priced the improved maize as much as 10 to 15% lower, especially if there was a bumper crop. The differential between the price of improved and local maize has diminished over time, however. Farmers' concern about the storage quality of the improved maize varieties arises from their inadequate husk cover, which allows insects and birds to damage maize in the field and leads to greater damage when maize is stored in the husk—the common practice.

Current priorities in maize breeding include improvement of grain qualities and husk cover; selection for nitrogen-use efficiency; screening for resistance or tolerance to *Striga,* drought, and stem borer; and development of quality protein maize (Asafo-Adjei and Soza 1993). Some effort is being made to develop the research capacity for hybrid maize, although a decision has not been taken to promote hybrid maize production.

Maize breeding research and efforts to develop new varieties have been complemented by research on crop management, especially fertilizer studies. The Soil Research Institute carried out several fertilizer response studies for maize in the 1960s and 1970s (e.g., Ofori 1963), and similar studies were done by a fertilizer promotion project in the early 1970s. At the inception of the Ghana Grains Development Project, however, farmers still used little fertilizer on maize. The project organized a series of on-farm trials aimed at developing fertilizer recommendations extension workers could pass to farmers. The recommendations ranged from no fertilizer application for Forest Zone plots with at least five years of fallow to roughly 90-40-40 (N-P-K) kg nutrient/ha for continuously cropped fields in the Transition Zone and the Guinea Savanna. Considerable care was taken to make these results easy for farmers to understand. Recommendations provided during demonstrations and field days were given in terms of bags of fertilizer per acre and the number of plants that could be fertilized from one condensed milk tin (farmers' most common application measure). Increases in fertilizer prices in recent years have required modifications in fertilizer recommendations.[4]

The adaptive research program also developed information about maize plant density and spatial arrangement. Farmers were advised to change from the traditional "random" planting of maize,[5] in which a relatively large number of seeds were planted in holes at least 1 m apart, to line planting, which allowed the establishment of densities more suitable to the short improved varieties. In addition, recommendations were made on plant stand establishment, such as early planting, deep planting, and firming of the soil.

Maize storage has also been addressed in adaptive research and extension. The insecticide Actellic is widely recommended and generally effective. Improved maize cribs are also promoted. Ghana now faces the problem of the larger grain borer, which is moving westward from Togo.[6]

Not all of the crop management research has resulted in useful crop management recommendations. Weed control is one of the principal constraints on maize production in Ghana. Farmers have difficulty marshaling the labor to control weeds; reduced fallow periods and the introduction of new weed species further aggravate the weed problem.[7] Surveys indicate that most farm credit is used to hire labor for clearing land and weeding (GGDP 1991; Bumb et al. 1994), and farmers describe weeding as their major constraint (Amanor 1994). On-farm experiments have examined various herbicides, but hand weeding is still more cost-effective than herbicides. Even though increased levels of hand weeding provide better weed control, most farmers already balance their labor supply as efficiently as they can, so exhortations to increase weeding are not a sound extension strategy. Weed control remains a priority for future research.

Technology Transfer

Ghana has experimented with a wide range of extension strategies. During the 1970s, the Focus and Concentrate program worked with selected farmers to introduce inputs and crop management advice (Atsu 1974). Other campaigns in the 1970s mobilized government loans for food production, especially Operation Feed Yourself under the Acheampong regime. Later in that decade, the World Bank's Training and Visit (T&V) extension program was included in two regional development projects.

The extension strategy of the 1980s drew on a countrywide network of extension agents from the Ministry of Agriculture and the Grains and Legumes Development Board who organized maize demonstrations. The diffusion of maize technology was given a significant boost in 1987 when the Sasakawa-Global (SG) 2000 food production project was launched in Ghana (Yudelman et al. 1991; Tripp 1993). The SG 2000 project is sponsored by the Sasakawa Africa Association, with participation from the Sasakawa Foundation in Japan and the Carter Center in the United States. SG 2000 operates in a number of African countries; Ghana is considered one of its principal successes.

The SG 2000 strategy in Ghana featured large demonstration plots, called "production test plots"; provided inputs to participating farmers at low rates of interest; and linked widespread coverage of farmers with the development of policy support for agriculture. The program expanded from around 1,600 production plots in 1987 to 16,000 in 1988 and 78,000 in 1989. SG 2000 scaled back its activities in 1989, however, because of the logistical problems posed by handling numerous loans and managing input distribution. Without question, the SG 2000 program achieved considerable coverage. A survey in 1990 indicated that almost one in four maize farmers had experienced some contact with SG 2000 (either through attending a demonstration or receiving a loan). Farmers who had participated in extension activities had significantly greater knowledge of the recommendations (GGDP 1991).

Input Policy

Both research and extension to develop and disseminate improved maize production technology have been supported by input policies to varying degrees. Seed production and distribution have always been problematic in Ghana. The Ghana Seed Company was established as a parastatal in 1979, but its 10-year life was plagued by both quality and supply problems. During the same period the Seed Company was active, the Grains Board also produced and distributed maize seed. The board has now been assigned the role of producing foundation seed for private seed producers using the breeder seed provided by the maize research program. All commercial maize seed is now produced by small private enterprises, most of which establish their own links with input dealers for marketing the seed (Bockari-Kugbei 1994). Ninety-four registered maize seed growers produced more than 500 t of seed in 1993 (less than 5% of total maize seed use) (ISSER 1994). The seed producers and the Ghana Seed Inspection Unit have received significant support from SG 2000 in an effort to develop a viable private maize seed sector.

All fertilizer in Ghana is imported, and more than half is used on maize. Fertilizer was subsidized from the mid-1960s until 1990. The fertilizer subsidy accounted for 60% of the cost of fertilizer at the farm level in 1985 (Dapaah and Otinkorang 1988) but was eliminated between 1988 and 1990 (Bumb et al. 1994).

Removal of the subsidy precipitated a sharp decline in fertilizer use on maize. Fertilizer imports reached a peak of 65,000 t in 1989, but imports in subsequent years have been less than half this figure and in some years have been negligible.

The ratio of fertilizer price to maize price was fairly attractive in the 1980s and was important in encouraging farmers to fertilize maize. Experimental data show response levels of improved maize varieties in the range of 15–20 kg grain for every kilogram of applied nitrogen for continuously cropped land under good management (Akposoe and Edmeades 1981, 1982). Taking account of labor costs, cost of capital, and risk, it is reasonable to expect that given this level of fertilizer response, farmers will not invest in fertilizer if the nitrogen-to-grain price ratio is higher than 7 or 8. Current ratios of nitrogen (ammonium sulfate) to grain price are higher than this.

Higher fertilizer prices make it imperative to develop efficient fertilizer management recommendations. Ghana recently changed its policy toward the importation of higher-analysis fertilizers and has moved toward replacing ammonium sulfate with urea. On-farm experiments indicate that urea is as effective a source of nitrogen as ammonium sulfate and a cheaper one, but farmers prefer ammonium sulfate because it can be broadcast. Although traders have been forced to lower prices for urea, demand remains low.

Marketing and Price Policy

As noted earlier, more than half of Ghana's maize production enters the market. Maize marketing has always been in the hands of the private sector, and the Ghanaian government has never been able to play an influential role in maize marketing. The Ghana Food Distribution Corporation was established to help implement government policy in food crop marketing and to offer floor prices to farmers, but it has handled only a small percentage of the maize that is marketed (Alderman 1991). Proposals in recent years to increase Ghana's public food storage capacity (Sarris 1992) have yielded little concrete action.

The available evidence indicates that the private sector is fairly efficient at marketing maize (Southworth, Jones, and Pearson 1979). Alderman and Shively (1991) concluded that most food crop markets in Ghana have functioned satisfactorily and that prices have been transmitted across commodities fairly well. They showed that the real wholesale and retail prices of maize (and many other food crops in Ghana) have been declining since the 1970s and that the ratio of retail to wholesale maize prices has also fallen during this period, indicating improvements in marketing efficiency.

But Ghana's maize marketing system is not without problems. A principal concern is the large seasonal swing in maize prices, which typically reach a peak in June and then fall sharply by September (Alderman and Shively 1991). In the past several years, two- or threefold differences have sometimes been observed between the highest and lowest monthly prices for maize. This price fluctuation is often blamed on hoarding by traders, but there is little evidence to support this hypothesis. Several studies show that a large proportion of maize is kept on the farm and marketed well after harvest (GGDP 1991; Alderman 1992a; Thornton 1973). Farmers stagger maize sales over the year, chiefly in anticipation of receiving a better price (Southworth, Jones, and Pearson 1979). Also, landowners who receive a harvest share from tenants often speculate with their maize, waiting for market prices to rise (Amanor 1994). Southworth, Jones, and Pearson (1979) concluded that the seasonal price fluctuation results mostly from the high cost of storage, but Alderman (1992b) has argued that "behavioral considerations are at the crux of the issue" and that "storage and interest costs cannot account for either the level or the variability of seasonal price rises."

Technology Adoption

Several adoption studies have documented considerable change in maize growing practices through the 1980s (Table 7.2). By 1990, 48% of the

Table 7.2 Ghana: Adoption of Maize Technology in Three Areas, Mid-1980s and 1990 (percentage)

	Zone, Area, and Year					
	Transition Zone		Transition Zone		Forest Zone	
Production Practice	Ejura, Ashanti, 1987[a]	Ejura, Ashanti, 1990[b]	Brong-Ahafo, 1986[c]	Kintampo, Brong-Ahafo, 1990[b]	Central Region, 1986[d]	Swedru, Central Region, 1990[b]
Farmers using improved varieties	38	78	81	75	17	34
Farmers row planting	52	81	68	61	3	22
Farmers using fertilizer	30	66	47	22	6	14

Sources: a. Dakurah and Arias (1987).
b. GGDP (1991).
c. Tripp et al. (1987).
d. Opoku-Apau et al. (1987).

maize area in Ghana was planted with improved varieties. Adoption rates vary considerably by zone, however, with higher rates in the Transition and Guinea Savanna and lower ones in the Forest and Coastal Savanna. Even where maize consumption is relatively high, there are contrasting adoption rates. In the Coastal Savanna, where maize consumption is based on fermented dough preparations, the adoption of improved varieties is generally low. In the Guinea Savanna, on the other hand, where maize flour is simply substituted for sorghum or millet flour in local foods, adoption of the new varieties is often very high.

Since all improved maize varieties released in Ghana are open-pollinated varieties (OPVs), farmers can save the seed to use the following season. Most farmers use their own saved seed, except when they change maize varieties. When farmers decide to switch varieties, 40% obtain the seed from another farmer, and 30% obtain it from an extension agent. Only 20% of the farmers purchase seed through the Ghana Seed Company or the Grains Board.

Fertilizer has been adopted on about half the area planted to improved varieties (Table 7.2). Although almost half of the maize farmers in the 1990 survey reported experience in using chemical fertilizer, only 26% of the maize fields received fertilizer that year. As noted earlier, fertilizer purchases by maize farmers dropped sharply in 1990 because the fertilizer subsidy was removed and also because of late, erratic rains. Several surveys have demonstrated a strong relationship between cropping history and fertilizer use: Maize fields that are continuously cropped are much more likely to receive fertilizer, as are fields that are monocropped. The use of an improved variety increases the probability of using fertilizer, but fields that are row planted (i.e., that have adequate plant populations) receive most of the fertilizer.

Changes in plant population management are more difficult to assess, but evidence from several surveys indicates that farmers who row plant maize are also following the recommendations of closer plant spacing and fewer seeds per hill. In the 1990 survey, 36% of the maize fields that were not ridged were row planted. Row planting is much more common with monocropped maize and on fields that have been continuously cropped and hence are free of stumps and other obstacles.

The adoption of these maize production practices during the 1980s has undoubtedly contributed to the growth in maize production and yields since the late 1980s (Table 7.1), but data are not adequate to quantify the impact of this technological change. With respect to yield gains, data from demonstrations in the Transition Zone indicate that adoption of an improved variety, fertilizer (90-40-40 N-P-K), and adequate plant population increases yields from 1.8 t/ha to 3.5 t/ha (Edmeades et al. 1991). Without question, increased maize production contributed to the decline in the real wholesale price of maize throughout the 1980s, with the sharpest drop coming after 1984 (Alderman and Shively 1991).[8] Perhaps most important, the introduction of new maize technology in the 1980s came at a time when Ghana was beginning to recover from the deep economic stagnation of the previous decade (Alderman 1991).

Increasing Maize Productivity in Ghana: The Challenges Ahead

Since the early 1980s, research and extension have given much attention to the challenge of increasing maize productivity in Ghana. Meeting that challenge has involved a considerable strengthening of the maize research capability, as well as the development of a number of government and donor-sponsored extension programs, with the SG 2000 effort the most prominent. The result has been the development and adoption of improved maize production practices on a significant area.

But the challenges that lie ahead for maize in Ghana are less straightforward. For one thing, it is not clear how to take advantage of increased maize production capacity in the future. In addition, the potential demand for maize for direct human consumption is limited. Expansion in maize demand may come from industrial opportunities, such as brewing, or from the animal feed sector, but these have yet to be developed. Thus to a fair degree, the future for maize demand will be closely linked to the nature and rate of Ghana's economic development.

The supply of maize will almost certainly continue to be provided by the small-farm sector, which has proven to be dynamic and flexible but is not without its own problems. Although land shortage per se is not yet a serious problem at the national level, intensification, shorter fallows, and disintegrating rural infrastructure are leading to a decline in areas that were previously important sources of commercial food production (Amanor 1994). Fallow periods are decreasing significantly in many areas of the Forest Zone (Vercruijsse 1988). As long as unexploited land remains, particularly in the Transition Zone, commercial maize production will be viable with current practices. But the sustainable management of land is a challenge that must be addressed as soon as possible.

Attention should also be paid to what may be a growing division between commercial (although small-scale) and more subsistence-oriented maize production. The former is most prevalent in the Transition and Savanna Zones, usually takes advantage of purchased inputs and often of mechanized land preparation, and has access to cheaper labor from the north of Ghana. The less commercially oriented maize producers, in the Forest Zone and parts of the Coastal Savanna, have been slow to adopt new maize technology. The 1990 survey found that about one-third of maize fields were prepared by slash and burn on land that had been fallow for at least a year; were intercropped (usually with cassava); and, for the most part, did not use new maize varieties, fertilizer, or recommended planting methods (GGDP 1991).

Another problem related to sustainable land management in Ghana is land tenure. In some maize growing areas in the south, maize is grown under various types of tenancy arrangements.[9] The patterns of leasing and sharecropping may be fairly complex (Vercruijsse 1988; Amanor 1994). In some areas, maize is planted under a sharecropping system developed for cocoa production and now applied to other cash crops. The sharecropper provides all of the inputs and labor and returns to the owner one-third of the harvest (or, in areas of land scarcity, one-half). This practice was probably efficient when land was relatively abundant (Robertson 1987), but when production practices require considerable investment in inputs and long-term land management strategies, sharecropping and similar tenurial arrangements present serious problems.

Structural adjustment and the rush to privatization have affected Ghana's maize economy and present some dilemmas for the future (Tripp 1993). The poor performance of the parastatal Ghana Seed Company was one of the factors that motivated the current reliance on private seed producers. The viability of the private seed sector is questionable, however, given that demand for seed of OPVs is low once farmers have adopted a variety. Maize hybrids are often discussed as an alternative, but considerable work remains to be done before hybrids become a viable option. Likewise, the privatization of fertilizer imports and distribution and the elimination of the subsidy were accompanied by a considerable reduction in fertilizer use on maize. Potential gains in efficiency of distribution offered by the private sector have not yet been realized.

Several important maize research priorities emerge from our analysis. Although more research is needed on developing new varieties suitable to Ghana's growing conditions and market demands, increased attention needs to be directed toward soil fertility and crop management. With fertilizer only marginally profitable in many fields without the fertilizer subsidy, efficient methods of combining chemical fertilizer with other management techniques need to be developed. Farmers are only beginning to come to terms with the implications of shortened fallow periods. Although green manuring or agroforestry systems offer possibilities for restoring soil fertility, they will require a strong research effort involving farmers. Another management challenge closely related to soil fertility is weed control. Weeds are the single most important limiting factor on many Ghanaian maize fields, but unless more effective weed control techniques are developed, improvements in soil fertility will not be realized.

The performance of Ghana's economy has improved significantly during the past decade, and maize technology development has made an important contribution to strengthening the agricultural sector. But maize research is entering a difficult period. Complaints are now heard that maize research has been overfunded, although the real problem is surely that research on other crops has been underfunded. Agricultural research in Ghana is being reorganized, with World Bank support, with the goal of setting priorities across a number of public institutions. Maize is included among the top-priority crops, and the high-priority themes for maize research cover both varietal development and crop management (NARP 1994). It remains to be seen whether this reorganization of research will be sufficient to attract the local and external resources necessary to address the challenges of food crop research. Whereas a scarcity of research and extension funds is leading to a retrenchment of the public sector from the countryside, the challenges of the future require more, not less, field research, extension presence, and location-specific research. Ghana's economic recovery and its recent success in developing maize technology may not be enough by themselves to increase maize productivity in the country over the next decade.

Notes

1. These figures are estimates based on monitoring methods dictated by the limited resources available to the Ministry of Food and Agriculture.

2. The Ghana Grains Development Project is executed by the CRI and the International Maize and Wheat Improvement Center, under the sponsorship of the governments of Ghana and Canada. The Grains and Legumes Development Board, the Ministry of Food and Agriculture, and the International Institute for Tropical Agriculture are cooperating institutions.

3. Some farmers maintain several local maize varieties, often distinguished by their time to maturity, but farmers in most maize growing areas seem less aware of which particular maize varieties they use than they are about specific varieties of other crops they grow, such as cassava.

4. Research has shown a significant response to zinc application in many of the soils of the Transition Zone and the Guinea Savanna (Akposoe and Edmeades 1982; Balasubramanian 1984), but this information has not been translated into recommendations for farmers, in part because zinc fertilizers are not available. Current research by the CRI includes examination of the management of various types of organic fertilizer.

5. This is the common practice, except in areas of the Guinea Savanna where maize is planted on hoed ridges.

6. Recent research has shown that the insecticide Actellic Super is useful for combating the larger grain borer, but supplies of the product are currently inadequate (Boxall 1994).

7. Amanor (1991) has reported on farmers' management innovations in response to changing weed populations.

8. The prices of many other foodstuffs in Ghana also declined during the same period (Alderman 1991).

9. Pearce (1992) has reported that 30% of agricultural plots are planted by tenants in the Eastern and Central Regions.

8

Fostering Sustainable Increases in Maize Productivity in Nigeria

Joyotee Smith, Georg Weber,
M. V. Manyong & M. A. B. Fakorede

The food security challenge in western Africa is well documented. In Nigeria, the average per capita supply of calories in 1988 was 12% less than the minimum requirement for a healthy life, despite increases in per capita food production between 1980 and 1990 (Cleaver 1993b). Since the 1970s, maize has achieved the highest growth rate of any major food crop in western Africa. Maize is also widely believed to have the greatest potential among food crops for attaining the technological breakthroughs that will improve food production in the region.

This chapter describes a general framework for analyzing the evolution of western African agricultural systems, traces recent changes in the Nigerian maize economy, and investigates the potential role of maize in Nigeria. The analysis shows that the green revolution approach, which focuses primarily on the development of high-yielding varieties and fertilizer application, has promoted rapid production increases but has not led to sustainable production systems. Although high-yielding, fertilizer-responsive maize can make a major contribution to food production in Nigeria and elsewhere in western Africa, it can do so only in a limited number of areas. An alternative holistic approach is proposed that explicitly recognizes and characterizes the heterogeneity and dynamics of western African agricultural systems in an attempt to design technology and policy interventions that will promote more sustainable systems.

A Conceptual Framework for Analyzing the Evolution of Agricultural Systems in Western Africa

The agricultural systems of western Africa can be broadly characterized by agroecological zone and by the evolutionary path each system has followed in each zone. The agroecological zones range from the subhumid north (the Northern Guinea Savanna and Southern Guinea Savanna) to the humid south (the Derived Savanna and the Humid Forest).[1] The humid zones generally possess better transport infrastructure than the subhumid zones, are located closer to seaports, and contain the major population centers. Input-output price ratios should be lower in the south, which would favor production of nutrient-responsive crops such as maize. Climatic and soil factors in the north, however, are more favorable for maize production. In particular, the high solar radiation and low night temperatures prevalent in the Northern Guinea Savanna are considered to give this agroecological zone a comparative advantage in maize production (Kassam et al. 1975). The length of the growing season in the Northern Guinea Savanna (150 to 180 days) closely matches the length of the maize growing cycle, which enables maize to use moisture and nutrient supplies more effectively in this zone than in the others. The longer growing season in the Southern Guinea Savanna favors day-length-sensitive sorghum or millet varieties and roots and tubers; further south, in the humid zones, the growing season favors semiperennial or perennial shrub and tree crops such as cassava, banana, and cocoa. In agroecologies characterized by longer growing periods, sole cropping of maize can severely leach nutrients from the soil unless maize is intercropped with semiperennial or perennial plants.[2] Thus from an agroecological point of view, only the Northern Guinea Savanna, which occupies 17% of western Africa, is appropriate for intensive maize production. In the more humid environments, maize is more appropriate as a niche crop[3] fitted into systems based on shrub or tree crops.

In each zone, agricultural systems follow two different evolutionary paths: a subsistence-oriented path, driven primarily by population growth, or a market-driven path, driven primarily by opportunities for cash cropping (Smith and Weber 1994). Preconditions for the market-driven path are investments in transport infrastructure and technologies for crops with a natural comparative advantage (Freeman and Smith 1996). Within each path, increasing population density and land-use intensity over time are accompanied by increasingly secure, privatized land rights. The early phase of this process is the *expansion phase,* in which soil fertility is restored through long fallows. During the later *intensification phase*, long fallows are replaced by bush fallow and permanent cropping. In West Africa, market-driven systems are largely in the intensification phase (Manyong et al. 1996).

Because market-driven systems are characterized by lower marketing margins for inputs and outputs, producers in these systems have a greater ability to adopt the land-saving, input-using technologies that foster profitable production of commodities for which the system has a natural comparative advantage.[4] Farmers in market-driven systems in the intensification phase (MDI) seek food and cash crops that can utilize purchased inputs to give high returns to land. In population-driven systems, in the expansion phase (PDE), labor-saving food crop production is farmers' chief objective. In the intensification phase of population-driven systems (PDI), farmers favor crops that give the highest caloric output per unit of land area with minimal use of purchased inputs.

As systems move from the expansion to the intensification phase and fallow periods become shorter, the fertility inherent in the soil declines. In PDI systems, characterized by minimal fertilizer use, declining soil fertility reduces the yield of maize relative to less nutrient-demanding crops, such as cassava and traditional sorghum and millet cultivars. In the absence of inorganic fertilizer, maize yields are higher in the humid zones than in the subhumid zones, most likely because organic matter mineralization is higher in the humid ecologies. With fertilizer application, as occurs in market-driven systems, the response of maize to applied nitrogen reaches about 30 kg grain per kilogram of nitrogen in the subhumid zones, whereas only half that response is achieved in the humid zones. Thus as fertilizer application increases, maize yields in the subhumid zones approach and eventually surpass those obtained in the humid zones at a nitrogen application rate of 80–100 kg/ha (Mughogho et al. 1986). In contrast, millet and sorghum achieve at most half the fertilizer response of maize in the subhumid areas (Mudahar 1986). In MDI, with the adoption of high-input technologies, maize is likely to yield more than competing crops in the subhumid zone and more than maize from the humid zone; eventually, maize can become a lead crop in the system. In PDI, however, the use of purchased inputs is generally low, and maize is likely to yield less than competing crops in all agroecologies.[5]

In summary, western Africa's heterogeneous agricultural systems span four major agroecological zones; in each zone, production systems fall into three major classes (PDE, PDI, and MDI), whose characteristics are listed in Table 8.1. Market-driven systems are more prevalent in the Northern Guinea Savanna and the Humid Forest than in the other two agroecologies (Manyong et al. 1996; Manyong, Smith, and Baker 1995). In the expansion phase, the importance of maize is determined by food preferences. As land use intensifies and bush fallow and continuous cultivation systems prevail, maize is likely to become a lead crop in market-driven areas of the subhumid zone, attaining greater importance in the Northern Guinea Savanna than in the Southern Guinea Savanna. Elsewhere, maize is likely to be a niche crop (where appropriate niches exist). Maize will be more prevalent in the population-driven systems of the south and less prevalent in such systems in the north. With improvements in transport infrastructure among agroecological zones and increased technical progress in the lead crops of all agroecologies, maize is likely to become more important in MDI systems of the Northern Guinea Savanna relative to other areas.

This concept of the role of maize in agricultural systems in western Africa has been confirmed by Manyong et al. (1996). Market-driven systems

Table 8.1 Western Africa: Distinguishing Characteristics of Agricultural Systems in Humid and Subhumid Areas of Nine Countries

Population-Driven Systems		Market-Driven Systems
Expansion Phase (PDE)	Intensification Phase (PDI)	Intensification Phase (MDI)
Shifting cultivation (Mean R=0.23)[a]	Bush fallow, continuous cropping (Mean R=0.58)[a]	Bush fallow, continuous cropping (Mean R=0.61)[a]
Land abundance	Land scarcity	Land scarcity
Labor scarcity	Abundant family labor	Abundant family labor and hired labor
Minimal purchased inputs (no fertilizer used in 79% of systems)	Minimal purchased inputs (no fertilizer used in 61% of systems)	High levels of purchased inputs (fertilizer used in 78% of systems)
Poor infrastructure	Poor infrastructure	Good infrastructure
Minimal cash cropping; labor-saving food crops (crops vary according to food preferences)	Minimal cash cropping; land-saving food crops using low levels of purchased inputs	Input-intensive food and cash crops (e.g., maize, cotton, cocoa)
Minimal crop-livestock interactions (nonexistent in 72% of systems)	Minimal crop-livestock interactions (nonexistent in 71% of systems)	Crop-livestock interactions in 61% of systems

Sources: Smith and Weber (1994); Manyong et al. (1996); Manyong, Smith, and Baker (1995).
Note: a. R = (Years of cultivation) ÷ (Years of cultivation + years of fallow).

are characteristic of a substantial part of the maize area in the subhumid zones (36 to 43%). In the humid zones, population-driven systems are far more common where maize is grown, whereas yams and tree crops dominate cropped area wherever market-driven systems are in place.

Heterogeneity and Dynamics of Nigerian Maize Production Systems

The variation in agricultural systems with respect to both agroecological and socioeconomic factors explains the rapid changes in the role of maize in different systems. An understanding of this variation is useful for envisioning the future role of maize and identifying entry points to realize that vision. We begin by looking at Nigeria's maize economy and the forces that have shaped it; we then examine how maize production systems, operating under diverse agroecological imperatives, have evolved over time.

Maize in the Nigerian Economy

Nigeria produces 43% of the maize grown in western Africa. The crop is most widely grown in the Northern Guinea Savanna, where it is one of the two major crops on 30 to 40% of the area under agricultural production. Staple foods in the north are sorghum, millet, and—increasingly—maize. Maize is least widely grown in the Humid Forest, where it is generally a minor crop (the traditional staples are cassava and yams). The growing season lasts five to seven months in subhumid areas and seven to nine months in the humid areas.

Although maize consumption at the national level is low (16 kg/yr), maize production and consumption have grown rapidly over the past two decades (Figure 8.1). Growth in area and production can be attributed to the successful development of high-yielding varieties combined with the provision of cheap fertilizer, improved infrastructure, and extension services. These inputs were provided largely in selected areas of the north through the Agricultural Development Projects, many of which were funded by World Bank loans, from the late 1970s to the late 1980s. Major changes in the regional shares of maize production were apparent by the mid-1980s. In 1972–1973, the Humid Forest and the Derived Savanna together accounted for 60% of total maize production, the Southern Guinea Savanna for 24%, and the Northern Guinea Savanna and the Sudan Savanna only a minor share of production (16%). By 1983–1984, the Northern Guinea Savanna and the Sudan Savanna combined were the largest maize-producing area (54%);[6] the Humid Forest and the Derived Savanna produced just 23% of Nigeria's maize (Lele et al. 1989).

Figure 8.1 Nigeria: Maize Area, Yield, and Production, 1969–1993

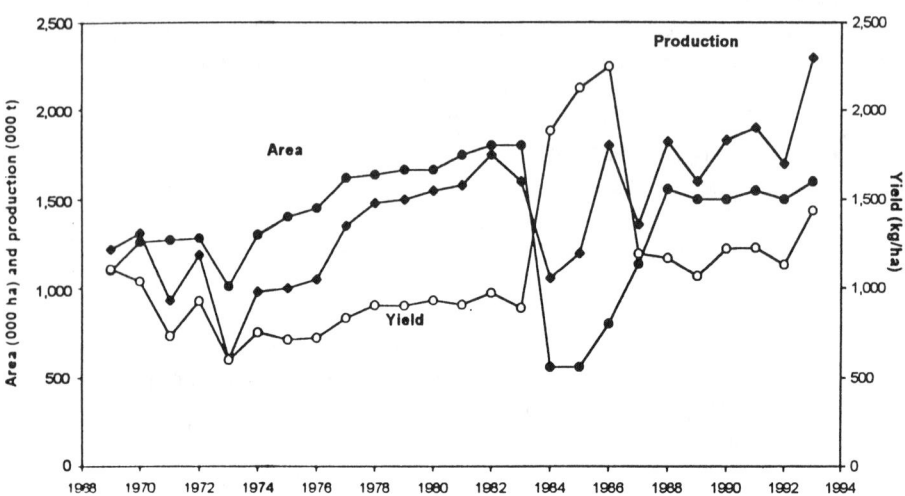

Source: FAO (various years).

The high-yielding varieties that catalyzed these changes were the product of decades of work, which began in the 1950s when the West African Rust Research Unit was established at Moor Plantation, Ibadan (1952), and the Federal Department of Agricultural Research was founded (1956). These programs developed two maize populations, Nigerian Composites A and B, which became important base populations for maize improvement by the International Institute of Tropical Agriculture (IITA), Ibadan, in 1970. The IITA maize improvement strategy emphasized increasing yield potential, breeding for durable pest resistance, and developing distinct varieties for each ecosystem. Among the varieties emerging from this program were TZB, developed for the subhumid zones, and TZPB, adapted to the humid zones. Since the early 1970s, the Nigerian breeding program and the IITA have also collaboratively developed and released a wide range of disease-resistant varieties.

Research on hybrid maize by the national research system began in the early 1970s, and research by the IITA started in 1979, with strong support from the Nigerian government. By 1984, the first hybrids were available for on-farm trials, and the IITA assisted the government in setting up the first private seed company in western Africa.

The potential for maize in the Northern Guinea Savanna was first identified in 1970 by the Institute of Agricultural Research in Zaria (Kassam et al. 1975), but no maize breeding has been done in the zone except for the evaluation of advanced populations. In 1982, following earlier work by Norman, Simmons, and Hays (1982), a farming systems research program was initiated at the Institute of Agricultural Research. This program linked experiment station and on-farm research with extension through monthly meetings with the Agricultural Development Projects. Maize was a key crop in this research program. The IITA's resource management work in the Northern Guinea Savanna started in 1989.

The release of TZB was a technological breakthrough for the Northern Guinea Savanna. The variety gave dramatically higher returns to land—six times as much as sorghum and millet, the traditional food crops, and seven times as much as cotton, the traditional cash crop (Balcet and Candler 1981). Maize also became a major food crop, particularly valued as a replacement for sorghum in the local dish (*tuwo*). The pure white grain of TZB, its improved husk cover, and its resistance to ear rot all contributed to its acceptance, but farmers stressed that maize owed its new importance largely to its dual role as a food and a cash crop. When maize prices were unattractive, maize could always be stored for home consumption.

Aside from the development of suitable maize varieties, another precondition for the expansion of maize in the north was the construction of transport systems. Surplus maize could be shipped from the north to ready markets in the population centers of the south. Road mileage within the northern states increased more than fivefold between 1967 and 1980, and major highways were improved. Improved rural infrastructure also promoted the increased use of agricultural inputs, especially fertilizers. The agricultural transformation of the Northern Guinea Savanna was undoubtedly aided by a fertilizer subsidy of around 85% of the real cost of delivering fertilizers to farmers. The fiscal cost of this subsidy has been high: 32% of federal government agricultural expenditure and 3.7% of total government expenditure (Lele, Christiansen, and Kadiresan 1989). In addition, the subsidy has provoked severe fertilizer shortages, because supplies have been smuggled out of the country. The net result has been that farmers bought fertilizer at prices well above the subsidized price but still well below the unsubsidized price. Because of the fertilizer subsidy, the nitrogen-to-maize price ratio was as low as 0.9–1.9. These favorable prices, together with the high fertilizer responsiveness of TZB, enabled maize to capitalize on its natural comparative advantage in the Northern Guinea Savanna (Smith et al. 1994a).

The performance of maize in Nigeria has been affected by the overall policy environment, which in turn has been dictated by oil revenues. During the "oil bonanza" from the mid-1970s to the early 1980s, intersectoral terms of trade moved strongly in favor of food crops in spite of an overvalued exchange rate and soaring food imports.[7] Maize prices fluctuated widely but remained well above the 1976 level throughout the period. Even with these incentives and the fertilizer subsidy, food availability per capita declined. The increasing gap between urban and rural wages drew large numbers of rural people away from their farming communities to urban centers. Maize production alone fell by an average annual rate of 6.7% between 1973 and 1982.[8]

In 1982, as oil revenues diminished, maize production responded strongly to the favorable terms of trade, policy incentives, and return migration to rural areas. Incentives for maize production were at their peak. By 1986, however, increased production and a currency devaluation led the real price of maize to plummet; in 1987, it was less than half the 1976 level (Lele, Christiansen, and Kadiresan 1989). Maize prices recovered at the end of the 1980s after the government banned cereal imports. The fertilizer subsidy remained in effect, maize once again became highly profitable, and production grew at an annual rate of 5.3%.

Evolution of Nigerian Maize Production Systems

Almost everywhere in Nigeria, maize production has entered the intensification phase: MDI systems cover 39% of the humid and subhumid zones, PDI systems occupy 59%, and the remaining area is dedicated to PDE systems.[9] All three types of production systems have been studied in each agroecology through group interviews and interviews with key informants.

In the Northern Guinea Savanna, group interviews in 27 randomly selected villages confirmed that maize, considered a minor backyard crop in the mid-1970s,[10] had emerged by 1989 as the lead crop in 90 to 100% of the villages surveyed in Kaduna and Katsina States. Virtually all farmers in every village were using improved maize varieties and fertilizers. Local maize varieties had been completely replaced by improved varieties in most of the survey villages (Table 8.2). Sorghum is grown almost entirely for home consumption, but maize is both consumed and sold. Sixty-one percent of farmers surveyed in 1991 claimed that maize was the major source of cash for purchasing crop production inputs. Thus maize is the linchpin of the market-driven system. Maize gives higher returns to land, labor, and cash than all competing crops except cotton, which in spite of its high returns remains a minor crop because of market uncertainties. The adoption of improved maize, and the consequent degree of change in the production system, are significantly lower in other areas of the Northern Guinea Savanna, such as Bauchi and Sokoto States. These areas are classified as PDI rather than MDI, because soils and rainfall are less favorable and feeder roads are less well developed.

Like many northern areas, the Derived Savanna in southern Nigeria has fairly good road connections to major markets, but agricultural intensification remains population-driven in this zone because of insufficient technical progress in the crops for which the zone has a comparative advantage. Survey data from three villages in the zone show that tree crops (cocoa and oilpalm) occupy half of the cultivated area. Other kinds of

Table 8.2 Northern Guinea Savanna, Nigeria: Importance of Maize in the Mid-1970s and 1989 (percentage of villages)

	All Villages (27)	Northern Kaduna (10)	Southern Katsina (5)	Bauchi (9)	Southeastern Sokoto (3)
Mid-1970s					
Maize is major food crop	33	30	20	33	67
Maize is major cash crop	0	0	0	0	0
1989 (group interviews in villages)					
Maize is major food crop	96	100	100	89	100
Maize is major cash crop	70	80	100	33	100
Maize is most important food or cash crop	59	90	100	22	0
Improved maize grown	100	100	100	100	100
Minimal or no local varieties grown	52	60	100	22	33
Fertilizer adoption by all or almost all farmers	81	100	100	56	67

Source: Smith et al. (1994a).
Note: Numbers in parentheses at the top of each column are Ns.

crops occupy just under one-third of the area, where the dominant enterprise is intercropped maize and cassava. Maize is not a major staple, yet farmers in the Derived Savanna say they appreciate it for providing food—often in the form of green cobs—during the "hungry period" when off-season prices can rise as much as two to three times higher than postharvest prices. Cassava is valued, however, because it provides food and cash throughout the year, so farmers' production practices emphasize cassava rather than maize. Although much of the maize area is sown to improved varieties with a significant yield advantage, maize is unlikely to become a major crop, because tree crops have a natural comparative advantage in the Derived Savanna. In 1988, the returns to land with established tree crops (intercropped cocoa and oilpalm) were 64% higher than returns to land with a maize-cassava intercrop.

Finally, the Southern Guinea Savanna represents an intermediate case between the Northern Guinea Savanna and the humid areas. Survey data from 16 villages in the zone indicated that yam was the lead crop, grown for food and cash. Maize was the major cash crop. Early maize was relayed with sorghum and planted before the yam harvest, and late maize was grown after the yam harvest. Farmers planted cassava on older plots of declining fertility, often in combination with cowpeas. Although the area had very poor feeder roads, farmers applied fertilizer to maize, although not to yam. This use of fertilizer probably occurred because the area was an important focus of the Agricultural Development Projects, and an input supply center had been set up to provide subsidized fertilizer to farmers (Oputa et al. 1983).

Results from all of these surveys are broadly consistent with recent IITA studies of key informants in all agroecologies in Nigeria. Maize area was reported to be expanding in most parts of the Northern Guinea Savanna, declining in the Derived Savanna, and showing no clear trend in the Southern Guinea Savanna. The majority of farmers throughout the Northern and Southern Guinea Savannas were reported to use improved maize varieties, whereas improved maize was used by the majority of farmers on 63% of the maize growing area in the Derived Savanna and 15% in the Humid Forest. Together, the various surveys basically confirm that maize has the greatest potential in the market-driven areas of the subhumid zones and that this potential is greater in the Northern than in the Southern Guinea Savanna. The relative importance of the Northern Guinea Savanna is likely to be greater where good transport infrastructure links it to other agroecologies and technological progress has been good. In the humid zones, maize is likely to remain a minor niche crop, serving a useful purpose but unlikely to contribute to solving the food crisis.

Sustainability of Maize-Based Systems

Even though maize production has expanded rapidly in some systems, it is by no means clear that this expansion is sustainable. The incidence and severity of threats to sustainability (summarized in Manyong et al. 1996, and Fakorede et al. 1995) are influenced by the physical environment, management practices, and evolutionary phenomena (Weber, Smith, and Manyong 1996). As land use intensifies, soil organic matter cannot be regenerated through fallowing. Soil organic matter has deteriorated severely in market-driven systems of intensified cereal cropping, where high levels of fertilizer use have led to high biomass removal, elimination of fallow periods, and reduced area in legumes. These MDI systems are also particularly susceptible to soil acidification and micronutrient deficiencies, owing to the use of high-concentration fertilizers that are cheaper to transport.

These potential problems are especially important in the Northern Guinea Savanna, where the fertilizer subsidy has encouraged high rates of fertilizer use (as high as 216 kg/ha) and reduced the incentive to use legumes as regenerating crops. The result has been a highly intensive cereal-dominated system. In the mid-1960s, legumes occupied 22% of the cultivated area, whereas cereals were planted on just over half of the area (Norman, Simmons, and Hays 1982). By 1991, the area planted to legumes had declined to 11%, and cereal area had increased to 77% of the cropped area. Intensive farm monitoring shows that indicators of the system's sustainability are closely linked to the dominance of cereal crops. Levels of organic matter, for instance, appear to decline with intensive, continuous cereal cropping and are lowest in Katsina State, where intensive maize production was initiated 15 to 20 years ago, and highest in Bauchi State, where maize is less important and short fallows still exist.

In spite of the high level of fertilizer application in the Northern Guinea Savanna, the nutrient balance for potassium and micronutrients is negative (in other words, the nutrients extracted in the form of grain and crop residues exceed the nutrients applied), indicating soil mining and a high probability of nutrient deficiencies in the medium term. Although these constraints pose no immediate threat to the cropping system's sustainability, their cumulative effect over time could be serious. Moreover, the impact of these problems on maize yields appears to be masked by the use of higher levels of fertilizer, because 61% of farmers claim yields have increased over the past five years, whereas 19% say yields have remained constant.

Weeds remain by far the most important biotic constraint, particularly in PDI systems in which land-use intensity is high and cash to pay laborers for weeding is scarce (Weber, Elemo, and Lagoke 1995). High humidity leads to severe problems with insect and disease pests in the humid zones (Bosque-Perez and Mareck 1990). In subhumid areas, the parasitic weed *Striga hermonthica* severely plagues PDI systems based on frequent sorghum cropping, whereas nematodes can be a serious problem in MDI systems based on maize cropping. On the positive side, intensively managed maize-based systems were found to suppress *Striga* and impede its adaptation to maize.

These results illustrate the complexity and dynamics of interactions between cropping system characteristics and the evolution of sustainability constraints. The implication is that it is not possible to make categorical statements about the impact of intensification and market opportunities on system sustainability. Certain constraints, such as *Striga* and nitrogen deficiency, are ameliorated in MDI systems, but others are exacerbated.

These threats to sustainability are also linked to recent changes in government policy. Recently, the Nigerian government removed the fertilizer subsidy and partially privatized fertilizer distribution. Preliminary results from group interviews with farmers in the Southern Guinea Savanna after the subsidy was removed show that maize area is declining in 39% of sample villages. Maize is being replaced by sorghum in the drier part of the Southern Guinea Savanna and by cassava and rice in the more humid areas. Farmers attribute the decline in maize cropping to the high cost and poor availability of fertilizers, as well as to *Striga* infestation.

Results from a simulation model (Freeman, Roe, and Smith, forthcoming) indicate that removing the fertilizer subsidy would lead to reduced maize production in the Northern Guinea Savanna, because maize cropping is a fertilizer-intensive enterprise. The marketed surplus of maize would decline by just over 20%, leading to decreased consumption of food, purchased commodities, and leisure. Production of all other crops would decline as well, because maize sales finance farm inputs. This decline would be sharpest among poorer farmers. If removal of the subsidy were accompanied by technical progress that increased maize yields by 10%, however, the reduction in marketed surplus would be substantially less. These results imply that increases in maize productivity can compensate for the negative effects of removing the fertilizer subsidy, as well as improve fiscal balance and reduce inefficiencies and rent-seeking activities stimulated by the subsidy.

The required technical progress in maize could be provided by hybrids. Currently, adoption of hybrids is minimal owing to uncertainties in fertilizer supply and farmers' perceptions (based on extension messages) that hybrids require higher rates of fertilizer than open-pollinated varieties (OPVs). Data on fertilizer response, however, show that the maximum yield advantage of hybrids over OPVs occurs at moderate rates of around 60 kg N/ha, which is substantially lower than the rates farmers used while the fertilizer subsidy was in effect (80–100 kg N/ha). At a rate of 60 kg N/ha, hybrids with a yield advantage of 25% over OPVs are capable of providing returns that should be acceptable to small-scale farmers. Current hybrids frequently outyield OPVs by 33 to 45%. If a consistent yield advantage is achieved and is accompanied by reliable fertilizer supplies and appropriate extension messages, hybrids could catalyze the expansion of MDI systems into other areas of the Northern Guinea Savanna where infrastructure is good. In areas where market-driven systems are already established, hybrids could compensate for the negative impact of subsidy removal on farmers' income and consumption (Smith et al. 1994b).

The adoption of hybrid maize alone would not lead to sustainable systems, however. The higher returns to hybrids would increase the dominance of cereals in the cropping system and undermine sustainability. The implication is that focusing only on the lead crop is unlikely to lead to sustainable increases in food production, because the agricultural systems of western Africa are highly complex and heterogeneous and require diversity to be sustainable. Policy and technological interventions for attaining sustainable systems must be targeted to appropriate systems based on compatibility with farmers' biophysical and socioeconomic environment.

Issues for Research

Clearly, the adoption of improved maize varieties and fertilizer has been widespread in Nigeria, and several other technologies have had a more limited impact (see the summary in Table 8.3). The classification of agricultural systems, described earlier, is now being used to identify high-risk

Table 8.3 Nigeria: Maize Production Technologies and an Assessment of their Adoption

Technology, Major Organizations Involved, and Time of Research on Technology	Problems of Adoption	Process of Adoption	Assessment of Impact
Rust-resistant maize varieties (Moor Plantation), 1950–1960	Varieties with reduced storage and processing quality; limited seed supply	Direct adoption minor; in 1950s, farmers selected moderately resistant material out of local materials and mixed crops of local and improved materials	Most farmers' maize now moderately resistant
Streak-resistant maize varieties (IITA, IAR&T, IAR), 1975–1985	Epidemics of disease are unpredictable	Adoption in areas of epidemics (the Humid Forest and Derived Savanna); spread in areas of increased yield potential and where maize was expanding as a new crop (Northern and Southern Guinea Savanna)	Resistance now widespread in maize grown by farmers; transport of maize from the Northern Guinea Savanna (mainly resistant) to markets in the Humid Forest has contributed to the spread of resistant materials in the Humid Forest
Maize varieties resistant to downy mildew disease (IITA, IAR&T), 1980–1990	Farmers in the Humid Forest zone rarely store seed; limited seed supply	Repeated multiplication and supply of seed to farmers in affected areas	Lasting effect limited, as farmers do not maintain variety
Inorganic fertilizers (IAR&T, IAR), 1970–1980	Inefficient distribution, poor infrastructure	Introduction of fertilizer-responsive improved maize in savanna zones; improved road infrastructure	Impact has been widespread
Less-acidifying nitrogen fertilizer (IAR), 1975–1985		Change in official fertilizer recommendations and in production, importation, and release	Ammonium sulfate replaced by urea and calcium ammonium nitrate
Micronutrient additions to fertilizers (IAR, IITA, IAR&T), 1980–1990	Farmers not aware of problems	Incorporation of zinc and sulfur into compound fertilizers	New fertilizer blends on market

(continues)

Table 8.3 continued

Technology, Major Organizations Involved, and Time of Research on Technology	Problems of Adoption	Process of Adoption	Assessment of Impact
Integration of legumes (many institutes and NGOs; work on soybeans by IITA, NCRI, and IAR), 1960–1990	Labor and land demands of technology reduce farmers' interest	Adoption occurred in areas where characteristics of the legume(s) fit farmers' needs	*Tephrosia* provides firewood and soil nitrogen, controls erosion (Mbila Plateau, Mid-altitude Savanna); soybeans provide cash from market (Benue State, Southern Guinea Savanna)
Animal traction (various development institutions)	Tsetse fly in the Humid Forest, the Derived Savanna, and part of the Southern Guinea Savanna restricts livestock production; fallow vegetation and trees in areas of extensive land use; lack of fodder in areas of intensive land use; sloping land in the Mid-altitude Savanna	Adoption in areas of moderate land-use intensity, often by settled pastoralists	Widespread in the Northern Guinea Savanna; market developed for renting service of oxen for ridging
Herbicides for weed control (IAR, IITA), 1965–1980	Uneconomical on small farms; selectivity of herbicides for intercropping		Restricted to large commercial farms
Minimum tillage (IITA), 1970–1980	Need for herbicides; indigenous no-till planting in some areas of the Northern Guinea Savanna		Little adoption of researcher-designed technology; indigenous method not widespread
Tied ridging (SAFGRAD, IAR), 1975–1985	Labor demand for tying ridges; risk of waterlogging	Indigenous method in areas of sloping land, sandy soils, and frequent drought stress	Common in certain areas (drier parts of the Mid-altitude Savanna; sandy soils in the Northern Guinea Savanna); little spread from areas of indigenous use

Note: IAR is the Institute for Agricultural Research, Zaria; IAR&T, Institute for Agricultural Research and Training, Moor Plantation, Ibadan; IITA, International Institute of Tropical Agriculture, Ibadan; NCRI, National Cereals Research Institute, Badeggi; NGO, nongovernmental organization (various sites); and SAFGRAD, Semi-Arid Food Grain Research and Development Project, Ouagadougou, Burkina Faso.

systems and guide the development of appropriate technological solutions. A legume database and a methodology for targeting legumes have been developed. It is encouraging that the adoption of legumes, which contribute to system sustainability, is now beginning to occur on a wider scale. The creation of a market for soybeans has stimulated widespread adoption of that crop in Benue State in the Southern Guinea Savanna (Smith, Woodworth, and Dashiell 1993), and soybeans are also moving into areas dominated by MDI systems in the Northern Guinea Savanna. Adoption of velvetbean (*Mucuna* spp.) to control weeds and improve soil fertility has occurred in PDI systems in Benin (Versteeg and Koudokpon 1993).

Progress is being made on controlling *Striga*. Seed treatments and *Striga*-tolerant maize materials are being developed for MDI systems (Kim and Winslow 1991), and sorghum transplanting is being tested in PDI systems, thus ensuring that control measures appropriate for western Africa's diverse high-risk systems are being developed. These examples illustrate that the analysis of dynamics and heterogeneity can be a powerful tool for orienting technology generation by identifying areas in which technological innovations can make a difference or in which some approaches, such as herbicides, should receive less emphasis because they are unsustainable or unlikely to be adopted.[11] These examples also indicate that new optimism about the successful introduction of resource management technologies, such as legumes and *Striga* control, may be warranted.

The important research issues for the future in the Northern Guinea Savanna are given in Table 8.4. Although MDI systems in the Northern Guinea Savanna have captured the imagination of maize technology developers, some important issues in PDI systems should not be neglected, because they have equity implications. These issues include maize varieties for planting in fields near household compounds. These fields may occupy a small area, but they could contribute to food security in PDI systems. Open-pollinated varieties that capture hybrid vigor through apomixis (nonsexual seed production) offer one approach to developing improved varieties for PDI systems. Another approach is to develop maize varieties with improved nitrogen-use efficiency that better synchronize their demand for nitrogen with the release of nitrogen from the soil.

A notable feature of Table 8.4 is that only a few items (maize with better nitrogen-use efficiency, *Striga*-tolerant maize, apomixis, and hybrids) relate specifically to maize breeding. The other potential research topics relate to other crops, resource management, and policies, but many of these topics nevertheless contribute to the productivity and sustainability of maize in the Northern Guinea Savanna. As noted earlier, research focused solely on varietal improvement for the lead crop would obviously be inadequate for strengthening the sustainability of the cropping system.

Table 8.4 Northern Guinea Savanna, Nigeria: Some Strategic Research Issues

Type of Intervention	Applicability		
	PDE Systems	PDI Systems	MDI Systems
Policies and technologies to stimulate conversion to market-driven systems (infrastructure, hybrid maize, fertilizer policy)	√	√	
Policies and technologies to stimulate land intensification in areas where land is abundant (stop penetration roads in virgin areas, improve infrastructure in established areas; improve fallow; tree crops; land titling; conversion to market-driven system)	√		
Nitrogen-efficient maize (for household compound fields)	√	√	√
Hybrid vigor captured through apomixis (in maize for household compound fields)		√	
Striga-resistant sorghum (appropriate grain quality and stover characteristics)		√	√
Striga-tolerant maize			√
Impediments to crop-livestock integration (particularly to livestock fattening)			√
Rotation with cash crop legumes to improve quality and quantity of organic matter; legume as trap crop for cereal pests			√
Rotation with food legumes to reduce nitrogen deficiency; legume as trap crop for cereal pests	√	√	
Intercropped nonfood legumes to reduce weed infestation and nitrogen deficiency		√	
Sweet cassava, sweet potato adapted to the Northern Guinea Savanna		√	
Alectra-resistant groundnut, cowpea		√	
Cereal seed treatments			√
Biological control: all agricultural systems	√	√	√
Sorghum transplanting for *Striga* control		√	

Source: Adapted from Weber and Smith (1995).

Conclusions

Our analysis of the heterogeneity and dynamics of agricultural systems in Nigeria shows that maize can make a major contribution to alleviating West Africa's food crisis but only in a limited area, primarily in the market-driven systems of the Northern Guinea Savanna. To a lesser extent, maize can also contribute to national food production in the market-driven systems of the Southern Guinea Savanna.

Most West African agriculture, however, is driven by population growth and oriented toward subsistence. Given the high cost of purchased food in areas with these characteristics, food security must be achieved by producing the major staples, which in the case of Nigeria are cassava and yam in the south and sorghum in the north. Even in population-driven systems where maize is an important staple, maize yield increases should be achieved through improved nitrogen-use efficiency (specifically, better synchrony between plant nutrient demand and soil nutrient release), early maturity, and pest and disease resistance rather than increased fertilizer response in view of the limited ability of farmers in these areas to purchase inputs.

Although maize production has increased rapidly in Nigeria over the past two decades, food requirements under alternative population and income scenarios indicate that a 4.5% annual increase in maize production will be required to meet the demand for maize for food. This increase will have to be even higher if nonfood uses of maize increase. Technologies that dramatically improve the yield potential of maize will play a central role in meeting this challenge, but as events in northern Nigeria illustrate, technologies should be targeted to appropriate areas—that is, areas where good transport infrastructure across agroecological zones exists and where the crop has a natural comparative advantage. Food crops like maize have a particular advantage in this respect relative to nonedible cash crops like cotton, as their dual role as food and cash crops is highly attractive to farmers in areas where prices of agricultural commodities tend to vary considerably.

The key elements in the maize transformation in the Northern Guinea Savanna have been improved varieties, fertilizer, and improved infrastructure. The fertilizer subsidies designed to jump-start the Nigerian maize revolution, however, have contributed to fiscal problems, inefficiencies, corruption, and an unsustainable predominance of cereal cropping. Without the subsidies, the agricultural transformation of the Northern Guinea Savanna would probably have been less spectacular but more sustainable. Government expenditure on agricultural research, extension, and infrastructure rather than on subsidized fertilizer is likely to yield better returns.

Our analysis has suggested that a maize seed-fertilizer strategy, although providing rapid gains in the short to medium term, will be inadequate by itself to sustain production in the long run, even in high-potential areas. We have shown, for example, that in the maize-based system in the Northern Guinea Savanna of Nigeria, greater dominance of cereal cropping could drive the system to become unsustainable. An increase in the fertilizer responsiveness of maize through the introduction of hybrid maize could worsen the situation. Accompanied by complementary technologies and policies derived from an understanding of system dynamics, however, hybrid maize could make a valuable contribution to sustainable production systems.

Even though we have uncovered potential threats to the sustainability of maize-based systems in the Northern Guinea Savanna, we believe West Africa's agricultural potential can be unlocked by using a very different approach from that which worked in the favorable and homogeneous areas of Asia. Production systems in western Africa, even intensive ones, require diversity to be sustainable. Technological priorities need to be developed within the context of a holistic vision of the agricultural systems in the mandate area. Clearly, success depends to a great extent on the ability to recognize and characterize the heterogeneity of systems, to understand their dynamics, and to develop an awareness of how these dynamics interact with the evolution of threats to sustainability.

Notes

1. Two zones are excluded from this analysis, which focuses primarily on the humid and subhumid areas: the dry Sudan Savanna, which lies beyond the Northern Guinea Savanna and produces relatively little maize, and the more temperate Mid-altitude Savanna, which occupies relatively little area (about 2%) in western Africa.

2. This problem could be overcome by the development of long-season maize materials if an appropriate, deep-rooted legume intercrop can be identified.

3. A niche crop exploits spatial or temporal niches in a cropping system to meet particular needs of farmers.

4. The average marketing margin (defined as the difference between farmgate prices and retail prices in rural markets) for cereals in a market-driven area of the Northern Guinea Savanna was 7 to 12%, whereas it was 34% in a population-driven area (Freeman, Roe, and Smith, forthcoming).

5. Even in PDI, however, maize can be grown on fields near the family compound to take advantage of the household refuse applied to these fields.

6. Most of this maize is grown in the Northern Guinea Savanna; as noted previously, the Sudan Savanna produces very little maize.

7. The price of food commodities relative to nonfood commodities was 80% higher in the early 1980s than in the mid-1970s (Lele, Christiansen, and Kadiresan 1989).

8. Production figures from Nigeria should be treated with caution because of their unreliability (Lele, Christiansen, and Kadiresan 1989).

9. Because market-driven systems in the expansion phase occupy only 3% of the area in western Africa and are virtually nonexistent in Nigeria, they are excluded from the rest of the analysis.

10. See Norman, Simmons, and Hays (1982).

11. For more detailed information see Smith and Weber (1994); Weber and Smith (1995); and Weber, Smith, and Manyong (1996).

Part 3
Technology, Institutions & Policy

9

The Technological Foundation of the Revolution

Derek Byerlee & David Jewell

The foundation for renewed growth in Africa's agricultural sector is investment in cost-effective, relevant research and extension systems that can generate rapid and broadly based technical change in staple food crops, including maize. The development of such systems is a long process, however, and focused research efforts on food crops, especially for smallholders, are a recent undertaking in most African countries. This chapter reviews the achievements of Africa's research and extension systems with respect to smallholder maize production and examines the challenges that remain to be addressed. We first summarize evidence on the adoption of maize technology, particularly improved seed, and discuss the technological constraints on increased maize production in Africa. Next, we review the progress that has been made over the past two decades in building research and extension systems. We conclude by discussing ways of fostering a research and development (R&D) capacity that will keep new technologies flowing to Africa's extension workers and small-scale farmers in the decades to come.

Evolution of Maize Research Programs in Africa

Maize research began in colonial times in Africa and focused mostly on improving maize for large-scale European farmers. Work on hybrid maize was initiated in 1932 in Zimbabwe; Kenya followed in 1955. Zaire and a few other countries established maize research programs for smallholders in the 1930s, but these efforts remained fairly modest until recently in most African countries in relation to research on cash crops and research for commercial farmers (Miracle 1966). At independence in 1960, research systems were generally small, and 90% of the researchers in Africa were still expatriates (Pardey, Roseboom, and Anderson 1991). Moreover, few scientists had training in, experience with, or orientation to smallholder agriculture.

In the decades since independence, investment in public national agricultural research systems (NARSs) has expanded more than fourfold (Pardey, Roseboom, and Beintema 1997). Expansion was particularly rapid until 1980, but research budgets have stagnated or declined ever since. The number of scientists has continued to rise, however, resulting in a sharply reduced expenditure per scientist (Pardey, Roseboom, and Beintema 1997).[1] Data from a 1991 survey show that about 272 maize scientists worked in public-sector maize research programs in Africa, and an additional 31 worked in the private sector (Table 9.1). The intensity of maize research in Africa, measured in scientists per million tonnes of maize, is still less than that in other regions of the developing world. The fact that most African countries have several maize production environments in a relatively small maize area compounds this resource constraint.[2]

The national maize research effort in Africa has been complemented by the efforts of the international agricultural research centers (IARCs), such as the International Institute of Tropical Agriculture (IITA), which launched a maize research program in western Africa in 1970. The International Maize and Wheat Improvement Center (CIMMYT) initiated maize research in eastern and southern Africa in the late 1970s and established its regional maize research station in Zimbabwe in 1985. In the mid-1980s, the IARCs spent an estimated 43% of their U.S.$11 million maize research budget in Africa (Gryseels and Anderson 1991).

Table 9.1 Africa, Asia, and Latin America: A Comparison of Maize Research Resources and Varietal Releases, 1991

	Africa	Asia[a]	Latin America[a]
Number of countries surveyed	20	9	12
Maize area per country, 1989–1991 (million ha)	0.67	1.86	0.96
Number of maize megaenvironments per country[b]	5.4	5.7	4.4
Number of public-sector maize researchers			
Breeders	86	175	182
Nonbreeders	186	240	87
Percentage of public-sector researchers holding M.S. or Ph.D. degree	51	60	39
Number of private-sector maize researchers	31	125	108
Percentage of maize breeders in the private sector	27	42	37
Total number of public and private maize breeders per million t of maize	6.7	10.7	13.0
Number of maize varieties/hybrids released per million ha maize, 1966–1990	25	11	30

Source: CIMMYT maize impacts data, 1992.

Notes: a. Excludes Brazil and China, the two largest maize producers in the developing world. Maize is used mostly as a commercial crop for livestock feed in Brazil and China.
  b. Breeding environments as defined by CIMMYT.

The Impact of Maize Research in Africa

Release and Diffusion of Improved Varieties

Maize research in Africa resulted in the release of over 300 improved varieties and hybrids by national maize research programs between 1966 and 1990. African farmers can now choose from about as many open-pollinated varieties (OPVs) and hybrids as farmers in other regions of the developing world (Table 9.1). The diversity of production environments in Africa, however, means that additional varieties still need to be developed for some agroecological niches (e.g., Hassan, forthcoming).

Although most countries have conducted maize breeding programs for at least 20 years, their success in delivering appropriate varieties and hybrids to a wide spectrum of small-scale farmers has varied. Improved maize is grown on as little as 20% of the maize area in Ethiopia and on nearly 100% of the area in Zimbabwe (Table 9.2). The relative emphasis on hybrids versus OPVs has varied as well, changing markedly from eastern and southern Africa to western and central Africa. The first breakthrough with hybrid maize for smallholders occurred in Kenya in the early 1970s (Gerhart 1975). By the mid-1970s, most small-scale farmers in Kenya's more favorable highland areas grew hybrids originally developed for commercial farmers. In the 1980s and 1990s, hybrids spread slowly to less favorable areas of the country (Hassan, forthcoming). Smallholders had similar success with hybrid maize in Zimbabwe, Lesotho, Swaziland, and Zambia in the 1980s and Malawi in the early 1990s. The use of hybrids in eastern and southern Africa reflects the fact that most maize research programs were established to serve commercial farmers.

In contrast, research in western and central Africa emphasized the development of improved OPVs of maize specifically tailored to smallholders' needs. Open-pollinated varieties are appealing as an appropriate technology for small-scale agriculture because farmers can save the seed from year to year. These varieties were rapidly adopted in the savanna of western Africa (Table 9.2), especially Nigeria and Ghana, as described in Chapters 7 and 8. Open-pollinated varieties are dominant in western and central Africa for two reasons. First, the tropical lowlands (usually defined as areas less than 900 m in altitude) constitute the major ecology for maize production in the region, and hybrids for this ecology became available only recently. Second, the main source of improved tropical germ plasm in this region has been the IARCs, which have bred mostly OPVs.

The overall record of adoption of improved maize varieties and hybrids in Africa is impressive. Almost half of the maize area in eastern and southern Africa and one-fifth of the area in western and central Africa is planted to improved OPVs and hybrids (Byerlee and Heisey 1996). The

Table 9.2 Sub-Saharan Africa: Maize Area Planted to Improved OPVs and Hybrids, 1990

Country	Total Maize Area (000 ha)	% Area Under Improved OPVs		% Area Under Hybrids	% Area Under OPVs and Hybrids	
		Minimum[a]	Maximum[a]		Minimum[a]	Maximum[a]
Tanzania	1,631	6	18	6	12	24
Nigeria	1,500	22	87	2	24	89
Kenya	1,500	8	8	62	70	70
Malawi	1,344	3	3	11	14	14
Zimbabwe	1,150	0	0	96	96	96
Ethiopia	1,050	8	24	5	13	29
Mozambique	1,015	17	17	1	18	18
Zambia	763	5	5	72	77	77
Côte d'Ivoire	691	14	42	4	18	46
Ghana	465	16	48	0	16	48
Benin	454	9	27	1	10	28
Uganda	389	30	70	10	40	80
Togo	296	7	18	3	10	21
Burkina Faso	216	15	70	2	17	72
Cameroon	200	20	67	1	21	68
Mali	170	36	50	0	36	50
Lesotho	145	12	12	70	82	82
Burundi	124	5	20	0	5	25
Senegal	117	100	100	0	100	100
Swaziland	84	0	0	90	90	90
Africa[b]	14,500	11	26	23	34	49

Source: CIMMYT maize impacts data.

Notes: a. Minimum = area usually based on seed sales; maximum = area based on surveys or breeders' estimates.

b. Excludes over 1 million ha of maize in Zaire, Angola, Somalia, and Namibia, which were not covered by the survey.

adoption of improved maize varieties and hybrids has progressed almost as quickly in Africa as in Asia and Latin America, especially if developing countries with large commercial or irrigated maize sectors are excluded (China, Argentina, and Brazil).

Despite these past successes, maize breeding programs have often taken insufficient account of the special circumstances of small-scale farmers when setting research priorities, resulting in low adoption. For example, smallholders in much of Africa plant maize late because of labor constraints, risk considerations, and crop rotations (Low and Waddington 1990). These farmers generally prefer maize varieties that mature earlier than the materials grown by commercial farmers. In other cases, farm households can improve their food security by planting early-maturing maize that can be consumed in the "hungry season" before the main harvest (Low 1988). Recent research has addressed this requirement, and early-maturing maize materials are being released and adopted in areas where they had previously been considered unnecessary. The productivity of late-planted maize can also be increased through crop management practices for late-planted fields. In the past, agronomic research has focused on developing practices, such as fertilizer doses and timing of application, for fields planted at the optimum time (Low and Waddington 1990).

Maize breeding programs have also given insufficient attention to postharvest issues such as grain quality, storability, small-scale processing requirements, and consumer preferences. For instance, farmers in many areas of southern Africa prefer the harder flint maize varieties because they have much lower on-farm storage and processing losses than the available soft dent hybrids (Chapter 5). Increasingly, maize breeders are responding to these preferences. Matuba, an OPV recently released in Mozambique, has gained wide acceptance because its home consumption characteristics and the length of its growing cycle meet farmers' requirements (Sperling et al. 1995). However, since the cost of developing a variety increases with the number of selected traits emphasized in selection, however, research priorities must be selected carefully (Chapter 10).

Adoption of Improved Crop Management Practices

Even in countries where improved varieties cover much of the maize area, only modest yield gains have been achieved. Although the use of improved maize can be a catalyst for increasing farmers' use of other inputs, especially fertilizer (Smith and Goldsmith 1995), such broad-based change has occurred only in some parts of Africa. For example, maize area and yields expanded rapidly in the west African savanna following the release of suitable varieties and associated improvements in infrastructure and input supply (Chapter 8). But the more common experience has been that most farmers fail to adopt the additional production practices needed for sustained improvements in maize yields. This is particularly true of practices for maintaining and enhancing soil fertility, despite the fact that the demise of the bush-fallow system has made poor soil fertility the major constraint on maize productivity in much of Africa (Chapter 11).

Farmers grow improved varieties without fertilizer in many areas of Africa—especially in marginal areas, such as the drier zones of Kenya and Zimbabwe, but also in some relatively favored areas, such as Ghana (GGDP 1991). Higher adoption rates for improved seed reflect the greater availability of maize seed compared to fertilizer and the fact that improved varieties can perform better than local varieties under low-input conditions (i.e., without fertilizer).[3] Also, research on fertilizer use has emphasized a single recommendation for wide areas, which does not account for the diversity of smallholder situations and the acute cash constraints under which they operate. In addition, for many years agricultural researchers emphasized chemical fertilizer almost exclusively, at the expense of research on alternative means of maintaining soil fertility from internal sources of nutrients such as green manure, alley cropping, and animal manure. It is important to seek a balanced approach to improving soil fertility that combines both organic and inorganic sources of nutrients.

Research has also been conducted on a number of other management practices, which farmers have adopted on a limited scale. Small-scale farmers often reject recommendations for labor-intensive practices such as precise plant spacing, frequent weeding, and separate operations to apply fertilizer; consequently, the profitability of other elements of the recommended package is also affected. Experience from many countries has shown that seasonal labor availability is an important constraint on the acceptance of improved management practices such as plant spacing and weeding.[4] Even where land is in short supply, seasonal labor shortages often decisively influence farmers' choice of technology for several reasons: Hand-hoe agriculture demands a great deal of labor, off-farm work is important in many areas, and a pool of landless rural laborers is not available when demand for labor is greatest (Low 1988). It is therefore critical to evaluate new technologies in terms of their effect on the returns to labor.

Reflecting this labor constraint, farmers in the savanna of western Africa and much of southern Africa and Ethiopia have adopted animal traction in maize-based systems. A seasonal draft power constraint often emerges, however, because animals are in short supply or poor condition during the peak demands for land preparation (Collinson 1987). This situation has led to efforts to develop technologies to reduce tillage. Extensive work on reduced tillage in Zimbabwe has demonstrated the potential of this approach. It is still too early to assess success in terms of farmers' adoption, however, because farmers must also use chemical weed control (Shumba, Waddington, and Rukuni 1992) and retain crop residues, which have considerable economic value as forage—especially in drier areas (McIntire, Bourzat, and Pingali 1992).

There is little doubt that research on crop and resource management to overcome seasonal labor constraint and maximize returns to cash inputs while conserving the soil base and enhancing soil fertility over the longer run will go a long way toward increasing productivity and sustainability of maize-based systems (Lynam and Blackie 1994). Research on these constraints is increasing, but success, measured in terms of adoption, remains rare. Even when such technologies become available, special efforts will be required to transfer and adapt them to local farmers' situations, given that many crop and resource management strategies are fairly knowledge-intensive. In short, progress with crop and resource management research has been and will continue to be slow, because technologies will have to be adapted to specific situations and will be influenced by agroclimatic circumstances, population pressure, labor availability, and the stage of infrastructural and institutional development.

Returns on Investment in Maize Research

The adoption of improved maize has had a significant impact on maize production in Africa. The use of hybrids has increased farmers' maize yields by at least 40% in favorable areas. The evidence suggests that even in dry areas and drought years (such as 1991–1992), hybrids possess at least a 30% yield advantage over traditional varieties (Byerlee and Heisey 1993). The yield advantage of improved OPVs over local traditional varieties is 15 to 25% in tropical areas (Morris, Clancy, and López-Pereira 1992). Adoption of improved varieties and hybrids with these yield gains on over 40% of the area may account for over half of the annual growth in maize yields in Africa since 1970. In addition, the availability of early-maturing varieties and hybrids has enabled maize area to expand, especially in the drier savanna areas.

Several recent studies have evaluated the returns on investments in maize research in Africa (Table 9.3). Five of the seven studies concluded that the investment in maize research had provided high annual rates of return on investment, usually in excess of 40%. Not surprisingly, Kenya and Zimbabwe, whose success with maize research is well-known, feature among the countries showing high returns on investment (Karanja 1990; Kupfuma 1994). One of the highest returns, however, occurred in Mali, a relatively small-scale maize producer where low-cost adaptive research to screen varieties combined with good agronomic practices, input supply systems, and markets helped foster rapid growth in maize production (Boughton and de Frahan 1994).

Table 9.3 Sub-Saharan Africa: Summary of Estimates of Returns to Investment in Maize Research

Region and Country	Time Period	Rate of Return (%)	Source
Eastern Africa			
Kenya	1955–1988	40–60	Karanja (1990)
Uganda	1985–1996	< 0–15	Laker-Ojok (1994)
	1985–2006	27–58	
Southern Africa			
Malawi	1953–1992	4	Smale and Heisey (1994)
	1977–1992	63	
Zambia	1978–1991	100[a]	Howard (1994)
	1978–1991	< 0[b]	
Zimbabwe	1932–1990	44	Kupfuma (1994)
Western Africa			
Ghana	1968–1991	74	Sanders, Bezuneh, and Shroeder (1994)
Mali	1969–1991	135	Boughton and de Frahan (1994)
	1962–1991	54	

Notes: a. Includes only research and extension costs.
b. All costs, including additional marketing costs.

The studies that break down returns to research by time period, however, demonstrate the sensitivity of the results to the time period used (Oehmke and Crawford 1996). Thus Uganda, which has experienced low returns to maize research to date, is projected to experience high returns in the future, given current adoption trends. On the other hand, the high returns in Kenya and Zimbabwe are based on research that was conducted in the 1960s. Many farmers continue growing hybrids developed in the 1960s, because recent research has failed to deliver new streams of better-performing hybrids.

Potential for Future Productivity Gains
Through Breeding Research: Strategic Issues

A strategic decision confronting all research programs for small-scale maize farmers is the relative emphasis to give to developing hybrids or OPVs. Recent evidence suggests that hybrids can perform as well as or better than local varieties and OPVs under low-input conditions (see Chapter 5). Nevertheless, many contend that hybrid maize is inappropriate for small-scale farmers, because hybrid seed must be purchased annually, requiring that farmers have cash at planting time and a reliable local supply of quality seed. Experience suggests that some countries in eastern and southern Africa have done a good job of providing hybrid seed at competitive prices *even to small-scale farmers,* although these countries generally have good infrastructure.[5]

On the other hand, OPVs have had less of an impact than expected, because low profit margins in producing OPV seed have discouraged private seed suppliers from investing in the industry. In most cases, seed of OPVs has been distributed to farmers through various donor-assisted projects, especially in Nigeria, Zaire, Mozambique, and Ghana. Most of these projects provided a one-time injection of new seed instead of developing a seed industry capable of regularly replenishing seed. Some countries are now experimenting with small private seed companies (often family owned), private nongovernmental organizations (NGOs), or farmers' cooperatives, combined with active support from the public sector to provide credit and training in seed production for OPVs (Cromwell 1996). It is too early to assess whether these experiments will result in viable small-scale seed industries for OPVs.

A second major issue related to future productivity gains is the extreme instability of maize yields in many maize-producing regions of Africa (Chapter 2). Unstable yields are largely the result of weather conditions, although disease and pest problems play a role as well. Given that maize production in some countries has shifted from large-scale to small-scale farms and to more marginal production environments, yield instability may be increasing. This raises the issue of whether breeders should place more emphasis on improving yield stability and less on increasing yield potential, especially since farmers' yields are so much lower than the genetic potential.

Considerable progress has already been achieved in stabilizing maize yields by breeding for resistance to disease and insect pests (especially maize streak virus). Work on drought tolerance also promises to alleviate (although not eliminate) the effects of the periodic severe droughts that affect even the relatively favorable maize-producing environments, especially in southern Africa (Edmeades, Bolaños, and Lafitte 1992).

Biotechnology has considerable potential to enhance the efficiency of conventional breeding programs and to enhance yield stability by incorporating genetic resistance to insects and specific herbicides in maize germ plasm and tolerance to the parasitic weed *Striga,* which is becoming a major problem in many maize production systems. The major issue for Africa is how to gain access to this technology through the private sector, the IARCs, or the development of local research capacity. Any successful strategy for benefiting from the emerging biotechnologies will probably involve all three sources, but only a very few countries, such as South Africa, have the resources to establish their own biotechnology laboratories. Such laboratories, working in collaboration with the IARCs and private companies, can become regional centers for adapting new biotechnologies. Even so, the promise of biotechnology is becoming a reality much more slowly than earlier forecasts suggested (McCalla 1994).

To summarize, conventional maize breeding should continue to provide high payoffs to farmers, particularly in terms of the development of more locally adapted materials with more stable yields. The good potential for continued progress in developing new varieties and hybrids suggests that investments in maize breeding should be sustained at current levels. At the same time, increased attention is needed for research on soil and water management, soil fertility, and weed control (Blackie 1994a). The potential long-term payoff to research on these constraints is undoubtedly high, but realizing that potential will require a long-term, well-focused multidisciplinary effort on specific crop and resource management problems.

Revitalizing Maize Research Capacity

It is a fact of life that the productivity increases that will occur 10 to 20 years from now will originate with today's research investments. It is also a fact of life that sustained and rapid technical progress will not occur without a strong local capacity for generating and promoting new technologies over the long term (Lynam and Blackie 1994). Much of the success of hybrid maize in Kenya and Zimbabwe can be attributed to the remarkable continuity of the maize breeding programs in these countries. Over a span of 56 years in Zimbabwe (1932–1988), four senior maize breeders led a low-cost research program on hybrid maize (Eicher 1990). In contrast, Malawi's Department of Agricultural Research has had four directors in the past 10 years (Rukuni 1996).

Problems with research continuity are just one symptom of the deepening financial crisis that has engulfed many public research systems in Africa over the past decade. Donors now support over 40% of the total research expenditures in Africa (not including research at the IARCs), and the proportion of research supported by donors is much higher in many individual countries (Pardey, Roseboom, and Beintema 1997). Most of this support, however, has been earmarked for specific projects, with the result that donor-supported activities at the periphery of the research system have burgeoned, often to the point of dominating core research activities (Spurling et al. 1992).

The funding crisis has been aggravated by a management crisis in many research systems, so that available resources have not been effectively used. Many research institutions have been unable to articulate research priorities, provide incentives for scientists, or develop appropriate external linkages with farmers, agribusiness, and political leaders (Eicher 1989). As a result, NARSs have often been marginalized from mainstream agricultural development.

A number of measures are being taken to revitalize NARSs in Africa, including the development of research strategies and priorities, greater autonomy for public research institutes outside of the civil service, "rationalization" of the number of research personnel and experiment stations (in many cases resulting in significant downsizing), and collaborative programs to conduct research on a regional basis and also to coordinate research closely with the IARCs. It is still too early to judge the success of these initiatives. A recent evaluation of programs in six countries suggests, however, that significant progress has been made in restructuring and reorienting NARSs, implying that policymakers are giving greater attention to agricultural research than in the recent past (SPAAR 1995). Both African governments and donors will need to make an extended commitment if they wish to build the kind of research establishment that can serve African smallholders into the next century.

Given the small size of most countries in Africa, the role of regional and international research is particularly important. Regional research programs established during the colonial period did not survive independence, although the IARCs—nearly all of which operate in Africa—have partially substituted for regional research institutes. By the 1980s, over three-quarters of the maize varieties released in Africa were based, at least in part, on germ plasm provided by the IARCs.

Because most countries lack sufficient maize area to justify a comprehensive maize research program, they can benefit from collaborating with neighboring countries, especially those with similar ecoregions, and by actively importing and screening available technologies from elsewhere. Fortunately, the role of regional research collaboration is being recognized, and several regional organizations with strong NARS participation are in operation.[6]

The acute scarcity of public funds for agricultural research has also elicited repeated calls for privatizing research and outsourcing research to universities and private research firms.[7] Because hybrid seed technology allows appropriation (through trade secrets) of the benefits of research on hybrid maize, the private sector has become a major global player in maize improvement research. It is estimated that the private sector accounts for about half of the maize R&D expenditures in Latin America and Asia. In Africa, private-sector involvement in maize research is still in its infancy, outside of South Africa and Zimbabwe. The global experience reveals that effective seed delivery systems for small-scale farmers have been developed through a combination of public-sector research and private-sector seed production and distribution (Byerlee and López-Pereira 1993). Once the seed market is well established, the applied research needed to develop new hybrids and varieties will gradually be provided by the private sector (Chapter 12).

The Evolving Role of Extension

Without doubt, maize farmers have been major beneficiaries of the expansion of national extension systems. Extension was a driving force behind the diffusion of improved maize technology in all of the country studies presented in this book. For example, during the 1960s and 1970s, tens of thousands of demonstrations were laid out in Kenyan farmers' fields to advertise the benefits of hybrid maize and associated management practices. This contact with extension was important in farmers' decisions to adopt hybrid maize technology (Moock 1981; Karanja 1990). Despite these successes, management problems arose as the number of extension staff increased and operating budgets for travel and farm visits decreased.[8] In addition, the messages and recommendations promoted by research and extension were often inappropriate for smallholders, especially resource-poor farmers lacking good access to markets.

General disenchantment with extension led to three major experiments in the 1980s and 1990s to strengthen both the management and the relevance of extension services: (1) the on-farm or farming systems approach to research and extension, (2) the Training and Visit (T&V) extension system, and (3) the Sasakawa-Global 2000 (SG 2000) extension demonstration approach.

On-Farm and Farming Systems Approaches

The farming systems approach to research and extension (FSR/E) was developed in the late 1970s to assist researchers and extension workers with developing technologies for African smallholders. The approach, which gained wide popularity in Africa in the 1980s, featured on-farm participatory surveys and experimental methods to involve farmers in technology development and diffusion. The FSR/E explicitly incorporated elements of smallholder farming systems that influence the acceptance of new technologies, such as intercropping, seasonal labor bottlenecks, draft power and cash constraints, and the management of family food security over the year (Collinson 1987). Another hallmark of FSR/E was the recognition that farmers, even within a given agroecological zone, differ considerably with respect to resource constraints, household objectives, and managerial capacity (Low 1994). Researchers sought to identify target groups of farmers ("recommendation domains") encompassing relatively homogeneous agroclimatic zones and socioeconomic circumstances for the purpose of developing more location-specific recommendations.

The 1980s saw a tremendous effort to build capacity in FSR/E within national research programs and universities. Much of the experience with FSR/E was based on work in maize-based farming systems (Collinson 1987). Training programs gave thousands of researchers and extension agents a valuable understanding of African smallholder farming systems and the need to develop strong linkages among commodity researchers (e.g., plant breeders), social scientists, extension workers, and farmers so technology could be better designed to fit the needs of target farmers.

The FSR/E approach did not meet early expectations that it would dramatically increase farmers' adoption of technology for major food crops (Heisey and Waddington 1993). The major constraints were on implementing the FSR/E approach, poor research-extension linkages, the lack of effective input delivery systems, and policy-induced price distortions. Where effective linkages were developed with extension, as in Ghana in the 1980s, adoption was impressive (Chapter 7). Finally, FSR/E put too much emphasis on developing technologies for the existing policy environment (e.g., lack of reliable input supply) at the expense of focused efforts to change that environment.

The Training and Visit System

The T&V extension system was implemented at the initiative of the World Bank to reform the *management* of extension systems, although in most cases adoption of the T&V approach implied some expansion in the number of extension workers. Described as "a hierarchical organized method of extension management designed to exclusively focus on technology and to deliver selected and timely messages to farmers with strict regularity" (World Bank 1994b), the T&V system trained village extension workers and provided the means for regular meetings with contact farmers. Contact farmers, in turn, were expected to relay specific crop production recommendations to farmers in their villages.

The T&V approach spread rapidly throughout Africa in the 1980s with the assistance of loans from the World Bank. To date, 27 countries in Africa have implemented the approach; in almost all cases, T&V projects have helped extension agencies to reach greater numbers of farmers (Cleaver 1993a). An evaluation of the T&V approach in Kenya and Burkina Faso concluded that it successfully encouraged the adoption of technology and enhanced farmers' productivity. In Kenya, farmers who had contact with extension workers obtained significant increases in maize yields, and the rate of return on the extension investment was found to be high (Bindlish and Evenson 1993). A more recent study concluded that the T&V approach had a favorable impact on farmers' adoption of new maize technologies in Kenya, especially small-scale farmers (Hassan, forthcoming).

Nonetheless, there has been considerable debate about the fiscal sustainability of the T&V approach after donor aid has been terminated. A World Bank evaluation found that at least half of the extension projects in Africa were rated "unsatisfactory" (World Bank 1994b). The study identified the following limitations with implementing the T&V approach.

- A rigid model was applied without sufficient attention to the variation in historical, cultural, economic, and institutional factors among and within countries
- Problems of financing recurrent costs threatened the long-term sustainability of extension reforms
- A top-down approach to delivering extension messages was reinforced, and messages were often based on standard packages of recommendations that ignored the heterogeneity among farmers
- Technologies appropriate to the circumstances of small-scale farmers were scarce, especially for resource-poor regions
- Linkages between research and extension were weak

A second generation of T&V extension projects in the late 1980s and the 1990s has sought to overcome these limitations by introducing greater flexibility to meet local needs. Obtaining sustained financing for extension programs remains a major challenge, however. As in Asia, where T&V was first introduced, many African countries have sought to develop more cost-effective and participatory approaches to extension. For example, in Malawi an extension manager recently reported that the T&V model was modified to suit local resource availability (Rukuni 1996). The World Bank, the major "donor" for extension, is currently reviewing its involvement in extension in Africa with a view to formulating a new strategy.

The SG 2000 Extension Approach

The extension program of SG 2000, an NGO, was launched in Ghana in 1986 to demonstrate that given access to available technology, small-scale farmers can dramatically increase yields of staple food crops. SG 2000 assists extension workers in the Ministry of Agriculture to conduct thousands of large (0.5 ha) demonstrations on farmers' fields to show the potential of a new technological package for raising crop yields (Borlaug and Dowswell 1995). SG 2000 also supplies credit on a revolving basis to ensure that components of the package, especially seed and fertilizer, are available to farmers. Recently, SG 2000 projects have broadened the range of interventions to include draft power and implements, on-farm grain storage, and agroprocessing technology. In 1996, SG 2000 projects were underway in 12 countries: Benin, Burkina Faso, Eritrea, Ethiopia, Ghana, Guinea, Mali, Mozambique, Nigeria, Tanzania, Togo, and Uganda.[9]

The decade-old SG 2000 project in Ghana has claimed the most success. The extensive coverage of on-farm demonstrations was undoubtedly a major factor in Ghanaian farmers' widespread adoption of maize seed-fertilizer technology (Chapter 7). The SG 2000 program has also experienced some success in extending maize technologies in Tanzania (Putterman 1995) and in convincing high-level African political leaders to promote smallholder agriculture more actively (Borlaug 1996). Although active dialogue with senior policymakers is a strength of the approach, SG 2000 has nonetheless found it difficult to mobilize ministries of agriculture to sustain the effort and ensure efficient input delivery after the demonstration phase is over (Eicher 1988; Tripp 1993; Farrington 1995; Putterman 1995; and Jiggins, Reijntjes, and Lightfoot 1996). Also, the dependence of earlier efforts on one or a few technological components was often inappropriate for meeting the diverse needs of African smallholders. Nonetheless, the SG 2000 country projects have demonstrated that there is considerable potential to increase maize yields and have served as a reminder that rapid adoption of new technologies is possible in Africa when relevant technology is combined with appropriate economic policies and markets.

Looking to the Future: Institutional Issues in Extension

The three extension initiatives discussed earlier have been valuable testing laboratories for improving the relevance of research and extension and increasing the rate of diffusion of new technology. Components of each model have been incorporated into most African extension services, but to date none of the models has been institutionalized within public extension services with financing by African governments. Rukuni (1996) recently reviewed the experience in institution building for research and extension in southern Africa and concluded that "prepackaged institutions" have not proven to be fiscally sustainable. He emphasized the importance of experimenting with pluralistic and demand-driven research and extension models for small-scale farmers rather than trying to import "successful models" from elsewhere.

The major issues now facing extension programs are how to sustain the gains that have been achieved in light of continuing budget shortfalls and how to introduce institutional reforms that make extension systems more cost-effective and demand driven. The widely lamented gap between research and extension is being addressed through the design of more integrated research and extension projects and more emphasis on farmer-led approaches to extension (Scarborough 1996). Farming systems diagnosis receives greater attention in T&V-based extension systems, and more attention is being given to reaching women, who are important but neglected clients of most extension systems.[10]

These approaches, however, although a step in the right direction, do not give rise to demand-driven systems. A major issue now is how to involve farmers in financing and governing pluralistic extension systems that involve collaboration among the traditional public-sector system, NGOs, and the private sector (e.g., seed and fertilizer dealers). For example, NGOs, which often have an advantage in articulating grassroots demands, are rapidly expanding their role in natural resource conservation. To this end, greater emphasis is being placed on strengthening local farmer organizations with the expectation that they will eventually be able to manage and finance at least some of the local costs of extension programs.

Conclusions

The evidence marshaled in this chapter reveals that investments in research and extension have generated some impressive achievements in maize production. Farmers are now growing improved varieties and hybrids on 40% of the maize area. Adoption of improved maize varieties in Africa compares favorably with Asia and Latin America, and rate of return studies show that public investment in maize research has produced high returns.

But Africa's maize success story has some important qualifiers. First, the use of improved maize varieties has been patchy, concentrated in fewer than 10 countries. Some large maize producing areas have scarcely benefited from improved maize varieties. Identical maize production packages have yielded different results in neighboring countries owing to variations in farmer support services (often because of infrastructure) and pricing and input supply policies. A second qualifier to the maize success story is that farmers have not adopted the complementary improvements in cultural practices that would enable them to exploit the potential of improved maize seed. As a result, widespread use of improved maize varieties has not caused national maize yields to rise as expected. Third, building sustained institutional capacity to conduct effective research and extension for smallholders remains an elusive goal in most countries. National technology systems have been unable to garner reliable domestic political and financial support, and dependence on foreign aid to support these systems is increasing.

It is now clear that improved maize seed alone cannot provide the impetus needed for Africa's emerging maize revolution to fulfill its promise in smallholders' fields. Concerted efforts are needed to improve crop management. The organization of research on critical crop management problems, such as declining soil fertility, will require multidisciplinary collaboration over many years by research teams closely linked to extension, NGOs, policymakers, and input delivery agencies. It is important to be realistic, however, about the long time frame needed to achieve payoffs from this research, given the complexity of the task and the management intensity of the technologies that are likely to be required.

Extension programs have been central to transferring maize technologies to African farmers. Many extension programs have favored maize over other crops, but the dearth of appropriate technologies for smallholders has meant that the impact of extension has been less widespread or comprehensive than it might have been.

Weakened by crises in management and funding, research and extension systems have reached an important crossroads. Ways must be found to improve operating budgets, enhance efficiency, and integrate the efforts of a wider array of entities from the public sector, the private sector, NGOs, and—especially—farmers into both technology development and dissemination. In recent years, African governments and donors have been experimenting with new institutional models for research and extension. These experiments have made important contributions but have yet to be politically and financially sustainable. Nor is a single model adequate for the diversity of situations in Africa.

Notes

1. Pardey, Roseboom, and Beintema (1997) estimated that agricultural research expenditures in Africa increased in real terms by 6.6% annually in the 1960s and by 3.7% in the 1970s and declined by 0.3% in the 1980s.

2. Twenty-eight African countries are significant maize producers (that is, production surpasses 100,000 t of maize each year), and maize production in these

countries averages 700,000 t per country. (Rice production in Asia, however, averages 10 million t per country, even when India and China are excluded from the calculations.)

3. Results from hundreds of on-farm demonstrations in Malawi revealed that hybrid maize without fertilizer yielded significantly better than local varieties, even in a severe drought year (Byerlee and Heisey 1993).

4. Increased plant density and line planting have been adopted by farmers, as long as these practices do not conflict with seasonal labor demands and intercropping systems (e.g., GGDP 1991).

5. In Tanzania, where only 10 to 15% of rural roads are passable throughout the year, a large number of farmers who adopted hybrid seed through the SG 2000 extension program had to recycle their seed the next season because of the lack of a reliable supply of hybrid seed.

6. For example, a maize improvement network was recently established in southern Africa to promote collaboration and flows of maize germ plasm among NARSs in the region. Another network is being formed to foster cross-country collaboration in research on soil fertility.

7. The Agricultural Research Council in South Africa was advised to subcontract one-third of its national research budget to private firms and universities by the year 2000 (Corbett and Coulter 1995).

8. The number of agricultural extension workers in sub-Saharan Africa (excluding South Africa) increased almost threefold, from 21,000 in 1959 to 57,000 in 1980 (Judd, Boyce, and Evenson 1987:13).

9. Projects in the Sudan and Zambia have been phased out.

10. Hassan, Ngure, and Njoroge (1994) found significant discrimination against female farmers in extension programs in Kenya.

10

Maize Research Priorities: The Role of Consumer Preferences

Lawrence Rubey, Richard W. Ward & David Tschirley

One criterion for evaluating the performance of the maize subsector is the degree to which farmers and marketing agents provide consumers with the products they desire. Maize breeding is one of many activities that contribute to providing maize products for consumers. A well-functioning maize production and marketing system will articulate users' preferences for product attributes to scientists who develop new maize cultivars.

Maize breeders in Africa have long recognized the importance of ensuring that cultivars satisfy an often complex set of needs. Traits such as early maturity and processing and storage characteristics can decisively influence the rate at which new varieties are adopted. It is not easy, however, to discover which traits are most desirable. Haugerud and Collinson (1990) have argued that few biological scientists "are trained in techniques to elicit and utilize knowledge from farmers." Furthermore, information flows between scientists and farmers must include other users in the food system. Despite this acknowledged need for better information, the preferences of rural and urban consumers are often neglected or misunderstood by researchers.

Because of the long lag in plant breeding between the initial concept of an improved variety and the variety's actual release, research priority setting requires a minimum time horizon of 7 to 10 years. A simple feedback system—in which products are developed by agricultural scientists, end users react to them, and the products are refined and reintroduced—can mean unacceptably long lags in technology development. To avoid such lags, the preferences of end users must be incorporated in the early stages of research planning.

To achieve this objective, two alternative approaches to designing and prioritizing crop breeding efforts have been pursued in Africa. The first approach, which we refer to as the "political approach," is based upon serving the most powerful interests in the system. Much of the early success in southern Africa with the white dent hybrid SR52 and its variants stems from the ability of well-defined interest groups (i.e., large-scale commercial farmers) to take a leading role in guiding public investment in maize research (Eicher 1995). Similar conditions in Zimbabwe following independence helped smallholders more than double their maize production between 1980 and 1986. On the other hand, Malawi's delayed maize revolution reflects the lack of formal organization and supporting institutions for smallholders (Chapter 5). Without such institutions, smallholders have difficulty articulating their needs to public research institutions.

A second approach, which we term the "presumptive approach," entails research institutions implicitly or explicitly making assumptions about what users (usually farmers) want. This approach offers scope for meeting the needs of dispersed, resource-poor smallholders who are not an organized political force. Yet there are limitations (and failures) when researchers' presumptions and the conventional wisdom about what farmers "need" do not match the actual circumstances of complex farming and marketing systems.

In this chapter, we assert that real (and potentially high) costs are associated with both the political and the presumptive approaches in designing maize breeding programs. Two kinds of errors can occur. The first kind of error ensues when a cultivar is developed and disseminated but is subsequently shown to be unacceptable to farmers. A second kind occurs when genetic material is excluded from a breeding effort because it possesses traits that are *erroneously* thought to be unacceptable to users. It may be fairly common for researchers to fail to use the entire range of appropriate genetic material because of incorrect notions of what users want, but such errors are naturally extremely difficult to observe and catalog.[1]

Because of the limitations of these approaches, we suggest an alternative method of incorporating users' preferences for maize attributes into breeding strategies. This approach makes use of survey data on consumer preferences. We shall use this approach to analyze two attributes of maize in southern Africa: grain color (white versus yellow) and endosperm texture (the continuum from flint to dent maize).

Maize Attributes and Consumer Preferences in Southern Africa

The goal of incorporating different combinations of attributes—including attributes consumers prefer—into a finished maize variety presents several challenges for a breeding program. Maize breeders essentially seek to rearrange genetic material to create a new combination that is "better" than existing cultivars. At one level, breeding objectives can be summarized simply as "beating what is out there" and doing so at a rate that allows an institution or a company to release its products in the market as quickly as possible. The rate at which cultivars are developed, however, depends on the resources deployed, the availability and nature of parental materials possessing the improved characteristics to be embodied in new cultivars, and the complexity of the breeding objectives. Generally, each trait that is added to a minimum set of breeding objectives will reduce the rate of genetic improvement if all other factors are constant. Adherence to an absolute objective, such as flintiness or white grain color, retards progress toward higher yield performance, because it reduces the value of alternative germ plasm by making that germ plasm more costly for the breeding program to exploit.[2] When more traits are added to the list of traits required in a new cultivar, the size of the breeding population must be increased to maintain the same probability of success. The research program must choose between accepting either the costs of slower rates of success within breeding populations or a reduction in the total number of breeding populations. Each trait targeted will also involve additional costs in terms of the procedures needed to select for the trait, and the costs of different selection procedures will vary. Selecting for grain color may entail simple visual inspection, but evaluating increased resistance to maize streak virus involves costly insect rearing and field trials.

This broader recognition of how alternative breeding objectives affect the pace and cost of cultivar development is essential for deciding which objectives a breeding program should pursue. An accurate assessment of consumer demand for maize attributes can help overcome preconceptions and broaden the germ-plasm base on which maize breeders can draw, which may have important long-term implications for the costs of research and the scope for achieving gains in productivity.

White Versus Yellow Maize

Virtually all maize grown by smallholders in southern Africa is white, and in much of the region white maize also dominates rural and urban consumption patterns. Yellow maize is grown primarily by large-scale settler farmers and is used as animal feed. In much of southern Africa, yellow maize is regarded as a "drought food," imported and consumed when domestic production is insufficient and white maize cannot be procured on international markets.[3] For example, extremely poor rainfall during the 1991–1992 growing season forced the countries of southern Africa to import 10 million t of yellow maize during 1992–1993. Following the economic liberalization of the late 1980s, yellow maize grain became increasingly available to urban (and some rural) consumers through open markets. Since the early 1990s, open markets have essentially been the only source of supply of all food staples for urban consumers.

Although the perception persists among policymakers and donors that consumers throughout the region strongly prefer white maize over yellow, evidence emerging from Zimbabwe and Mozambique suggests that widespread consumption of white maize reflects a complex set of historical precedents, marketing restrictions, and policy-related constraints that have effectively limited consumers' options regarding maize color (Rubey 1993).

In Zimbabwe, the reliance on white dent maize suggests some degree of path dependence in technology choice. At the time of the first colonial penetration in the 1890s, a "low-yielding but hardy" flint maize of variable color was cultivated in the region (Weinmann 1972). Although the first settler farmers obtained maize seed locally, they soon sought improved cultivars from overseas. In 1910, the Gwebi experimental farm conducted a large-scale field trial with seven white cultivars and two yellow cultivars from the United States. The yellow cultivars were "not tested further because of the lack of a yield advantage over the white cultivars, uncertain local demand, and because the yellow cultivars were less resistant to maize blight than the white cultivars (Weinmann 1972).

By 1920, both smallholder and commercial farmers in Zimbabwe had largely replaced their white flint cultivars with improved white dent cultivars, which had higher yields (Weinmann 1975). As white became the predominant maize grain color, regulations that reinforced the dominance of white maize were promulgated. Aside from the desire to provide export markets with a uniform, high-quality product and establish the pseudo-branded status of "Southern Rhodesian maize," there were fears that cross-pollination of white maize with yellow maize grown on nearby plots would result in a lower-quality "mixed" grain. To protect growers of white maize from losses of uniform color through cross-pollination, the Maize Act of 1925 was enacted. This legislation permitted growers to petition the governor of Southern Rhodesia to restrict the color of maize grown in their area to any color of maize. Although this regulation, which remained in force until 1970, was originally intended for commercial farmers, it was also applied in areas where smallholders produced maize.

From these observations, it appears that the dominance of white maize in Zimbabwe may simply reflect a series of historical events fortuitous for white maize. Because of a slight initial advantage at the turn of the century, white quickly became the maize grain color of choice and remained so for the next 75 years. The average area planted to white maize by commercial farmers during the period 1980–1984 was more than 230,000 ha, whereas the area planted to yellow maize was less than 5,000 ha.

In 1985, however, the popularity of newly introduced, late-maturing yellow maize hybrids such as ZS206 began to rise dramatically in Zimbabwe, where yellow maize is grown almost exclusively by large-scale commercial farmers. Between 1985 and 1989, the area planted to yellow maize by commercial farmers grew to 61,000 ha, whereas the area devoted to white maize fell from 176,000 to 110,000 ha. The reason for this expansion in yellow maize area is readily apparent: From 1985 to 1989, commercial farmers realized yields of yellow maize that averaged 13% more than yields of white maize, whereas government-set producer prices for the two types of maize remained identical.

The superior yields of yellow cultivars in Zimbabwe in the late 1980s and early 1990s were probably a temporary phenomenon, similar to the better performance of white maize cultivars in the early part of the century. Evidence is emerging that new late-maturing white cultivars equal the yield performance of ZS206.

Flint Versus Dent Maize

Maize cultivars grown throughout eastern and southern Africa differ not only in color but also in texture. The grain of dent cultivars has a high proportion of soft endosperm, whereas the grain of flint cultivars has a higher proportion of hard endosperm.[4]

A century ago, when settler farmers first began to grow maize, most local landraces were flint maize types. Since then, most improved cultivars introduced in the region have been dent types. The reliance on dent maize types in much of the region is partially explained by arguments of path dependence similar to those we have just used to explain changing preferences for white and yellow maize grain. In colonial Rhodesia (now Zimbabwe), early settler farmers were interested in growing maize for export; consequently, export demand largely determined their choice of maize cultivars (Weinmann 1972). Dent maize types were preferred in the British starch market, because dent maize was easier to process in industrial roller mills (Kydd 1989). By contrast, countries such as Malawi and Mozambique, which did not have large communities of settler farmers and did not export large quantities of maize, were much less likely to devote resources to developing dent hybrids (Smale 1995).

Today, dent hybrids dominate the cropping systems of smallholders in Zimbabwe and Zambia. Virtually all maize destined for human consumption in these countries is processed mechanically using small hammer mills or (in urban areas) large roller mills. Yet in countries such as Malawi and Mozambique, where household maize processing practices include pounding by hand, flint maize types are normally grown by smallholders. Although both flint and dent maize can be ground by the small hammer mills that are common in both countries, households in Malawi and Mozambique most often consume a refined maize meal (*ufa* or *ushwa*) whose processing requires numerous steps, including hand pounding, soaking, washing, drying, and milling (Ninje and Weaver 1986; Smale 1995; Weber et al. 1992). When maize is pounded by hand, losses are less for flint maize types than for dent types, because the germ separates more easily from the bran. Flint maize cultivars are also said to store better than dents, since their harder grain discourages weevils (Smale 1995).

Prior to 1990, all of the cultivars released by the Malawian national research program were dent types, largely because of perceptions that dent cultivars had higher yield potential (Smale and Heisey 1994). Kydd (1989) contends that the maize breeders failed to recognize that smallholders, who are both producers and consumers of maize, demanded flintier cultivars to meet their processing and home storage requirements. According to Kydd, the low adoption of high-yielding dent cultivars stemmed from this apparent "dent bias" among national maize breeders. But Smale and Heisey (1994) report that Malawi's maize breeding program has always been concerned about the grain texture of the materials it develops but that the lack of suitable germ plasm, and staff and funding discontinuities during the pre- and postindependence turmoil, created delays in developing suitable varieties for smallholders. In 1990, researchers released two semi-flint hybrids (MH17 and MH18) adapted especially for smallholders' agronomic conditions. Both hybrids had yields similar to those of dent hybrids, and they also possessed the processing and storage characteristics smallholders desired. We assume that the new semiflints are partially responsible for the increase in the area planted to hybrids by smallholders, from 7% in 1988 to 24% in 1992 (Smale 1995).

Eliciting Information on Preferences: Contingent Valuation Approaches

Maize breeders have long recognized the need to develop products that meet the needs of farmers and processors. Researchers have often sought farmers' input in developing maize cultivars, and insofar as most farmers in Africa consume all or part of their maize production, consumer preferences have at least implicitly been a factor in setting research priorities. At times, preferences of urban consumers have also influenced the development of maize technology. For example, the apparent preference for white maize among urban workers in Zimbabwe is one of many factors that ensured the dominance of white maize cultivars.

In this section, we demonstrate how contingent valuation techniques can be used to assess the willingness of potential users to accept a new or nonmarket product (specifically, yellow maize products or dent maize materials) and to help provide an empirical basis for setting agricultural research priorities. These techniques rely on stated preference data, which is simply information on what a survey respondent *would* do in a given situation (unlike revealed preference data, which is information on what a respondent *actually did* in a given situation). Contingent valuation techniques have been used extensively to value the willingness to pay for nonmarket goods, such as environmental assets, but they have not been widely applied to market goods.[5]

Evidence from Zimbabwe

To gauge the potential demand for yellow maize in Zimbabwe, a survey was initiated in June 1993. Data were collected from a random sample of 512 households in the country's three largest urban centers. The primary food purchaser for each household was given a hypothetical scenario and asked if he or she would purchase a specified commodity at a particular price. In this market simulation, the commodity under investigation (yellow roller-milled maize meal) was referenced against the existing government-set price of white roller-milled maize meal.[6] Such techniques offer an avenue for consumers and other end users to articulate their preferences, even for "new," unavailable products (Rubey and Lupi 1995).

Survey results generally supported the conventional wisdom that strong consumer preferences for white maize meal products exist among all segments of the urban population. Assuming prices were the same, 89% of respondents said they "strongly prefer" white roller meal to yellow roller meal. Yet the critical question is whether a subset of consumers would purchase yellow maize meal if it were offered at a price discount relative to white maize meal. The survey results suggest that with a small price differential, a significant proportion of consumers would switch from white to yellow roller meal. Consumers in the lowest income class are more likely to switch from white to yellow meal at a specified price differential. When yellow roller meal is priced 20% lower than white roller meal, 49% of the poorest fifth of consumers would switch, whereas only 30% of the richest fifth would switch (Table 10.1). Lower-income consumers are likely to switch from white to yellow roller meal at a smaller price differential than higher-income consumers. These results, in conjunction with data on yield differentials between white and yellow maize, can be used to calculate the percentage of consumers that would be willing to accept yellow maize at price differentials that might be expected to prevail in an environment where the government did not set maize producer prices.

On the basis of yield data from the late 1980s, we assume a 13% differential in farmers' yields of yellow over white maize, which translates into a 10% difference in consumer prices of white and yellow straight-run meal. Assuming consumers' willingness to pay is independent of the level of processing, the consumer responses given previously can be used to predict the percentage of the urban population that would be willing to switch to yellow maize at a 10% price differential. A 10% price discount on yellow maize was found to be sufficient to induce 21% of urban consumers to switch the kind of meal they use. Although this analysis relies on data from consumer responses to market simulations and incorporates several simplifying assumptions, it does suggest that there is scope for consumers to accept yellow maize in Zimbabwe at the price levels that might be expected to exist in a liberalized market.

Table 10.1 Zimbabwe: Percentage of Consumers by Income Quintile Switching from White Roller Meal to Yellow Roller Meal at a Specified Price Discount, 1993

Income Class	Percentage of Households That Would Switch to Yellow Roller Meal at	
	10% Discount	20% Discount
Quintile 1 (bottom 20%)	25	49
Quintile 2	23	38
Quintile 3	21	37
Quintile 4	17	28
Quintile 5 (top 20%)	20	30
All consumers	21	36

Source: Rubey (1995).

Evidence from Mozambique

Mozambique is the only country in southern Africa where historical circumstances permit three kinds of data to be used to examine consumers' willingness to substitute between white and yellow maize grain: time-series data on prices of white and yellow maize, revealed preference data on consumers' actual white and yellow maize choices, and stated preference data from contingent evaluation methods. All three types of evidence support the hypothesis that white and yellow maize grains are substitutes in consumption and that relatively modest price discounts are sufficient to cause a significant proportion of urban consumers, especially the poorest, to switch to yellow maize.

The time-series price data show that market prices of white and yellow maize grain in Maputo have moved together, very closely, prior to and after the 1992 drought in southern Africa.[7] It is noteworthy that the correlations between yellow maize and rice (which are logically expected to be weaker substitutes) are uniformly small and are significant in only one of the four nondrought cases, whereas all four nondrought cases are significant for yellow and white maize.

This strong correlation over 29 months prior to the drought and 21 months after the end of the drought is striking. Although common supply movements may explain some of the comovement of the two series in the period after the drought, the decrease in both white and yellow maize prices in late 1994 was coincident with the arrival of more commercial food aid. With the arrival of a food aid shipment, white maize prices fell (and then stabilized) along with yellow maize prices. The evidence is compelling that yellow and white maize are substitutes in consumption.[8]

Revealed and stated preference data also suggest that white and yellow maize grains are substitutes in consumption. During April 1994, 400 randomly selected households in Maputo were surveyed regarding their maize supply sources and processing and consumption habits. Revealed preferences were obtained by determining the actual maize purchasing behavior of households, and this information was complemented by stated preference data obtained using a contingent valuation technique similar to that employed in Zimbabwe. Even though 96% of the survey respondents said they would choose white maize over yellow if prices of the two types of grain were identical, during the year preceding the survey nearly 70% of the respondents had actually purchased either yellow grain or whole yellow meal.[9] When asked how they would behave as the price of yellow maize grain was discounted relative to white, consumers showed a marked willingness to switch to yellow maize. This tendency was especially evident among, but not limited to, the poorest consumers. At a discount of only 14%, one-quarter of all households indicated that they would switch to yellow maize, and these had household incomes well below all others.

Findings from the revealed preference data are consistent with the stated preference results. Among the earliest switchers, more than 90% had in fact purchased some yellow maize product during the past year. Among the 25% of consumers who indicated they would never switch to yellow maize products at any conceivable price discount, more than half had in fact purchased either yellow grain or whole yellow meal during the past year, although in small quantities.

In Mozambique as in Malawi, the vast majority of maize produced domestically has been white flint maize. Maize received as food aid, on the other hand, is yellow dent maize. Hand pounding is the most common means of processing grain, even in urban areas. Of the households in Maputo purchasing white or yellow maize grain, 81% pounded it by hand to produce a refined flour. Another 13% hand pounded the grain only to remove the germ and pericarp and then sent the grain to a custom hammer mill. The remaining 5–6% sent their grain directly to the mill without any hand pounding. Nevertheless, many consumers have chosen to consume the yellow dent maize received as food aid instead of white flint maize, because yellow maize has a lower price. In fact, the presence of many hammer mills in Maputo has facilitated this switch. Of those consumers purchasing primarily yellow maize grain, one-third used hammer mills at some point in processing the grain, whereas only 15% of those purchasing primarily white grain used these milling services. Taken together, these results suggest that a significant market exists in Mozambique for lower-priced yellow maize. Although a firm conclusion awaits further research, consumers may also be willing to switch to white dent maize if it is priced lower than white flint maize.

Implications for the Design of Maize Research Strategies

In an era of declining resources for research, there is a need to develop sound national and commodity-specific research priorities. Although our analysis has focused on only two maize attributes—grain color and endosperm texture—because of their relevance in southern Africa, several findings emerge that may be significant for breeding programs in other regions and for different commodities.

First, alternative maize breeding strategies have differing economic implications. Erroneous notions of "what farmers or consumers want" artificially limit the size of the pool from which genetic material is drawn and exclude potentially valuable genetic material from consideration. As a result, research costs are higher and progress is slower than might otherwise be the case. Although researchers cannot respond effectively to the needs of all potential users, our analysis suggests that lower costs and faster progress can result from a more complete consideration of the costs and benefits of alternative breeding strategies.

Second, the application of contingent valuation techniques can give breeding programs a better understanding of users' preferences and acceptable trade-offs between alternative attributes. Such knowledge can help researchers develop more effective criteria to screen for needed traits and help guide resource allocation. By allowing consumers and other end users to articulate their preferences, even for "new," unavailable products, contingent valuation techniques offer a way of expanding participatory research beyond farmers' agronomic needs.

Third, the scope for consumers to accept products with different attributes may be greater than supposed, particularly when the alternative products are available at lower relative prices. In southern Africa, the conventional wisdom that consumers are unwilling to accept variation in grain color (i.e., yellow maize) is misleading. Evidence from Zimbabwe and Mozambique suggests that consumers will switch from white maize to yellow maize at modest price discounts. Results from Mozambique also suggest that consumers are more willing to switch from flint to dent maize than typically thought, even where flint cultivars have dominated production and consumption patterns for decades. Given consumers' willingness to consider relative prices when evaluating alternative combinations of attributes, narrowly focused breeding objectives may be misplaced. By confusing what consumers "prefer" when prices are equal instead of examining consumers' choices under differing prices, maize breeders may adopt false notions about the acceptability of particular attributes.

Finally, the direction of maize research strategies has important implications for national food security and access to and dependence on world markets. In Zimbabwe, consumer demand for yellow maize (when it is priced below white maize) offers scope for resolving a chronic agricultural policy dilemma that has been exacerbated by exclusive reliance on white maize. A recognition of consumers' acceptance of yellow maize would give policymakers greater flexibility. Instead of holding large stocks of white maize from year to year and incurring the costs associated with maintaining sufficiently high producer prices to spur domestic self-sufficiency, a cheaper option for covering expected deficits might be to purchase yellow maize from world markets.

Notes

1. Haugerud and Collinson (1990) suggest that scientists often reject potentially useful germ plasm because they do not correctly gauge its potential acceptability to farmers.
2. This argument is not based on the direct impact of flintiness, color, or other traits on yield performance. It is based on the principle that the progress a breeding program can make with a fixed budget is diminished by expanding the minimum number of traits that define a successful cultivar.
3. For example, Zimbabwe imported yellow maize for food during the droughts of 1967, 1984, and 1992. Mozambique is the only nation in the region in which yellow maize has been consumed in nondrought years.
4. Maize grain texture is best viewed as a continuum that ranges from extreme dent to extreme flint, and a cultivar's place on the continuum is determined by the relative quantities of both hard and soft endosperm contained in the grain. Thus "flint," "dent," and the voguish "semiflint" are to some extent subjective terms.
5. In the limited studies that have been done on market goods, data gathered from contingent valuation techniques and used to estimate potential demand have been shown to provide fairly accurate estimates of actual demand (Dickie, Fisher, and Gerking 1987).
6. Yellow meal was offered to consumers as a binomial choice in an iterative bidding process. If the offer was refused at a particular price, the price of yellow meal was systematically lowered until the respondent stated that he or she would purchase it. If the respondent agreed to purchase yellow meal at the initial price, the price was raised until the respondent refused to purchase it. In either case, the maximum willingness to pay for the product was obtained for each respondent.
7. Nominal prices were used in this analysis but were linearly detrended to remove the effects of exogenous factors that could contribute to comovement. Deflated prices were not used because of serious reservations about data quality of the Consumer Price Index.
8. The lack of correlation during the drought is to be expected, given the extreme scarcity of white maize and the overabundance of yellow. Thin markets are widely known to be unstable and to provide opportunities for sellers to exercise market power (Marion 1986; Hayenga 1979). Tomek (1980) found evidence that prices in thinly traded markets tend to become disassociated from prices in related markets.
9. Yellow roller meal was not generally available in the market during this time.

11

Soil Fertility Management in Southern Africa

*John D. T. Kumwenda, Stephen R. Waddington,
Sieglinde S. Snapp, Richard B. Jones & Malcolm J. Blackie*

The decline in soil fertility is the most widespread limitation on both yield improvement and the sustainability of the maize-based production systems of southern and eastern Africa, particularly in the wetter areas where yield potential is higher. Traditional agricultural systems relied largely on extended bush fallows and the harvesting of nutrients stored in woody plants to maintain soil fertility (see, for example, Blackie and Jones 1993; Blackie 1994a). But in most arable areas of Malawi, Zimbabwe, and Kenya, fallowing has almost disappeared, and continuous cropping is the norm. At the same time, in Zambia, Mozambique, and Tanzania, where fallowing is still widely practiced, the length of the fallow period is decreasing and is often insufficient to maintain soil fertility. The consequent downward spiral of soil fertility has led to a corresponding decline in crop yields and an increase in soil erosion (Araki 1993).[1]

Buddenhagen (1992) has estimated that in the tropics the weathering of minerals and biological nitrogen fixation will enable, at most, yields of 1 t/ha on a sustainable basis. Loss of soil through erosion, which is particularly common on soils cultivated with annual crops in the upland tropics, reduces this level considerably. Smaling (1993) estimated annual net nutrient depletion exceeding 30 kg of nitrogen (N) and 20 kg of potassium (K) per ha of arable land in Ethiopia, Kenya, Malawi, Nigeria, Rwanda, and Zimbabwe. The densely populated, erosion-prone countries of southern Africa, such as Malawi, where maize is a major food crop, are those with the greatest aggregate nutrient loss. In some areas, such as the communal lands with higher rainfall in northern Zimbabwe, soil depletion is so severe that maize will yield virtually no grain without fertilizer.

The central issue for improving agricultural productivity in southern and eastern Africa is how to build up and maintain soil fertility despite the low incomes of smallholders and the increasing land and labor constraints they face. The success of the green revolution in Asia, led by the introduction of high-yielding varieties that thrived on Asia's more fertile and uniform soils, has biased the research agenda in Africa toward plant breeding. We contend that maize varietal improvement will have only a transitory impact on smallholder farming in Africa unless the widespread decline in soil fertility is reversed by researchers, policymakers, and farmers.

The analysis shown in Figure 11.1 underlines the urgency of the task. Three possible scenarios for Malawi, based on "best bet" estimates of technology adoption, are depicted. For each scenario, a maize deficit or surplus is calculated as the projected balance between maize production and consumption in a given year. Assumptions include a 3.2% annual population growth rate, a per capita annual maize consumption of 230 kg, and a constant area of 1.4 million ha planted to maize.[2] In scenario one, the yield of improved maize is held constant at 2.5 t/ha, and the area planted to improved maize does not change from its current 20% share. In this scenario, the maize deficit widens rapidly to surpass current national maize consumption. In scenario two, the adoption of hybrid maize is assumed to increase rapidly at 20% annually, but hybrid yields remain at 2.5 t/ha. This scenario provides a small national surplus at first, but within about a decade the deficit reappears and soon widens. Finally, scenario three represents a combination of the adoption of improved maize (from scenario two) and efficiency gains from better crop management (see note 4), which increase hybrid yields to close to 4 t/ha. Under this scenario, Malawi would maintain current levels of per capita maize consumption without resorting to large quantities of imports. Concurrent adoption of improved varieties and better management practices appears to be essential for sustainable and long-term productivity gains.

Figure 11.1 Malawi: Scenarios for Maize Surplus/Deficit

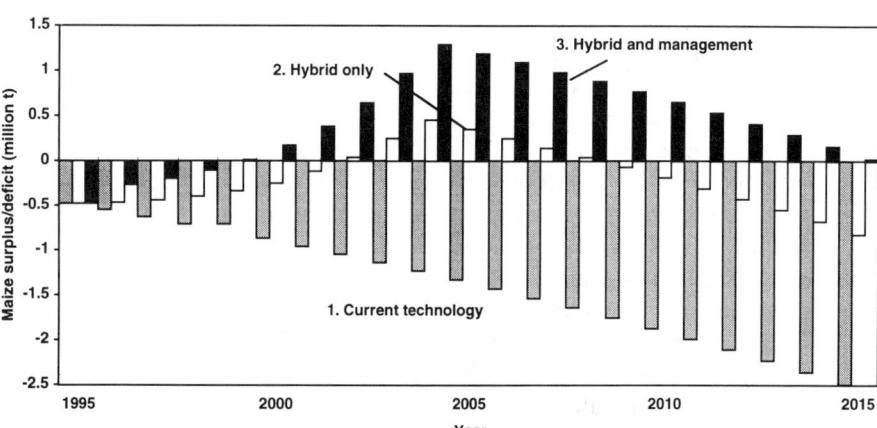

To address the soil fertility problem, we propose a research and extension paradigm that combines organic with inorganic sources of soil fertility and actively involves farmers and other clients in an integrated, long-term process. First, we review existing technologies for enhancing soil fertility to suggest how they might be made more attractive to farmers, giving particular attention to the circumstances of smallholder farmers. We then outline, based on examples drawn from practice, improvements in the technology development and transfer process that are required to facilitate the adoption of better soil fertility management practices.

**Improving Soil Fertility and Productivity:
Feasible Options for Farmers**

Although there is little argument about the need for more sustainable soil fertility management, there is less clarity on how such management might be achieved. In a review of past impacts of and future prospects for maize research in Africa, Byerlee et al. (1994) have contrasted the relatively high adoption of improved maize varieties and hybrids against the relatively limited adoption of resource management technology for maintaining soil fertility and increasing labor productivity.

Many technologies for addressing poor soil fertility are potentially available, but few are easy for farmers to adopt. To be successful, a technology must be appropriate to the cash and labor situation of farm households: Cash and labor are interlinked factors that dominate smallholders' decisionmaking. Labor may be provided by the family, bought from other farmers with food or cash, and sold to others for food or cash.

Often the farm household is headed by a woman, living with her small children on land that has been cultivated many times before. Given a hectare or more of land, she may be self-sufficient in food if her health is good and the weather favorable, but frequently seasonal illness will prevent the family from planting maize on time. If the rains are poor, the woman's crop may fail. The odds are that in some, if not many, years, she will not produce enough food for her family's needs. This failure requires her to work on a daily basis for neighboring farmers who will feed or pay her to plant, weed, or fertilize their fields. Her field is left unplanted, unweeded, and unfertilized until later in the season. The resulting poor harvest leaves her without food before the next crop can be harvested. This cycle must be broken. In Malawi, about 60% of rural households produce less than they require to feed themselves. The total annual cash income of these households is less than the amount of cash needed to pay for the recommended dosage of inorganic fertilizer (HIID/EPD 1994).[3]

Three kinds of technological options are available to improve soil fertility. First, both the type of inorganic fertilizer and its use could be tailored to smallholders' conditions to improve fertilizer-use efficiency and profitability. Second, the quantity and quality of locally produced organic materials could be increased to reduce the cash cost of fertilizer to the farmer and increase the efficiency of inorganic fertilizer use. Third, other practices that interact with soil fertility management practices, such as the use of nutrient-efficient varieties and timely weeding, could be improved. We discuss each of these options in turn.

Increasing the Use of Fertilizers

In most parts of the world, chemical fertilizers play a major role in maintaining or increasing soil fertility. Governments in southern Africa have made a considerable effort to promote the use of chemical fertilizer through research, extension, fertilizer subsidies, and credit programs, but the use of chemical fertilizer on smallholder farms remains very low. Despite the relatively low dosage, fertilizer is the most costly cash input used by the typical smallholder in southern Africa.

Moreover, the efficiency of fertilizer use (as measured by the grain yield response to the addition of chemical fertilizers) is poor, which reduces its profitability (see Jones and Wendt 1995 for an example from Malawi). Benefit-cost ratios in Malawi are 1.8 for fertilizer use on hybrid maize and 1.3 for unimproved maize (Conroy and Kumwenda 1995), substantially below the ratio of 2.0 usually assumed necessary for widespread adoption of fertilizer by smallholders (Chapter 13). An economic analysis of fertilizer policy in Malawi (HIID/EPD 1994) concluded, however, that improvements in fertilizer-use efficiency could make fertilizer financially attractive to smallholders.[4]

Low fertilizer-use efficiency reflects inappropriate fertilizer recommendations promoted by research and extension services. For 50 years, research on inorganic fertilizers in Zimbabwe was geared toward farmers who could afford relatively large quantities of fertilizer: commercial farmers and growers of cash crops. The recommendations emerging from such research have often ignored soil and climatic variation in smallholder farming areas, have been incompatible with farmers' resources, and have been inefficient in terms of nutrient balances. For instance, although Zimbabwean farmers are advised to apply basal fertilizer to maize at planting, they almost always apply it just after crop emergence. The farmers' practice is not only easier but is also less risky, resulting in a negligible loss of yield under farm conditions (Shumba 1989). A response to potassium is rare on the granite soils common in smallholder areas of Zimbabwe (Mashiringwani 1983; Hikwa and Mukurumbira 1995), but the compound fertilizer recommended contains potassium. Although much work has been done since independence on the appropriate types, amounts, timing, and placement of inorganic fertilizers for food crops produced by smallholders (see Grant 1981; Metelerkamp 1988; and a recent summary in Hikwa and Mukurumbira 1995), fertilizer recommendations in Zimbabwe still fail to take sufficient account of the cash constraints and risks affecting resource-poor farmers in marginal areas.

Recent research suggests that some relatively low-cost methods are available for increasing fertilizer-use efficiency. Some potential for cost saving exists by applying nitrogen just after planting and using cheaper forms of phosphorus at other times. The addition of micronutrients can improve the yield response to macronutrients on deficient soils. Micronutrients can often be included relatively cheaply in existing fertilizer blends and, when targeted to deficient soils, can dramatically improve fertilizer-use efficiency and crop profitability (see the example from Malawi later in this chapter).

A lack of moisture is a frequent constraint on maize yield and yield response to fertilizer. The efficiency (measured through grain production) of using both water and nitrogen is raised when supplies of both inputs are adequate, but the high risk of poor response to fertilizer in dry years is a major reason most farmers in semiarid areas use little or no fertilizer. Existing recommendations are often too risky for lower rainfall areas and need to be adjusted downward to be profitable. Adjusting fertilizer use to the evolving rainfall pattern in any one season can significantly increase the profitability of using fertilizer, resulting in increased maize yields and profitability of 20 to 40% (Piha 1993). Fertilizer use can also be refined to suit specific soil conditions through the use of crop simulation models to predict outcomes under variable water and nitrogen conditions. When crop simulation models are used in conjunction with geographic information systems, the resulting information can help delineate target agroecological areas or groups of farmers for which a particular input level is appropriate (e.g., Dent and Thornton 1988; Keating, Wafula, and Watiki 1992). This recent work has shown that productive and profitable agriculture is possible on poor soils and in semiarid conditions with the judicious use of inorganic fertilizers.

Evidence is mounting that the most promising route to improving the efficiency of inorganic fertilizer in smallholder cropping systems is through the addition of small amounts of high-quality organic matter to tropical soils (Ladd and Amato 1985; Snapp 1995). Increased levels of organic matter will increase soil microbial activity and nutrient cycling, as well as reduce nutrient loss from leaching and denitrification (Doran et al. 1987). Long-term cropping studies in Kenya and Nigeria indicate that organic plus inorganic inputs sustain fertility at a higher level than the expected additive effects of either input alone (Dennison 1961). The additional nutrients supplied with organic fertilizers are only part of the benefits of using these fertilizers. Many farmers already recognize other beneficial effects of organic manures and combine, where possible, small amounts of high-quality organic inputs with inorganic nutrients.

Increasing Availability and Use of Organic Sources of Fertility

The maintenance and management of soil organic matter are central to sustaining soil fertility in the tropics (Swift and Woomer 1993; Woomer et al. 1994). The more efficient use of inorganic fertilizers will solve only part of the problem of soil fertility management, and only those farmers who use fertilizer or have the potential to do so will be affected. Current inputs of organic materials, such as animal manures and green manures, however, are insufficient to maintain organic matter levels in the soils of tropical Africa. The continuous cropping of land and associated tillage practices provoke rapid declines in organic matter, which eventually stabilizes at a low level.

Conventional mechanisms for addressing the loss of organic matter are rotations (especially rotations involving legumes), the addition of animal manures, and forms of intercropping. Animal manures and crop residues are widely used by farmers. In areas where cattle are common, smallholders often apply manure to fields that will be planted to maize, although the amount of manure applied provides only a small proportion of the nutrient requirements for the maize crop.[5] Agricultural intensification in southern Africa, however, is often associated with a decline in the availability of animal manures as livestock production becomes more difficult.[6] As pressure on arable land rises, cropping encroaches on areas previously used for grazing. Manure from cattle and other animals is very important for most farmers in Zimbabwe but is rarely available in Malawi, where animals are scarce. Even in the best areas for crop production, however, both the supply of manure and, as important, its quality are inadequate for manure alone to maintain soil fertility.

Where animals are scarce, farmers must turn to other sources of organic matter. Although composted crop residues are used in wetter areas and where crop biomass production is relatively high, composts are usually sufficient for only a modest part of the cultivated area. Careful management of such residues is also required, since nitrogen immobilization can result in poor crop growth (Nandwa, Anderson, and Seward 1995), a fact well-known to farmers. In areas of unimodal rainfall, where stover tends to break down slowly in the soil, and on sandy soils, where the supply of soil nitrogen is very low, maize stover is usually fed to livestock. This practice serves not only to sustain animals but to cycle the residues in a way that makes them more beneficial to the crop. Like animal manures, these technologies require a substantial labor commitment on the part of farmers (see, for example, Huchu and Sithole 1994; Carter 1993; McIntire, Bourzat, and Pingali 1992).

The reality is that since organic materials are rarely sufficient to maintain soil organic matter levels, more organic materials must be introduced into the cropping system. Options include rotations, green manures, animal manures, intercropping, strip cropping, relay cropping, and agroforestry; of these, many of the most promising options involve legumes.

Legumes have been the focus of well-meaning efforts to transform smallholder agriculture in Africa since the turn of the century.[7] Annual legumes are used as sole crops in rotation with cereals, are intercropped, and occasionally are used as green manures. Perennial legumes are sometimes retained in farmers' fields and are just beginning to be incorporated as hedgerow intercrop or alley crop systems.[8]

Despite their promise, legumes remain a marginal component of many maize-based farming systems in southern and eastern Africa. Much of the work underlying legume-based technologies has been done on research stations with little attention to tailoring these technologies to farming circumstances where seasonal labor is in short supply. The fertilizer, especially phosphorus fertilizer, needed to jump-start the system may be too costly or unavailable, and legume seeds are often difficult to obtain (Giller, McDonagh, and Cadisch 1994).

It may also be difficult to release land from staple food crop production to produce legumes that are efficient in fixing nitrogen but have low food value. Thus there is often a direct conflict between the more immediate need to assure today's food supply and the longer-term need to build up soil fertility to assure tomorrow's food supply. Broadly speaking, the larger the potential soil fertility benefit from a legume technology, the larger the initial investment required in labor and land, and the fewer short-term food or fodder benefits the technology has.

Grain legumes have the fewest barriers to adoption and are widely grown by farmers, mainly for home consumption of the seed and sometimes leaves. But the more productive grain legumes add relatively little organic matter and nitrogen to the soil, because most of the above-ground dry matter and almost all of the nitrogen are removed from the field in the grain (see Giller, McDonagh, and Cadisch 1994). Species that combine some grain with high biomass, such as pigeon pea and dolichos bean, offer a useful compromise in promoting farmer adoption and improving soil fertility (K. E. Giller, personal communication).

On severely depleted soils in Malawi, soybeans produce more calories per unit of land than unfertilized maize, in addition to fixing nitrogen from the atmosphere (Carr 1994). Regional soybean development, however, is biased toward varieties that offer high grain yields, which are favored by large-scale producers whose interest in improving soil fertility is secondary. Self-nodulating soybeans that yield less grain are often attractive to smallholders, because they do not have to be inoculated with *Rhizobium* bacteria to fix nitrogen, and they also have high biomass. One self-nodulating variety, Magoye, has been grown widely in Zambia and is now grown by thousands of smallholders in Malawi (although Malawi has not yet approved it for release).

Intercropping with legumes is a widespread traditional practice in African agriculture (Andrews and Kassam 1976) that contributes some residual nitrogen to the subsequent crop (Willey 1979). In densely populated regions, intercrops are recognized for their stabilizing effect on food security and for improving the efficiency of land use. Low-growing legumes, however, are often shaded by taller cereals (Dalal 1974; Chang and Shibles 1985; Manson, Leighner, and Vorst 1986). Under smallholder management and conditions of low soil fertility, a common problem is poor emergence and growth of the legume in the intercrop, especially in semiarid areas. This problem limits the nitrogen and organic matter contribution of the legume to levels well below the potentials found on research stations (Kumwenda, Kabambe, and Sakala 1993; Kumwenda 1995).

Late-maturing pigeon peas would appear to be one of the more promising intercrops. Even though early growth of legumes is reduced when they are intercropped with maize, pigeon peas compensate by continuing to grow after the maize harvest to produce large quantities of biomass (Sakala 1994). Pigeon peas are easily intercropped with cereals, and even if the seed is harvested for food, the leaf fall is sufficient to make a significant contribution to nitrogen accumulation. The disadvantage is that pigeon peas are highly attractive to livestock. It is rarely practical to grow pigeon peas in smallholder systems where livestock are left to roam in the fields after harvest.

Cereal-legume rotations appear to offer greater prospects than intercrops for raising the yield of cereals (Natarajan and Shumba 1990). The amount of nitrogen returned from rotations depends upon whether the legume is harvested for seed, used for forage, or incorporated as a green manure.[9] The yield response of a cereal crop following a legume such as pigeon peas, groundnuts, bambara nuts, and sunnhemp can be substantial (MacColl 1989; Mukurumbira 1985; Temu 1982). Legume rotations, however, are a more important practice for restoring soil fertility on larger landholdings than on smallholdings.

Legumes may also be included in the fallow cycle. Natural short fallowing of continuously cropped land in Zimbabwe results in little or no improvement in soil fertility (Grant 1981), but improved fallows, using a legume such as *Sesbania sesban*, are a way of adding significant amounts of nitrogen and organic matter to soil. Improved fallows have been evaluated in Eastern Province, Zambia, where short fallows of two to three years are common. When inorganic fertilizer was not used (a common practice of farmers in this part of Zambia), it was profitable for farmers to invest in the improved fallow, which doubled maize yields (Place, Mwanza, and Kwesiga 1995). But where land is limiting, the feasibility of fallow systems is yet to be proved.

Finally, various agroforestry systems have been tested in which woody perennials, usually a legume, are grown in association with annual crops or pastures in a spatial arrangement, a rotation, or both. One such system is alley cropping, sometimes referred to as hedgerow intercropping, developed in the late 1970s at the International Institute of Tropical Agriculture (Kang, Reynolds, and Atta-Krah 1990). A leguminous tree crop is planted in rows some meters apart and pruned to supply nutrients to the soil; the nutrients are used by shallow-rooted crops planted between the rows of trees. Important aspects of the technology remain problematic, however. Trees and associated crops can compete for moisture, nutrients, and light (Mbekeani 1991; Ong 1994), and the technology is labor-intensive and management sensitive.[10] *Leucaena leucocephala* (the most widely researched hedgerow species) is susceptible to termite attack at the seedling stage, to defoliation by insects, and to poor biomass production under low soil fertility levels and on acidic soils.

A promising tree species for agroforestry is *Faidherbia albida,* a vigorous leguminous tree long used by African farmers for improving crop yields in areas where the tree is naturally abundant. Saka et al. (1994) have described how natural stands of *F. albida* have been used in maize production systems in Malawi. Although farmers recognize the value of the tree, *F. albida* has not been integrated systematically into maize growing areas.

Exploiting Interactions Between
Soil Fertility Technologies and Other Factors

By themselves, organic sources of soil fertility will only rarely provide the productivity boost needed by smallholders, and they will need to be combined with the judicious use of chemical fertilizers, improved pest and weed management techniques, and high-yielding crop varieties. Soil fertility management interacts with weed control practices, moisture availability, fertilizer-responsive and fertilizer-efficient maize varieties, labor, and draft power. Because of the long dry season, many farm operations—particularly planting, weeding, and fertilizing—are concentrated in the critical early weeks of the growing season. Thus the interaction of soil fertility technology with other, possibly more readily adoptable farmer inputs and management practices needs particular consideration.

Changes in these other practices will often aid efforts to improve soil fertility management. For example, in Malawi farmers who weed twice at the critical periods for maize can achieve a higher yield with half the amount of fertilizer than farmers who only weed once (Kabambe and Kumwenda 1995). Likewise, in Zambia the combination of basal and topdress fertilizer, applied when weeding the maize at a height of 20 cm, saved six person-days/ha during the peak demand period for family labor and resulted in a 19% yield increase compared with the standard farmer practice (a basal fertilizer application just after planting, followed by late weeding and topdressings) (Low and Waddington 1990, 1991).

A decline in soil fertility may cause weeds and other pests and diseases to build up and may thereby indirectly affect crop production. A well-known example of such an interaction is the parasitic weed *Striga,* which becomes a more severe problem when soils are depleted (once established, however, this weed is hard to control through improved soil fertility alone). Data from long-term trials in Kenya have shown that incorporated crop residues play an important role in reducing *Striga* (Odhiambo and Ransom 1995).

Finally, there is scope for partially alleviating the soil fertility constraint through crop breeding. Some improvement in nitrogen-use efficiency has been achieved through breeding by the International Maize and Wheat Improvement Center in Zimbabwe (Short and Edmeades 1991), although recent progress has been inconclusive (Pixley 1995). So far, most of these gains have come from improved nitrogen utilization to produce more biomass and grain yield with no increase in total nitrogen uptake (Lafitte and Edmeades 1994). Nevertheless, varieties bred for conditions of low soil fertility may gain their advantage from extracting more nutrients from the soil. The outcome of such a strategy is uncertain.

Research for Improved Technology Development and Dissemination

Improved soil fertility management requires not only adaptation of experiences from other regions but also a much greater understanding of the processes through which fertility can be managed under African conditions. This implies that basic research is needed to understand soil fertility issues and that such research should be well integrated with applied and adaptive research.

The methods for each type of research are different, and it is unusual for scientists involved in basic research to have a complete grasp of the skills needed for adaptive research (and vice versa). Developing a relevant agenda for basic research is particularly challenging. Changes in soil fertility have long-term effects on productivity that are typically incremental rather than spectacular. Consequently, understanding such changes requires a consistent, long-term effort by scientists. But it is important that scientists conducting basic, applied, and adaptive research share a vision of the problems faced by farmers so the research agenda remains fixed on important problems and does not get sidetracked into challenging but less relevant avenues.

Many of the key characteristics of applied and adaptive research are well described in the literature on adaptive on-farm research (e.g., Tripp 1992). The challenge now is for this kind of research to embrace a longer-term perspective so that investigation, review, and interaction with producers and other concerned parties lead to a coherent research strategy over time. In the following section, we give an example of such a process, which is already underway. We then show how progress in applied or adaptive research requires careful underpinning by high-quality, basic studies that are carefully prioritized.

A Case Study of Adaptive Research:
Overcoming Micronutrient Deficiencies on Maize in Malawi

Research on micronutrient deficiencies in maize in Malawi exemplifies the commitment needed for "successful" soil fertility research.[11] A study by Conroy (1993) over three seasons from 1990 to 1992 showed that smallholders following existing fertilizer recommendations[12] obtained only about 3 t/ha maize grain, even though farmers planted hybrids with a yield potential of at least 6 t/ha. A series of trials was initiated in smallholders' fields in 1989 to evaluate nutrient deficiencies at several sites throughout Malawi. The treatments and treatment methods evolved from year to year, based on experience from previous years. By the 1991–1992 season, regional deficiencies of potassium and several micronutrients had been detected. In deficient regions, average yields improved by 40% over the existing nitrogen and phosphorus application when these deficiencies were satisfied. Lime applications were necessary at some sites. Data from soil and plant analyses showed that phosphorus application was unnecessary at some sites, whereas at other sites the recommended phosphorus fertilizer rate was insufficient.

To refine these observations, the research team worked with a fertilizer demonstration project run by the extension service and the United Nations Food and Agriculture Organization. In 1992–1993, micronutrient supplements were included in selected fertilizer demonstrations throughout the country. From the analysis of soil samples from 400 smallholder sites, preliminary maps of areas with specific nutrient deficiencies for maize in Malawi were developed (Wendt, Jones, and Itimu 1994).

The next step was to work with a local fertilizer company to produce practical fertilizer blends that addressed these regional deficiencies. At the same time, the effort to characterize nutrient deficiencies was greatly expanded to refine the nutrient deficiency maps. More than 3,000 soil samples were collected and analyzed. The sample locations were geo-referenced with respect to elevation, longitude, and latitude to develop a computer database that could be used with a geographic information system. Verification trials were conducted in cooperation with the extension service and farmers to review the new fertilizer formulations under farmers' conditions and demonstrate more efficient methods for using fertilizer. Work was initiated to define new response curves for nitrogen and phosphorus when micronutrients were added.

This example of moving from blanket recommendations to targeted recommendations illustrates how progress is possible with only modest resources but a clear sense of direction. By exploiting the potential of fertilizer blending systems to produce small "runs" of fertilizer to a given specification, this focused research and verification program was able to provide fertilizer recommendations for localized areas at a reasonable cost.

Basic Research on Soil Fertility Processes

To be successful, the kind of applied research described in the preceding section must be underpinned by high-quality research aimed at understanding the basic processes of nutrient flows in tropical soils. The challenge is complicated by the fact that detectable changes in soil organic matter and other soil fertility parameters occur very slowly. To further improve technologies for managing organic matter, the first step is to quantify nutrient losses and inputs at different scales.[13] New methods of soil analysis and modeling are promising tools for assessing nutrient flows across the heterogeneous landscape of smallholder farms in tropical Africa. Research must be conducted on representative sites in well-characterized agroecosystems. This work requires a continuing commitment to building bridges among disciplines such as agronomy, ecology, soil biology, chemistry, and physics.

Nutrient budgeting can be used to develop improved ways of using available nutrients. For example, recent findings suggest that small additions of high-quality organic material can increase nutrient cycling efficiency.[14] Synchronizing the release of nutrients from organic materials with the crop requirement for nutrients could also increase nutrient cycling efficiency (Myers et al. 1994). Organic inputs of varying quality can be used to manipulate nutrient supply.

A vigorous, effective crop-rooting system is essential for efficient nutrient acquisition. Major nitrogen losses are thought to occur early in the cropping season because of high initial mineralization rates, inorganic nitrogen accumulation during the dry season, and limited root growth of young crop plants (Myers et al. 1994). Understanding root behavior is also essential to minimize competition associated with intercropping technologies. Several important agroforestry systems are based on the assumption that trees reduce nutrient losses by scavenging nutrients from below the depth of crop roots (Young 1989). Reducing negative interactions between tree roots and annual crops, and enhancing benefits from tree roots, may be fundamental to the success of the technology (Giller, Itimu, and Masamba 1996).

These are a few examples of the kind of basic research that is needed to provide a stronger foundation for applied and adaptive research on soil fertility issues.

Conclusions

Smallholders in parts of Africa have adopted high-yielding maize varieties with notable success, but increases in maize yields have still been disappointing, largely as a result of declining soil fertility. This widespread problem is becoming worse. In most production systems, mineral nutrient losses from soil far exceed nutrient inputs. Little fertilizer is applied, and the efficiency of fertilizer use is often low and declining because the level of soil organic matter is diminishing. Greater concentrations of people competing for smaller shares of cropland have made shifting cultivation obsolete in the dominant smallholder cropping systems of southern Africa, which are based on maize. Maize is now grown in continuous cultivation; as a result, the soil resource base is increasingly depleted, and yields are lower. The question that needs to be addressed is how to build up and maintain soil fertility under the poverty and other constraints faced by smallholders.

The technologies needed to manage soil fertility in southern Africa do not differ from those developed for other parts of the world. The relative emphasis on the kinds and amounts of technology used in southern Africa, however, does depart significantly from practices followed elsewhere. Inorganic fertilizers are a key element of fertility management. The need for added external nutrient inputs is inescapable. Inorganic fertilizers are expensive, however, and their use is often unprofitable, especially because blanket fertilizer applications are recommended even in semiarid areas. An urgent need exists to improve the profitability of using fertilizer. One way of doing so is to increase fertilizer-use efficiency by better targeting recommendations to site- and season-specific conditions; another way is to develop improved fertilizer management techniques appropriate for smallholders.

But the solution to the soil fertility problem will not consist of inorganic sources of fertility alone. More attention must be given to exploiting sources of organic fertility, especially the better integration of legumes into cropping systems. For many households, the lack of cash dominates decisionmaking. These households are often forced to sell their labor in return for food or cash, which in turn compromises the management of their own crops. For this group of farmers, organic sources of nutrients—especially legumes, which capitalize on the freely available nitrogen in the atmosphere—offer the best means of increasing fertility, at least from the point of view of nitrogen. Production of organic manures is highly variable, however, and they are usually in short supply. Mixing high-quality organic sources of nutrients with inorganic fertilizer can substantially improve nutrient-use efficiency and crop productivity and shows potential for wider adoption.

Legumes and their multiple uses are not new to farming systems. Grain legumes, legume intercropping and rotations, green manures, improved fallowing, agroforestry, as well as cereal residues and animal manures, are all techniques that can enhance soil fertility and sustain the resource base. As noted earlier, however, the larger the potential soil fertility benefit from a legume technology, the larger the initial investment required in labor and land, and the smaller the short-term benefits to farm households in increased food production. Although the potential of such technologies is rarely realized on farmers' fields, we see opportunities for combinations of low rates of several inputs, especially those that combine inorganic and organic fertilizer. In particular, the phosphorus requirements needed to produce legumes will have to be supplied from inorganic sources.

The proportion of locally produced organic materials must be increased not only to maintain soil organic matter and halt the downward spiral of fertility evident in many systems but also to improve fertilizer-use efficiency and consolidate and expand the base of fertilizer users. For both organic and inorganic sources of nutrients, important practical questions for smallholders remain unanswered. Information on (1) the optimum use of small amounts of inorganic fertilizers, (2) the best combinations of organic and inorganic fertilizers, (3) how to produce sufficient amounts of organic manures under low fertility conditions, (4) the best management compromises in the use of labor among critical seasonal tasks, and (5) adjusting fertility management to seasonal and other external factors is largely lacking despite some attempts to resolve these issues through adaptive research. Yet these are the very questions to which smallholders are seeking answers. Typically, farmers are looking for ways of combining these inputs and employing them in ways to most efficiently utilize their cash, labor, and land resources. There is little past or current experimentation, and little counsel from extension, to guide farmers' choices.

Finally, the paradigm for research and development of improved crop husbandry has to change. Emphasis in both research and extension needs to move away from a rigid and prescriptive approach to a flexible, problem-solving format. This change in emphasis will enable a process of technology development driven by smallholders' needs to evolve. Past research has tended to distill results into a few recommendations that ignore the important interactions in the system and fail to address the widespread diversity among farmers and agroclimatic zones. Flexible soil fertility recommendations are required that better address actual nutrient deficiencies, take advantage of cropping system opportunities, are efficient under the highly variable rainfall regimes faced by most smallholders, and are compatible with farmers' socioeconomic circumstances. It is also important to recognize that applied and adaptive research will not be enough. Additional basic research has to provide the foundation for extrapolating from site-specific trials to agronomic recommendations for specific agroecological zones and farmer groups. Without concerted action in this direction, the consequence will be a weakening natural resource base and a continuing decline in the standard of living.

Notes

1. Nutrient levels of the old and already highly leached soils in Africa's humid and subhumid zones are inherently low. Sandy and sandy loam soils derived from granite, with low organic matter, are common in southern Africa. Nitrogen (N) deficiency is ubiquitous on these soils, whereas deficiencies of phosphorus (P) and micronutrients such as sulfur (S), magnesium (Mg), and zinc (Zn) are common (Grant 1981; Wendt, Jones, and Itimu 1994).

2. Almost all of the arable land available to smallholders in Malawi is under continuous maize cultivation, so opportunities for expanding maize area are limited. At the start of each scenario, 20% of maize area is planted to improved maize—roughly the current situation. In all scenarios, unimproved maize yields 1,000 kg/ha.

3. In Zimbabwe, of the 32% of farmers applying the recommended package of fertilizer to their maize crop in the 1990–1991 cropping season (a near-average season), 48% failed to recover the value of the fertilizer (Page and Chonyera 1994).

4. Research on farmers' fields in Malawi shows that at farmers' levels of fertilizer application with improved timing and application methods, the response to nitrogen can increase from 15.0 to 20.0 kg grain per kg N applied for unimproved maize and from 17.4 to 25.0 kg grain per kg N for hybrid maize.

5. Cattle manure is applied in a dried, aerobically decomposed form, often with a high sand content and a nitrogen content that is frequently less than 1.2% (Mugwira and Mukurumbira 1984). Survey discussions with farmers and informal assessments reveal that the amounts applied are 8 to 20 t/ha, with applications every three to five years (Mugwira and Shumba 1986).

6. This problem is more common in the unimodal rainfall areas of southern Africa, where the long dry season makes zero-grazing techniques difficult or impossible for smallholders, than in the bimodal rainfall areas of eastern Africa.

7. Green manures were heavily researched from the 1920s to the 1940s (Metelerkamp 1988). They were widely used by large commercial farmers in Zimbabwe until the 1950s, when the real price of inorganic fertilizers fell, making green manures uneconomic. With the rise in real prices of inorganic fertilizers and concern over the sustainability of current cropping systems, green manures have attracted new research interest (see Hikwa and Mukurumbira 1995).

8. Giller, McDonagh, and Cadisch (1994) concluded that biological nitrogen fixation from legumes can sustain tropical agriculture at moderate levels of output, often double those currently achieved. Under favorable conditions, green manure crops generate large amounts of organic matter and can accumulate 100 to 200 kg N/ha in 100 to 150 days in the tropics.

9. In Malawi, MacColl (1989) estimated net nitrogen of 23 to 110 kg/ha from pigeon pea, 23 to 50 kg/ha from dolichos bean, and 25 kg/ha from groundnuts. In Nigeria, Jones (1974) and Giri and De (1980) estimated 60 kg N/ha from groundnuts.

10. Research has identified alternative species that are better adapted and produce more biomass than *L. leucocephala* (Bunderson 1994). But without a better understanding of the competitive effects between trees and crops, especially belowground, the potential of alley cropping is unproven at the farmer level.

11. The work synthesized here is the result of efforts by many persons and organizations in Malawi. In particular, credit for much of this research goes to John Wendt and Richard Jones, Rockefeller Foundation postdoctoral scientists attached to the Department of Agricultural Research in Malawi from 1989 to 1993.

12. A single blanket recommendation of 92 kg N and 40 kg P_2O_5 per ha was used for all farming areas of Malawi.

13. Typical scales include the plot level, farm level, watershed level, and regional level (Fresco and Kroonenberg 1992). To date, regional assessments of soil nutrient status and losses have necessarily been based almost entirely on extrapolation of plot estimates (Smaling 1993), but unfortunately results have been unreliable.

14. High-quality organic materials are high in available nitrogen and provide a source of energy (available carbon) to soil microorganisms. Enhanced activity of soil microbes can increase nutrient turnover rates, improving nutrient availability to crops while minimizing nutrient losses from leaching and volatilization (Doran et al. 1987; Ladd and Amato 1985).

12

Institutional Innovations in the Maize Seed Industry

Joseph Rusike & Carl K. Eicher

A large gap exists in the African development literature on the impact of structural adjustment programs on the performance of farmer support institutions, such as public and private research and seed and fertilizer delivery systems.[1] Research is urgently needed to guide public and private investments in seed improvement, because seed industry interventions in Africa have made so many false starts. For example, Pioneer Hi-Bred International recently wrote off U.S.$54 million in seed distribution and oilseed processing investments in Nigeria, Morocco, Côte d'Ivoire, Ethiopia, Sudan, Cameroon, Egypt, and Zambia because of political and institutional barriers and the lack of an adequate market to justify the company's level of investment (Pioneer Hi-Bred International 1993).

In this chapter, we investigate the technological, institutional, and organizational innovations in the maize seed industry in eastern and southern Africa. We begin by presenting a conceptual framework for analyzing seed supply systems. Next, we review the evolution of the maize seed industry in Africa, especially eastern and southern Africa, giving special attention to the industry's organization and performance in six countries in the region since the early 1980s. Finally, we examine the major institutional innovations that have occurred in the seed industry from 1980 to 1995 and discuss options for improving the seed industry's performance in the region.

Conceptual Framework for Analyzing Seed Supply Systems

Seed Supply

The development of a science-based seed industry begins with breeding, selecting, and testing genetic materials on different soils, in different climatic zones, and under various management applications to produce open-pollinated varieties (OPVs) and hybrids adapted to the conditions of different farmers. Because maize seed can become genetically contaminated, seed producers maintain quality by inspecting seed fields and conducting grow-out tests to guarantee that varieties remain pure and true to type. Seed production is followed by seed conditioning, which ensures that harvested seed contains no impurities that could affect the yield and quality of the subsequent crop. Finally, seed is distributed to farmers. Seed companies provide information to farmers on the relative advantages of different varieties and hybrids to demonstrate that they can earn a competitive return on their investment in high-quality seed.

A Life Cycle Model of Seed Industry Development

A number of analysts have suggested that science-based industries such as the seed and fertilizer industries follow a life cycle pattern of evolution. A typical life cycle pattern of development for the seed industry is shown in Table 12.1, along with a number of common market problems and a range of technological, organizational, and institutional innovations that are introduced to overcome these problems during the life of the seed industry.

During the first stage of the cycle, farmers save their own seed and exchange seed with a few other farmers. The emergence stage is marked by increased specialization, greater division of labor, and the emergence of a scientific community (Grant 1991). New knowledge in the form of product innovation leads to the development of a specialized industry. Because there are only a few pioneering firms and customers are not knowledgeable about their products, market penetration is initially slow. During the growth stage, as customers become more knowledgeable and experienced, there is increased growth in demand and market penetration. In the maturity stage, information is widely diffused, customers are influenced by suppliers' reputations, production technology is sophisticated, and the market becomes saturated. Normally, private organizations are the dominant actors in seed production and marketing in the fourth stage (Douglas 1980).

Transaction cost theory predicts that production costs will decline over the life cycle of an industry because of advances in technology (North 1990), but transaction costs will increase over time because of the increasing complexity of plant varieties and seed products and the growing size of the market. Rising transaction costs eventually erode the productivity gains from improvements in technology and economies of size and scope. This framework suggests that growth in the seed industry is shaped by technological innovations that reduce transaction costs; by institutional innovations, such as political, legal, and contractual structures; and by norms of behavior regarding contract fulfillment, honesty, and other efforts that hold down transaction costs.

At the farm level, transaction costs are caused by imperfect information, as well as the cost of transportation, negotiation, motivation, monitoring and

Table 12.1 Key Market Problems and Innovations During the Life Cycle of the Maize Seed Industry

	Stage of the Life Cycle			
	Farmer Seed Exchange	Emergence	Growth	Maturity
Key Market Problem	Uncertainty of Seed Quality	Technological Uncertainty	Demand and Financial Uncertainty	Logistics and Intellectual Property Rights Uncertainty
Innovations				
Technological	Testing and selection of varieties	Government research, seed certification, and testing of foreign varieties	Generic technologies	Proprietary technologies
Organizational	Farmer organizations	Government departments	Government departments and public- and private-sector partnerships	Public- and private-sector partnerships and commercial seed companies
Institutional	Informal habits and customary law	Political, legislative, administrative, and judicial structures; property ownership rights, laws, and regulations	Business and commercial contract law; trade secrets law	Plant breeders' rights, plant utility patent law, trademark law, reputation effect mechanisms

Source: Rusike (1995).

supervision, and contract enforcement in agricultural input and output markets. Transaction costs create price bands between the effective price farmers receive for products they sell and the price they pay for purchased inputs (de Janvry and Sadoulet 1994). Transaction costs can be reduced by establishing competitive markets, improving information, lowering transportation costs, and improving banking and legal services, education and extension, and farmer organizations that can achieve economies of scale (Fafchamps, de Janvry, and Sadoulet 1995). These activities will speed up the adoption of yield-raising genetic products and agronomic practices.

The most efficient transaction and production cost–minimizing organizational structures in the seed industry will vary across the different stages of the life cycle depending upon asset specificity, technological and market uncertainty, and the frequency of exchanges (Williamson 1990). The distinctive competencies to develop new crop varieties and to produce and market superior seed will shift from farm households to public organizations and then to private organizations over the life cycle of the industry because of changing technologies and resources, technological and market uncertainty, breadth of seed markets, demographic factors, and political and legal rules.

Overview of the Maize Seed Industry in Africa

Africa's maize area is concentrated in eastern and southern Africa, where maize is the major staple food crop and occupies around 78% of the cereal area (this figure rises to more than 95% in Malawi and Swaziland).[2] Because hybrids cover most of the maize area in southern Africa, commercial seed sales are high, accounting for 74% of maize seed sales in Africa. By contrast, much of the maize area in western and central Africa is planted to improved OPVs, and most maize area in central and eastern Africa is under local varieties.

Despite the high proportion of farmer-saved seed that is planted in western, central, and eastern Africa, these regions have a larger number of public research stations and public researchers per million ha of cultivated maize than does southern Africa (Table 12.2). Public seed companies dominate the market in western Africa because much of the maize area is planted to OPVs (Table 12.3, p. 178). In many countries in western, central, and eastern Africa, the government produces and distributes seed, seed prices are subsidized, and the price margins between OPVs and public and proprietary hybrid seed are high. By contrast, the margins are narrower in southern Africa, where the seed industry is dominated by the private sector and seed prices are not generally subsidized. Because hybrids account for two-thirds of all of the maize seed planted in southern Africa, it is understandable why private seed companies are so dominant in this part of Africa.

Table 12.2 Sub-Saharan Africa: Public and Private Investment Share of Commercial Maize Seed Sales and Seed Prices

Region/ Country	Public-Sector Breeders (per million ha of maize)	Seed Sales, 1993 (000 t)	Share of Commercial Seed Sales (%)		Ratio of Price		
			Publicly Bred Hybrids	Privately Bred Hybrids	OPV Seed to Grain Price	Public Hybrids to Grain Price	Proprietary Hybrids to Grain Price
Eastern Africa							
Burundi	16.1	0.13	100	0	1.3	n.a.	n.a.
Ethiopia	22.0	2.80	87	13	2.5	3.5	11.4
Kenya	7.3	21.61	100	1	6.3	6.3	6.3
Rwanda	12.5	0.03	100	0	3.7	n.a.	n.a.
Tanzania	3.1	2.67	74	26	5.5	6.1	7.3
Uganda	4.5	1.04	100	0	6.1	13.0	7.0
Subtotal	10.9	28.28	94	6	4.3	7.2	8.0
Southern Africa							
Lesotho	0.0	1.88	3	97	5.7	n.a.	n.a.
Malawi	2.2	8.50	42	58	n.a.	8.1	8.7
Mozambique	4.8	13.19	96	5	7.1	n.a.	10.0
South Africa	6.1	37.50	7	93	5.5	n.a.	13.2
Zambia	7.8	10.81	96	4	3.0	3.9	4.7
Zimbabwe	4.5	30.00	66	34	n.a.	n.a.	4.2
Subtotal	4.2	102.91	51	49	5.4	6.1	9.1
Western Africa							
Benin	6.8	0.22	100	0.0	2.1	n.a.	n.a.
Burkina Faso	10.0	0.19	100	0.0	4.0	n.a.	n.a.
Cameroon	38.1	0.46	55	45.2	6.5	10.0	13.4
Côte d'Ivoire	3.1	3.25	98	2.3	8.9	16.0	19.0
Ghana	8.0	0.45	100	0.0	4.2	n.a.	n.a.
Mali	5.7	0.15	100	0.0	2.5	n.a.	n.a.
Nigeria	3.3	2.50	88	12.0	2.7	8.8	n.a.
Senegal	19.0	0.23	100	0.0	2.7	n.a.	n.a.
Togo	7.7	0.20	100	0.0	3.9	4.1	n.a.
Subtotal	11.3	7.65	93	6.6	4.2	9.7	16.2
Total	7.9	138.84	62	38	4.5	8.0	10.0

Source: CIMMYT (1994).
Note: n.a. = data not available.

Evolution of the Maize Seed Industry in Six Countries of Eastern and Southern Africa

Until the early 1900s, farmers in eastern and southern Africa saved their own seed and exchanged seed with other farmers. The establishment of colonial research stations at the turn of the century initiated a chain of events that led to the emergence of commercial seed trade and science-based seed industries. The following discussion traces the evolution of the maize seed industry in six countries in eastern and southern Africa: South Africa, Zimbabwe, Kenya, Zambia, Malawi, and Tanzania (see summary in Table 12.4, p. 179).

Table 12.3 Sub-Saharan Africa: Number of Companies Engaged in the Maize Seed Industry by Type of Company, 1992

Region/ Country	Public	Private	National	Multinational Cooperative	Nongovernmental Organization	Total
Eastern Africa						
Burundi	1	0	0	0	0	1
Ethiopia	1	0	1	0	0	2
Kenya	0	1	0	0	0	1
Rwanda	1	0	0	0	0	1
Tanzania	1	1	1	2	0	5
Uganda	1	0	0	0	0	1
Subtotal	5	2	2	2	0	11
Southern Africa						
Lesotho	0	2	2	0	0	4
Malawi	0	0	2	0	0	2
Mozambique	2	0	0	0	0	2
South Africa	1	3	1	1	0	6
Zambia	1	0	1	0	0	2
Zimbabwe	0	0	3	1	0	4
Subtotal	4	5	9	2	0	20
Western Africa						
Benin	3	0	0	25	8	36
Burkina Faso	7	0	0	4	5	16
Cameroon	1	1	1[a]	1	1	5
Côte d'Ivoire	1	1	1[a]	0	0	3
Ghana	1	3	0	0	0	4
Mali	5	0	0	0	0	5
Nigeria	33	2	1[a]	0	0	36
Senegal	1	1	0	1	0	3
Togo	3	0	0	1	0	4
Subtotal	55	8	3	32	14	112

Source: CIMMYT (1994).
Note: a. Pioneer Hi-Bred International withdrew in 1994.

South Africa

The South African seed industry was in the emergence phase from 1900 to around 1960. At the turn of the twentieth century, commercial farmers encouraged the government to organize agriculture along scientific lines. The Department of Agriculture established experiment stations to select and test foreign and local varieties and multiply varieties for distribution to commercial farmers. The government and commercial farmer associations imported seed from established seed growers in the United States, and the accidental hybridization of two of these varieties produced Potchefstroom Pearl, which became the anchor of South Africa's hybrid maize seed industry (Saunders, undated).

A number of institutional innovations in the public sector helped lay the foundation for a science-based seed industry. The Fertilizer and Seeds Acts, introduced in 1907, prohibited the sale of adulterated inputs. A few

Table 12.4 Eastern and Southern Africa: Stages of the Life Cycle of the Maize Seed Industry in Six Countries, 1996

Country	Stage					
	Emergence		Growth		Maturity	
	Period	Major Events	Period	Major Events	Period	Major Events
South Africa	1900–1960	Government research stations; new varietal introductions and selections; importation of seed; Fertilizers and Seeds Acts; Maize Seed Associations.	1960–1980	Release of hybrids SA4, SA11, SA13, and SA200. Seed Act; transfer of seed production and marketing to private sector; private breeding programs; Plant Breeders' Rights Act. Entry of multinationals.	1980–1996	100% hybrid coverage on commercial farms. Agricultural reforms and desegregation; consolidation of industry; transfer of seed certification, laboratory testing, and phytosanitary regulations to private sector. Entry of Pioneer Hi-Bred International and Pacific Seeds.
Zimbabwe	1900–1960	Colonial research stations; new varieties; seed certification; Maize Breeders' Association; hybrid breeding program; Southern Rhodesia Seed Maize Association.	1960–1985	Release of hybrids SR52, R200, R201, and R215. Tripartite Agreements; Plant Breeders' Rights Act; private hybrid program.	1985–1996	100% hybrid coverage. Economic reforms; transfer of seed certification and seed testing to the private sector. Entry of Pioneer, Pannar Seed Company, Cargill Hybrid Seeds, and Africa Pacific Seeds.
Kenya	1900–1970	Colonial research stations; new varietal introductions; hybrid maize program; Kitale Synthetic II. Release of H611, H622, and H632.	1970–1996	Release of H614, H625, and H680. Transfer of seed production and marketing to Kenya Seed Company; seed law; economic reforms. Entry of Pannar, Cargill, and Pioneer Hi-Bred International.		

(continues)

Table 12.4 continued

Country	Stage					
	Emergence		Growth		Maturity	
	Period	Major Events	Period	Major Events	Period	Major Events
Zambia	1900–1970	Colonial research stations; new varietal introductions; importation of seed from South Africa, Federation of Rhodesia and Nyasaland; seed testing. Initiation of national maize breeding program.	1970–1996	Hybrid breeding; seed inspection; seeds act; Zamseed; economic reforms. Entry of Pioneer Hi-Bred International, Pannar, Cargill, and Carnia Seed International. Transfer of breeding and seed certification to the private sector.		
Malawi	1900–1985	Colonial experiment stations; varietal introduction; government breeding; seed production and marketing; seed certification; seed testing laboratory.	1985–1996	Economic reforms; entry of Cargill, Lever Brothers, and Pannar. Transfer of breeding, seed certification, and laboratory testing to private sector.		
Tanzania	1900–1996	Colonial experiment stations; maize breeding; new varietal introduction and testing; USAID-funded Seed Multiplication and Distribution Project; economic reforms. Entry of Cargill and Pannar.				

years later, European settler farmers began organizing maize seed shows to spread the word about new varieties. The formation of a maize breeders' association and a seedsmen and nurserymen association in 1917 paved the way for the diffusion of knowledge about hybrid seed production. The government initiated a hybrid breeding program in 1925, established the Maize Control Board in 1944, and organized a seed inspection service and an official seed testing laboratory that same year. The Maize Board facilitated the seed industry's transition to the growth phase by directly undertaking hybrid seed production and marketing, as well as nurturing cooperatives and private seed companies.

The growth phase of South Africa's seed industry lasted about 20 years, from 1960 to 1980. A number of research breakthroughs led to the release of superior hybrids. During this phase, the Seeds and Foundation Seed Acts and Plant Breeders' Rights legislation were passed, and plant breeding and variety development, foundation seed production, and seed growing and marketing were transferred from the public sector to domestic private and multinational seed companies. By 1980, virtually all of the commercial maize area in South Africa was planted to hybrids.

The seed industry entered the maturity phase around 1980. Seed companies adopted biotechnology, computer, and information technologies; financial and exchange rate reforms were introduced; private companies assumed responsibility for seed certification, laboratory seed testing, and the management of national trials; and the South African National Seed Organization and the Agricultural Research Council were established (Van der Walt 1990).

Average maize yields on commercial farms doubled between 1950 and 1994. The increase in yields and production on commercial farms between 1950 and 1970 can be attributed almost entirely to the adoption of hybrid seed. A second growth spurt in commercial maize yields in the 1970s and 1980s resulted from a combination of genetic improvements with intensive fertilization, crop protection, and more timely crop operations (Gevers 1988). During this second period, about 40% of the increase in yields came from genetic advances and 60% from the increased use of fertilizer, pesticides, and better agronomic practices. The number of registered maize cultivars on the list of official varieties increased from 16 hybrids and 37 OPVs in 1964 to 284 hybrids and 19 OPVs in 1993 (Van der Walt 1990). Smallholders do not use these improved varieties extensively, however, because improved agronomic practices and input delivery systems for smallholders have not yet been developed.

The South African seed industry in 1996 was dominated by five firms: Pannar-Saffola, Sensako-DeKalb, Carnia-Asgrow Seed, Pioneer Hi-Bred International, and Pacific Seeds. The challenge facing the industry is to develop maize cultivars and agronomic technologies for smallholders, who cultivate 20% of South Africa's maize area but produce only about 15% of South Africa's maize.

Zimbabwe

Zimbabwe's seed industry, like South Africa's, was in the emergence phase from 1900 to 1960. Like their counterparts in South Africa, European farmers in Zimbabwe successfully lobbied the government to organize agriculture along scientific lines, and the Department of Agriculture was established in the late 1890s (Smith 1979). The department's experiment stations were responsible for introducing, selecting, and testing foreign and local varieties and distributing seed of suitable varieties to commercial farmers. Even though government and farmer associations imported maize seed of superior varieties from seed growers in the United States in the early 1900s, it was a local variety, Salisbury White, developed through an accidental hybridization, that became the anchor of Zimbabwe's seed industry (Weinmann 1972; Smith 1979). An important institutional innovation was the introduction of a voluntary seed certification scheme in 1913. Under this scheme, farmers and seed merchants submitted samples of and information about the names, origin, breeding history, and price of varieties available for sale to the Department of Agriculture, which then acted as a clearinghouse between buyers and sellers. In the 1910s, European farmers organized maize seed shows to help others gain access to superior strains of established varieties. A maize breeders' association, established in 1919, encouraged the production of varietally pure maize seed and paved the way for hybrid seed production (Mainwaring 1922).

The government prepared the transition to the growth phase by initiating a hybrid maize breeding program in 1933. The Southern Rhodesia Seed Maize Association was set up by commercial farmers in 1940, and an official seed testing laboratory was established in 1950, followed by enactment of the Fertilizers, Farm Feeds, Seeds, and Remedies Act in 1952. This act required commercial seed to conform to minimum standards of genetic purity, germination, and weed-seed content.

Zimbabwe's maize seed industry was in the growth phase for about 25 years, 1960 to 1985 (Tattersfield and Havazvidi 1994). This phase was dominated by a breeding breakthrough in 1960 that produced the world's first single-cross hybrid, Southern Rhodesia 52 (SR52). Without question, the popularity of SR52 seed—widely adopted at home, exported to neighboring Zambia and South Africa, and grown as far north as Ethiopia and as far west as Cameroon—helped the Seed Co-op of Zimbabwe to become the leading seed company in Africa. Further advances in plant breeding in Zimbabwe led to the release of several three-way hybrids (R200, R201, and R215) in the mid-1960s; these hybrids were well adapted to areas where rainfall was less reliable.[3]

In 1970, the Tripartite Agreement among the Seed Maize Association (later renamed the Seed Co-op), the Commercial Farmers' Union, and the government gave the Seed Maize Association exclusive marketing rights to publicly bred germ plasm. In return, the association agreed to maintain a strategic reserve of 20% of Zimbabwe's annual seed requirements. In addition, the government enacted a Plant Breeders' Rights Act in 1973, which induced the Seed Maize Association to initiate private breeding of hybrids in 1974. The accession to power of the first democratically elected government in 1980 spurred the growth of the maize seed industry as well: Smallholder maize area expanded following the expansion of agricultural extension, improvements in infrastructure, and favorable producer prices (Rohrbach 1989). By 1985, more than 90% of all maize area in Zimbabwe was planted to hybrids.

The Zimbabwean seed industry entered the maturity phase around 1985. Economic reforms from 1985 to 1995 encouraged four multinational seed companies to enter the national market: Pannar Seed Company (South Africa), Pioneer Hi-Bred International, Cargill Hybrid Seeds, and Africa Pacific Seeds. In addition, the Seed Co-operative Company of Zimbabwe, which controls 90% of the seed market, recently formed strategic alliances with seed companies in South Africa, the United States, Mozambique, Zambia, and Kenya. The government is gradually transferring responsibility for plant breeding, seed certification, and laboratory seed testing to the private sector (De Woronin 1993). Eight companies sold maize seed in Zimbabwe in 1996, compared with one in 1986.

The development, production, and adoption of hybrids have been described as the greatest single contribution of public research to the agricultural industry of Zimbabwe (Weinmann 1975). The 35 hybrids released by the government breeding program between 1950 and 1990 fueled Zimbabwe's maize-based green revolution (Eicher 1995). Although virtually all smallholders now plant hybrids, their yields still lag behind yields obtained on commercial farms. This yield gap persists for several reasons: Most smallholders plant old hybrids, and the genetic advances offered by breeding research have not been matched by agronomic practices and efficient support services for smallholders, many of whom are located in marginal areas.

Kenya

The Kenyan seed industry remained in the emergence phase for 70 years, from 1900 to 1970. The Department of Agriculture secured U.S. maize varieties through South Africa during the early part of the twentieth century. A few farmer-breeders developed the Kenya Flat White variety, subsequently the anchor of the hybrid seed industry. The government launched a maize improvement program in 1930, a public hybrid breeding program in 1954, and a maize agronomy research program in the 1960s. The Kenya Seed Company was established in 1956 as a cooperative of commercial growers. A superior-performing Kitale Synthetic II was released in 1961, followed by hybrids in 1964; seed sales accelerated around 1970 after smallholders rapidly began adopting hybrids. The government enacted and enforced a seed law governing the testing and release of new varieties and hybrids, seed certification, and quality control.

Kenya's seed industry has been locked into the growth phase since 1970. The Kenya Seed Company, which originally started as a private company but is currently owned by the government (51%) and private investors (49%), has a monopoly on all seed trade in Kenya. The government recently committed to opening up the seed industry, however, by allowing domestic and multinational seed companies to engage in the breeding, production, marketing, and distribution of agricultural and horticultural seeds. Cargill Hybrid Seeds, Pannar Seed Company, and Pioneer Hi-Bred International have tested their hybrids in Kenya since 1992; in 1994–1995, these multinationals started to sell their proprietary hybrids in Kenya. The Kenya Seed Company has responded to this growing international competition by entering into a strategic alliance with the Seed Cooperative Company of Zimbabwe and Zamseed of Zambia, which covers research, production, and marketing. These innovations have set the stage for Kenya's seed industry to move from the growth to the maturity phase.

The Kenyan maize seed industry has helped to double national average maize yields and increase maize output fivefold from 1955 to 1988 (Karanja 1996). By helping to achieve maize self-sufficiency and exports during most of the past 45 years, maize research in Kenya has benefited both large- and small-scale farmers, as well as consumers. Some of this success can be attributed to effective donor support, including foreign assistance spread over 22 years (1955–1977). But despite this impressive continuity of assistance, Kenya was unable to develop the scientific, organizational, and financial capability to sustain its maize research program after foreign assistance was withdrawn in 1977. In 1987, the government asked the U.S. Agency for International Development to resume technical assistance to its maize research program. Kenya's inability to develop a sustainable public research program in maize after 30 years of independence raises questions about why government financial support for maize research is lacking, why scientists in government research stations have few career incentives, and how much time it will take to develop dynamic, productive, and adequately financed national agricultural research systems in Africa. Because Kenya has been unable to develop and sustain a national government maize improvement program with its own financial resources, it also seems appropriate to ask why the government has been slow in encouraging private maize seed companies to take over more of the research activities from the government.

Zambia

In Zambia, the maize seed industry was in the emergence phase from 1900 until 1970. The Department of Agriculture set up experimental gardens in

1913 to test and introduce new maize varieties from South Africa, the United States, and Zimbabwe. Zambia imported SR52 maize seed from Zimbabwe in the early 1960s, and almost all of the commercial farmers switched from OPVs to hybrids in the 12 years from 1955 to 1967. Commercial yields quadrupled and production increased sevenfold between 1950 and 1994. Commercial farmers, however, account for only a small proportion of the maize produced in Zambia.

After independence in 1964, the government launched a hybrid breeding program and seed production scheme, enacted a seed law governing the testing and release of new varieties and hybrids, established seed certification and quality control by the Seed Services, and marketed seed to farmers through the parastatal National Agricultural Marketing Board. In 1980, the government set up the Zambia Seed Company (Zamseed). Recent economic reforms have stimulated several multinational seed companies—including Pioneer Hi-Bred International, Cargill Hybrid Seeds, Pannar Seed Company, Sempro, and Carnia Seed International—to enter the Zambian seed market and invest in seed production, conditioning, and marketing facilities and personnel. The government introduced mandatory seed certification for hybrids and required private companies to submit samples of their inbreds to the Seed Control and Certification Unit. Since Zambia lacks Plant Breeders' Rights legislation, Pioneer Hi-Bred International withdrew from the Zambian seed market in the early 1990s rather than release samples of its inbred lines. Because of increasing global competition, Zamseed recently forged a strategic alliance with the Seed Co-operative Company of Zimbabwe (Kabaghe 1992).

During the 1980s, smallholders rapidly adopted improved hybrids, but the removal of fertilizer subsidies in 1989 and the government's failure to deliver inputs and pay for produce resulted in declining maize area and production and led to substantial maize imports in the 1990s (Howard 1994). Even so, the institutional reforms of the 1980s opened up the seed industry and prepared it to evolve in much the same manner as the industry has evolved in Zimbabwe over the past decade. Zambia's maize seed industry is now prepared to move from the growth to the maturity phase.

Malawi

In Malawi, the Department of Agriculture set up an experiment station in 1909 to introduce, test, and select new maize varieties from South Africa and the United States. It was not until 1950, following the severe 1949 famine, however, that the government launched a hybrid breeding program. By the time Malawi joined the Federation of Rhodesia and Nyasaland in 1953, maize researchers had developed an improved, locally selected semiflint variety. In 1958, two flint varieties were released, as well as the first dent hybrid. But the government did not develop seed certification and laboratory seed testing procedures to control seed quality. Also, because varieties were not maintained and field inspection and seed testing services were lacking, the new varieties and hybrids became contaminated, and the rate of adoption declined (Gausi 1970). The lack of demand for the varieties and hybrids led the government to suspend the hybrid breeding program and discourage the use of hybrids from Zimbabwe.

With financial assistance from the UK, the government implemented a maize and groundnut development project in 1948 to expand the production of superior varieties and increase the availability of high-quality seed to farmers. The government later introduced a seed certification scheme, revived the hybrid program, and established the National Seed and Cotton Milling Company of Malawi (NSCM) to take over foundation and commercial seed production, conditioning, and storage. In 1989, Cargill, a multinational seed company, entered the market by acquiring 55% of the equity of the NSCM and reorganizing the company into a profit-driven firm. Lever Brothers, another multinational, entered the seed market in 1991, and Pannar from South Africa forged a strategic alliance with Lever Brothers in 1993. Recently, the government announced plans to deregulate the seed industry as part of its structural adjustment program. Malawi's seed industry is now poised to enter the maturity phase.

Although Malawi achieved self-sufficiency in maize production in the 1950s and 1960s, maize was imported in the 1970s and early 1980s because of stagnant yields and a large influx of Mozambican refugees. Smallholders dramatically increased the area planted to hybrids, however, from 7% in 1988 to 24% in 1992 following the release of two high-yielding semiflint hybrids with good processing and home storage characteristics (Heisey 1990; Smale 1995). Although hybrid maize is planted on only 25% of the maize area in Malawi, it accounts for more than 50% of national production.

Tanzania

In Tanzania, the Department of Agriculture initiated a maize improvement program before World War I. During the 1950s, the government organized research stations to develop new varieties, but the breeding work had little impact because the government failed to establish seed laws, seed certification programs, quality control services, and distribution systems (Experience Incorporated 1969). In the early 1970s, with assistance from the U.S. Agency for International Development, the government implemented a Seed Multiplication and Distribution Project and established foundation seed farms, conditioning plants, seed testing laboratories, and a seed law (which the Tanzania Official Seed Certification Agency was charged to enforce) (USAID 1985). The government also established two parastatals—the Tanzania Seed Company (TanSeed), which assumed responsibility for seed production and distribution, and TanWatt, which took over the production of foundation seed. TanSeed and TanWatt were poorly managed, however, and lacked the trained staff and organizational capabilities needed to produce high-quality seed (Danagro 1987).

In 1990, the government launched economic reforms and deregulated the seed industry, decontrolled seed and commercial grain prices, privatized TanSeed, and created incentives to attract foreign companies. Cargill Hybrid Seeds entered the seed industry in 1991 with South African–bred hybrids. Pioneer Hi-Bred International and Pannar began testing their hybrids in official variety performance trials in 1993. Pannar now sells seed in Tanzania that it imports from its operations in Zimbabwe and South Africa. Pioneer Hi-Bred International has registered its products in Tanzania but is not selling them because distribution channels and effective demand are lacking. For many reasons—unfavorable government policies; low producer prices; unavailability of high-quality seed, fertilizer, and agrochemical inputs; poor marketing infrastructure; and weak institutions—maize yields have been erratic in Tanzania over the past 30 years. Tanzania remains in the emergence phase of the life cycle of the seed industry. To date, most of the increase in maize production has come from area expansion.

Stages of the Life Cycle of the Maize Seed Industry

The thesis advanced here is that the seed industry progresses through different stages of a life cycle—the farmer-to-farmer stage, emergence, growth, and maturity—by an incremental and path-dependent learning process for governments, agribusiness firms, and farmers. This learning-by-doing process implies that government policies, foreign investment, and donor assistance must be carefully tailored to the specific stage of the life cycle of the seed industry in each country. For this reason, country-specific research is needed on the seed industry and other input delivery institutions (De Capitani and North 1994). Having reviewed the evolution of maize seed industries in six countries of eastern and southern Africa, we now examine the stages themselves in greater detail. As we describe each stage, we note which countries in our study had reached that stage by 1996 and discuss the steps that need to be taken to move the industry in a particular country to the next part of the life cycle. Table 12.5 displays the stage of growth of the seed industry and the number of seed companies in these six countries in 1996.

Farmer-to-Farmer Phase

The farmer-to-farmer phase of the life cycle is important in many parts of Africa where farmers save their own seed from the open-pollinated maize

Table 12.5 Eastern and Southern Africa: Public and Private Companies Distributing Maize Seed in Six Countries, 1996

Country	Stage of Life Cycle	Public	Public/Private	Private
Kenya	Growth	—	Kenya Seed Company	Pannar, Cargill, Pioneer Hi-Bred International
Tanzania	Emergence	TanSeed	—	Pannar, Cargill, Pioneer Hi-Bred International, Bondeni Seeds
Malawi	Growth	—	National Seed and Cotton Milling Company of Malawi	Pannar, Lever Brothers, Cargill
Zambia	Growth	—	Zamseed	Cargill, Pannar, Sempro, Carnia
Zimbabwe	Maturity	—	—	Seed Co-op of Zimbabwe,[a] Cargill, National Tested Seeds, Pannar, Pioneer Hi-Bred International, Pacific Seeds, Carnia, Agri Seeds and Services
South Africa	Maturity	—	—	Pannar-Saffola, Sensako-DeKalb, Pioneer Hi-Bred International, Carnia-Asgrow Seed, Pacific Seeds

Note: a. The Seed Co-op became a private company in 1995.

varieties and other staple food crops they grow. Even if improved seed is available in a few parts of a country, farmers in marginal areas usually lack the complementary resources—such as draft power, labor, fertilizer, and cash—for adopting improved varieties. As a result, the market for improved varieties is small, and private seed companies cannot recover their investments in research and development, seed plant and equipmental, and sales promotion. In some African countries, various governmental and nongovernmental organizations distribute free seed to farmers following a drought. These seed distribution schemes enable farmers to see the value of using improved varieties and help lay the foundation for commercial seed trade.

The transition from farmer-to-farmer seed exchange to the emergence phase of the life cycle requires an elementary science base to enable technological and institutional innovations to moderate the costs of obtaining information and negotiating and enforcing contracts.[4] The critical element in moving the seed industry to the emergence phase is government research, because long lags occur between the initiation of research and availability of research products and the risky nature of research (Pray and

Ramaswami 1993). In most cases, private firms cannot finance research over the 10 to 15 years that are typically needed to develop, test, and release a new variety. Once public researchers have developed an improved variety or large-scale farmers or traders have imported an improved variety that is well adapted to local conditions, the demand for improved seed will increase significantly. This increase in demand is crucial in providing opportunities for the private sector to sell seed.

Emergence Phase

The government is the dominant actor during the emergence phase, in part because of the small, uncertain market for seed and in part because of the lack of legal protection and economic certainty for private firms to invest in specialized research stations, seed farms, and conditioning and marketing facilities. As a result, the government underwrites virtually the total cost of plant breeding, variety trials, production of foundation seed, multiplication, conditioning, certification, and laboratory testing. Since the market is small, private firms are unable to obtain competitive returns on their investments in seed production and marketing.

Once improved seed is available from the public sector, seed certification is the next step in promoting the transition from the farmer-to-farmer to the emergence phase of the seed industry life cycle. The government certification label informs farmers of the quality of seed offered for sale. As farmers increasingly adopt certified seed of improved varieties, agricultural productivity increases; resources become available for buying seed, fertilizer, and other scientific inputs; farmers increase their demand for proprietary hybrids; and the seed industry enters the growth phase. The driving force that moves the industry to the growth phase is the profits private firms make from producing and distributing seed. New generic technologies are required, such as hybridization, detasseling, and cytoplasmic male sterility. Infratechnologies (seed certification, sampling, laboratory testing, and health and pest control trials) are also needed to reduce buyer uncertainties and achieve rapid market penetration.[5]

Of the six countries studied here, Tanzania is the only one currently in the emergence phase. The government seed company, TanSeed, dominates the industry. Four small private companies sell seed in Tanzania, but the overall legal and policy environment is not conducive to the private sector in Tanzania. Until private firms become more profitable, Tanzania is unlikely to move to the next stage of the life cycle.

Growth Phase

Three of the six countries studied (Kenya, Malawi, and Zambia) are in the growth phase of the seed industry life cycle (Table 12.5). When seed markets become increasingly commercialized, private seed companies assume an important role in plant breeding, variety development, seed testing, and marketing. During the growth phase, legal codes of conduct are developed to regulate the seed trade, plant breeding, seed certification, and the preservation of trade secrets. These codes provide protection for private seed companies. Gradually, private firms begin to certify seed and advertise that their seed exceeds the minimum government standards. They use a combination of government certification labels and brand names to develop a reputation for product quality. The growth of the private seed industry increases the array of high-quality seed of superior varieties and hybrids.

As farmers adopt higher-yielding seed, they adopt improved agronomic practices and increase the use and efficiency of fertilizer and pesticides. Market competition helps assure seed quality, because farmers do not patronize firms that market inferior seed products. During the growth phase, government seed services continue to enforce quality and phytosanitary standards for seed imports, but the government transfers the responsibility for seed certification and laboratory testing to the private sector.

What can Kenya, Malawi, and Zambia do to move from the growth to the maturity phase of the life cycle? The answer is to tighten the legal framework for private investors, remove controls over seed prices, and allow private firms to make sufficient profits to invest in plant breeding programs and improved facilities. The transition to the maturity phase will require investments in intellectual property rights, such as patents, trademarks, plant varietal protection certificates, plant patents, formulas, confidential information, and trade secrets (López-Pereira and Filippello 1995).

Maturity Phase

Currently, two of the six countries studied here (Zimbabwe and South Africa) have entered the maturity phase of the life cycle. During this phase, private firms "do most of the research on plant breeding, produce most of the breeder seed, and produce virtually all of the foundation or commercial seed" (Pray and Ramaswami 1993:318). In many countries, seed companies pay royalties to the government for new varieties that are developed by government research programs. Seed companies initially focus on hybrid crops such as maize, sunflower, and sorghum and then gradually shift to other crops as market competition intensifies and drives companies to broaden their product portfolios.

As competition for market share intensifies, the seed industry moves through a period of consolidation through mergers, acquisitions, and strategic alliances among seed companies and agrochemical firms. Intellectual property rights are protected through laws conferring plant breeders' rights and plant patents. This protection, in turn, stimulates investments in seed enhancement processes, biotechnology, and information technology. Private seed companies use a mix of brand names, logos, trademarks, advertising, field demonstration, personal selling, and agronomic signals to inform farmers about the characteristics of their products. Because of the intense competition among seed companies and the need to gain and retain farmers' loyalty, there is a continuous flow of superior-performing, high-quality genetic products to farmers, which contributes to an increase in the national average maize yield.

To summarize, the private sector dominates the seed industry in the maturity phase, but the government continues to play a major role as a watchdog over the private sector. The government enforces the seed laws and regulations and carries out breeding programs for OPVs and for farmers in resource-poor areas.

Conclusions

The six case studies of maize seed development examined here support the view that the seed industry evolves through four stages of a life cycle, which begins with the traditional, farmer-to-farmer seed exchange phase and moves on through the emergence, growth, and maturity phases. In some countries, the initiative for the seed industry originated with a group of farmers; in others, the initiative originated with the government. The government plays the dominant role in the early stages of seed industry development by initiating research to develop well-adapted, superior varieties or hybrids and establishing a reputable seed inspection and certification service.

As revealed by the six case studies described here, however, the roles of the public and private sectors in seed industry development change over time. In mature seed industries, such as those in South Africa and Zimbabwe, the private sector replaces the government as the dominant actor, although the government plays a continuing role even in countries with a mature seed industry. In South Africa, for example, the government is responsible for germ-plasm conservation, breeding for minor crops and marginal areas, and maintaining an official register of varieties.

Throughout Africa, structural adjustment programs are encouraging reduced roles for government parastatals, such as seed and fertilizer companies, and wider roles for the private sector in maize improvement research and seed production, conditioning, and marketing. As a result, several government seed companies in eastern and southern Africa have entered into strategic partnerships with private regional and multinational seed companies. A number of institutional innovations have been introduced to provide the legal protection and seed quality assurances required to induce private investments in the seed industry. Revisions in tax laws have encouraged new private seed companies to enter the seed business in eastern and southern Africa. Even so, seed industry conditions vary from one country to another. For example, private investors remain hesitant to invest in the seed industry of Tanzania, because they lack legal protection and a stable economic environment.

Despite these differences among countries, the maize seed industry is clearly undergoing rapid change across the region. From 1980 to 1996, the number of public, public–private, and private seed companies in five countries in southern Africa (Kenya, Malawi, Tanzania, Zambia, and Zimbabwe) tripled, growing from 5 to 15. The combination of new private seed companies and new forms of government–private seed partnerships has helped increase the number of new maize materials offered to farmers, which, in turn, has led to higher yields and a greater percentage of maize area planted to hybrids in the region. Because structural adjustment programs are still being implemented, however, at present it is impossible to quantify their overall impact on the seed industry in the region.

The varied patterns of seed industry development in the study countries provide some important insights for future donor assistance in seed industry development. Donors have initiated seed projects in many African countries under the false assumption that the legal and institutional framework for a viable seed industry was already in place, with the result that many of these projects were ineffective. In fact, the great diversity among and within African countries—all of which are at different stages of institutional, legal, and scientific development—means considerable care must be given to designing and tailoring seed assistance programs to the specific needs of each country. It behooves analysts to eschew cookbook approaches to seed industry development in Africa. Country-specific research is needed to assess the political, legal, and institutional environment of each country; determine the present stage of seed industry development; identify problems facing the industry; and outline the specific policy, legal, and institutional innovations that are needed to help the industry improve its performance.

Notes

1. Cromwell (1996), Friis-Hansen (1992), Venkatesan (1994), and Wiggins and Cromwell (1995) are among the few studies of the seed industry in Africa.

2. See Byerlee et al. (1994) and CIMMYT (1981, 1990, 1992a, 1992b).

3. Rohrbach (1989) has reported that 20 to 30% of the smallholder maize area was planted to hybrids before 1980.

4. The necessary institutional preconditions are a coherent political, legislative, administrative, and judicial system for resolving the question of whose interests count; property ownership rights; and laws and regulations.

5. Generic technologies are product concepts that have been shown to work in a laboratory environment. Infratechnologies consist of measurement methods, quality standards, and quality assurance procedures that reduce buyer uncertainty and facilitate market penetration.

13

Fertilizer Use and Maize Production

Paul W. Heisey & Wilfred Mwangi

One of the most crucial issues in African agricultural development today is how to manage soil fertility to improve food crop production. A strategy that seeks to meet Africa's future food needs will have to feature the use of mineral fertilizer. Because maize has a relatively high response to fertilizer compared with other staple food crops in Africa, maize production and fertilizer use will likely be linked even more closely in the future than they have been in the immediate past.

Although the appropriateness of seed-fertilizer technology will continue to be debated, it is clear that Africa can no longer be regarded as primarily a land-abundant region where farmers can increase crop production simply by bringing more land into production. This has been one of the major arguments propounded in the past against reliance on a seed-fertilizer strategy as a principal component of agricultural development. Although conditions across the region are extremely variable, land for agriculture is scarce in many countries (Binswanger and Pingali 1988), and yield increases rather than area expansion will become more important as a source of growth in crop production.

The declining real price of fertilizer over much of the twentieth century enabled fertilizer to contribute greatly to increasing world crop production (Tomich, Kilby, and Johnston 1995). In developing countries outside Africa, 50 to 75% of the increases in crop yields from the mid-1960s onward have been attributed to fertilizers (Viyas 1983). Fertilizers are also complementary to the other major inputs (improved seed, better water control) that have boosted yields the most.

The potential for mineral fertilizer to have an impact on agriculture in Africa is great: Most African agriculture has suffered as soil nutrients have been depleted over time (Smaling 1993). Nutrients from sources other than inorganic fertilizers can help to restore the balance, and improved organic techniques for supplying soil nutrients will undoubtedly contribute to future soil health and productivity (see Chapter 12). If farmers rely only on the efficient recycling of nutrients available in depleted soils, however, food production will not grow to the extent that will be necessary (Janssen 1993). For the foreseeable future, "the environmental consequences of continued low use of fertilizers" through nutrient mining and increasing use of marginal lands "are more inevitable and devastating than those anticipated from increased fertilizer use" (Dudal and Byrnes 1993:152).

Based on these considerations, many observers have called for increases in fertilizer consumption in Africa of at least 15% annually (Mellor, Delgado, and Blackie 1987). Even based on the experience of non-African countries in the developing world, where fertilizer consumption in the aggregate has increased far more rapidly than that in Africa, achieving such rates of expansion in an economically efficient fashion is not easy.

Some of the characteristics of inorganic fertilizer itself help to explain why devising an optimal fertilizer policy is difficult and why the elements of such a policy are likely to vary in time and space. Fertilizer, like improved seed, is a divisible input; thus in theory it is likely to diffuse rapidly, even among small farmers, wherever agronomic responses and price ratios combine to favor its use. In practice, fertilizer costs are a large part of cash outlays for crop production, so the purchase of fertilizer is likely to subject the farmer to greater financial risk than the purchase of improved seed. After fertilizers are first adopted, the informational requirements of determining the optimal fertilizer types, application rates, and application times pose considerable challenges to agricultural research systems.

Two other characteristics of inorganic fertilizer underline how challenging it is to develop an optimal fertilizer policy: Demand for fertilizer is seasonal, and fertilizer is a bulky product. Turnover in fertilizer stocks is relatively low, and storage requirements are considerable, which in turn results in higher financing charges. Distributors are faced with their own credit requirements, as well as with the need to offer credit to end users or to work closely with credit agencies (Shepherd 1989).

In addition, fertilizer production is characterized by considerable economies of scale. For example, to operate at maximum efficiency, an ammonia-urea plant needs to produce about 500,000 t of urea per year.[1] Such a plant takes three to four years from initiation of construction to come on-line and another two years to reach full capacity utilization. The economics of investment in fertilizer production depend heavily on the size of the potential demand, the availability of local feedstocks, the cost of capital, and the manufacturer's pricing policy (Segura, Shetty, and Nishimizu 1986). Most fertilizer-producing countries tend to plan for output to meet domestic demand, with exports occurring as a residual. As a result, world fertilizer prices have tended to be more volatile than prices for other commodities, making the decision to rely on the world market or to initiate domestic production particularly difficult, even for countries where potential demand is large (Ahmed, Falcon, and Timmer 1989).[2]

Many knowledgeable observers have concluded that "perhaps more than any other important component of the rural economy, fertilizer use in developing countries is in a continuous state of disequilibrium" (Ahmed, Falcon, and Timmer 1989:26), and supply-side constraints are often more important than demand factors in limiting growth in consumption. These disequilibrating features of the fertilizer economy are more problematic in Africa than in the rest of the developing world. First, the control of water, a major factor that complements the use of fertilizer, is more costly to achieve. Second, production environments in Africa may be somewhat more variable in time and space than environments elsewhere, which makes it more costly to develop the information base about agronomic potential and disseminate this knowledge to farmers. Third, many African countries have very small markets for fertilizer, making reliance on the world market through trade or aid inevitable. Fourth, infrastructure is less developed in much of Africa than in other regions of the world, which raises real fertilizer distribution costs, reduces product prices at the farm level, and in general leads to large price variability for agricultural inputs and outputs (Kelly et al. 1996).

Fertilizer Use on Maize

Fertilizer Use in Africa

Fertilizer application rates are lower in Africa (10 kg/ha in 1993) than in the developing world as a whole (83 kg/ha in 1993). Nonetheless, fertilizer consumption in Africa has increased substantially over the past 30 years. Recent growth in the use of fertilizers on cereal crops, particularly maize, has contributed substantially to the increased use of fertilizer. Current application rates remain low because fertilizer use started from a low base, and increases in fertilizer use have been relatively small—slowing to near zero in the period 1981–1993.

Fertilizer Use in Maize Production

Analysts of fertilizer use in Africa (e.g., Tshibaka and Baanante 1988) often contend that export or plantation crops (primarily cotton, sugar, tea, and coffee) receive most of the fertilizer used in the region. Although this is still true in francophone western Africa, where cotton is important, and in Kenya, for Africa as a whole fertilizer use has shifted to cereal crops, particularly maize (Table 13.1). Maize drives fertilizer consumption in much of southern Africa. In eastern Africa, cereal crops also dominate fertilizer consumption, although the pattern is more diverse—with a large proportion of fertilizer going to teff in Ethiopia, wheat in Sudan, and noncereal export crops in Kenya. Even in western Africa, maize and other cereals now account for the bulk of fertilizer consumption, although these results are driven by apparently large increases in Nigerian consumption in the 1980s. In most

Table 13.1 Sub-Saharan Africa: Fertilizer Use by Crop, 1991–1992

Country	Year of Estimate	Maize Area, 1990–1992 (000 ha)	Percentage of Fertilizer[a] Applied				Maize Area as a Percentage of Total Cereal Area	Fertilizer Applied to Maize as a Percentage of Fertilizer Applied to Cereals
			Maize	Other Cereal Crops	Export/Plantation Crops	Other Crops		
Tanzania	1990	1,796	77	2	21	0	61	97
Nigeria	1990	1,517	43	35	13	9	14	55
Kenya	1983	1,428	20	3	61	15	80	87
Malawi	1990	1,365	74	0	26	0	95	100
Zimbabwe	1990	1,044	39	16	34	11	73	70
Ethiopia	1988	1,000	9	81	10	0	19	10
Angola	1991	750	82	5	0	18	83	100
Zambia	1987	681	90	5	5	0	84	95
Côte d'Ivoire	1989	675	6	12	68	14	49	33
Ghana	1991	525	64	16	5	15	52	79
Benin	1989	413	6	2	92	0	70	75
Togo	1990	273	23	10	66	1	44	70
Cameroon	1985	220	4	11	67	18	27	n.c.
Burkina Faso	1991	201	3	40	57	<1	7	n.c.
Total	—	11,888	41	24	27	8	36	63

Sources: IFA/IFDC/FAO (1992); figures for Nigeria calculated from data in Smith et al. (see Chapter 8 in this book); figures for Kenya and Cameroon from Lele, Christiansen, and Kadiresan (1989); figures for Zambia from Tshibaka and Baanante (1988); figures for Ghana from Bumb et al. (1994); figures for Malawi are authors' estimates.

Notes: n.c. = not calculated.
a. Measured in nutrients.

maize-producing countries in Africa, the proportion of maize fertilizer consumption in total fertilizer consumption by cereals tends to equal or surpass the proportion of maize area in total cereal area. In contrast, in other developing countries maize shows a slight tendency to receive less than its share of total nutrient consumption (Heisey and Mwangi 1996).

Despite the increasing importance of maize in African fertilizer consumption, application rates at the aggregate level generally remain low compared with application rates on maize in other developing countries (Table 13.2). Five countries—Nigeria, Kenya, Malawi, Zimbabwe, and Zambia—are exceptional in that a significant portion of their maize area, about half or more, was fertilized by around 1990, and application rates on fertilized areas in these countries ranged from two-thirds to over two times the nutrient application rates on rain-fed maize in India. The area fertilized in Africa was lower in general than in Asia or Latin America, however, and fertilizer use has stagnated or declined in recent years. For example, fertilizer use on maize in Malawi was curtailed following changes in policy at the micro- and macrolevel. In Zambia, fertilizer use has been declining for a decade or more and may fall further with recent policy shifts. Consumption in Kenya, Zimbabwe, and Nigeria has been erratic over the same period, with little discernable trend in Kenya and Zimbabwe.

Technical Response to Fertilizer

Several factors complicate the estimation of a crop's response to fertilizer. An important issue in estimating fertilizer response is to what degree the conditions under which the response is measured are representative of conditions in farmers' fields. Many have contended that the agronomic responses to fertilizer on experiment stations are considerably higher than those on farmers' fields, because the use of complementary inputs such as timely weeding is greater on experiment stations. Others have argued that under certain conditions responses are higher on farmers' fields, because the nutrient being tested is not a limiting factor on experiment stations, where soil fertility levels are high. Either argument may be correct, depending upon the empirical situation.

The cropping system or rotation may also play a role in determining fertilizer response. For example, in Ghana the response of maize to nitrogen can be twice as high on depleted soils that have been continuously cropped as on soils with high natural fertility that have lain fallow for a number of years (Edmeades et al. 1991). Variability in weather may mean moisture is limiting in some years and not others, and response may differ markedly as a result.[3]

Nonetheless, under rain-fed conditions, maize in Africa tends to be more responsive to fertilizer than other cereals are, with the possible exception of rice. Response data for maize in Africa are not markedly different from those observed in rain-fed areas of Latin America and Asia. This is un-

Table 13.2 Sub-Saharan Africa and Other Regions of the Developing World: Fertilizer Applied to Maize, 1989–1993 period

Region	Percentage of Maize Area Receiving N-P-K	Fertilizer Application Rate (kg N-P-K/ha)		Maize Yield, 1989–1993 (t/ha)
		Area Receiving Fertilizer	Total Maize Area	
Africa				
Tanzania	20	71	14	1.39
Nigeria	64	176	113	1.27
Kenya[a]	50	37	18	1.62
Malawi	50	52	26	1.02
Zimbabwe	45	122	55	1.47
Ethiopia	15	33	5	1.69
Angola	8	137	11	0.34
Zambia	65	108	70	1.64
Côte d'Ivoire	5	103	5	0.77
Ghana	26	52	14	1.34
Benin	n.a.	n.a.	1	0.96
Togo	16	63	10	1.02
Cameroon	n.a.	n.a.	3	1.84
Burkina Faso	n.a.	n.a.	3	1.43
Africa[b]	37	96	36	n.a.
Africa[c]	n.a.	n.a.	33	n.a.
Developing world				
Asia[d]	74	80	59	n.a.
Latin America[e]	65	124	80	n.a.

Sources: IFA/IFDC/FAO (1992); authors' calculations (Tanzania, Nigeria, Kenya, Malawi, Ethiopia, Zambia, Ghana, Benin, Cameroon); D. Byerlee (personal communication), India. Sources consulted for making authors' calculations include FAO (various years); Smith et al. (1994a) (see Chapter 8 in this book); Lele, Christiansen, and Kadiresan (1989); Tshibaka and Baanante (1988); Howard (1994); Bumb et al. (1994); GGDP (1991).

Notes: n.a. = data not available.
a. Microlevel studies (e.g., the Kenyan study described in Chapter 6) suggest application rates may be somewhat higher.
b. Includes all African countries in the table with complete information.
c. Includes all African countries in the table.
d. Includes India, Nepal, Pakistan, the Philippines, and Thailand. Overall fertilizer application rates for maize in Indonesia may also be about 60 kg/ha. China is excluded.
e. Includes Colombia, Ecuador, Honduras, Mexico, Nicaragua, Paraguay, Peru, and Venezuela. Brazil and Argentina are excluded.

doubtedly one of the reasons maize production appears to be positively linked to fertilizer consumption.

Factors Influencing Farmers' Adoption and Intensity of Fertilizer Use

It is difficult to separate demand from supply factors in determining farmers' initial adoption of fertilizer and subsequent choice of application rates. For example, many key influences discussed in the adoption literature (farm size, access to credit, membership in cooperatives, contact with extension, access to outside information, availability of inputs, distance to markets) may be related at least as much to supply-side constraints as to farmer demand factors (Mwangi 1995). In this section, we focus on the demand for fertilizer; in the next section, we turn our attention to problems of supply.

Prices

Theoretically, the demand for farmers' initial adoption of fertilizer is created by the interaction between agronomic response and the nutrient-to-grain price ratio. Agronomic response, in turn, is determined by soil characteristics and climatic factors. If the marginal agronomic response at a level of 0 kg/ha applied nutrient is greater than the nutrient-to-grain price ratio, in theory the farmer should adopt some fertilizer. In practice, other considerations are important, including the cost of operating capital over the cropping season, information and learning costs, plus perhaps the effects of risk aversion (risk is considered separately later) (CIMMYT 1988). Many observers contend that the marginal agronomic response must be at least twice the nutrient-to-grain price ratio (in other words, the marginal rate of return on working capital invested in fertilizer must be at least 100%) for significant adoption to occur.

These considerations can be used to examine some implications of observed technical responses under African conditions. Let us assume that agronomic responses have been measured under conditions similar enough to those faced by farmers, so that an assumed reduction of 20% indicates the response farmers will actually achieve (CIMMYT 1988). This combined with the assumption of a marginal rate of return of at least 100% implies that where 25 kg of grain can be obtained with the application of 1 kg of nutrient, farmers will adopt fertilizer as long as the nutrient-to-grain price ratio is 10 or lower. In areas with an intermediate response, the nutrient-to-grain price ratio must be 6 or lower for widespread adoption to occur. Where response is low (5 kg of grain for 1 kg of nutrient), the threshold price ratio is 2.

In Africa, nitrogen-maize price ratios in the absence of subsidies are considerably higher than the median for the rest of the developing world (Table 13.3).[4] Nonetheless, there is a considerable area of high- and medium-potential maize where some fertilizer use is profitable (Heisey and Mwangi 1996). Since calculations of the farmer profitability of fertilizer use can be fairly sensitive to assumptions about both input and output prices, however, it is important to consider both of these assumptions, as well as the prospects for improving price ratios (Table 13.4).

Assumed input prices are affected by whether the effects of subsidies and exchange rate overvaluation are taken into account (Martin and Lele 1992) and whether the nutrient source is low or high analysis (HIID/EPD

Table 13.3 Sub-Saharan Africa and Other Regions of the Developing World: Nitrogen-Maize Price Ratios

Country/Region and Time Period	Nitrogen-Maize Price Ratio (median)
Tanzania	
1980–1985	2.6
1995	7.0
Nigeria	
1985–1992	2.0
Kenya	
1980–1995	7.3
Malawi	
1977–1987	10.7
1988–1994	7.7
Zimbabwe	
1980–1994	6.4
Ethiopia	
1983	6.4
1992	1.9
Zambia	
1971–1989	3.3
1990–1994	5.4
Côte d'Ivoire	
1980–1992	5.4
Ghana	
1982–1987	2.2
1991–1994	10.2
Asia	
1980–1992	2.7
Latin America	
1980–1992	3.8

Sources: CIMMYT *World Maize Facts and Trends,* various issues; Smith et al. (1994a) (see Chapter 8 in this book); Tripp and Marfo (see Chapter 7 in this book); J. Howard (personal communication); S. Waddington (personal communication); and Heisey and Smale (1995).

1994). Assumed output prices can be affected by subsidies and exchange rate overvaluation as well, but maize prices may also be lowered by implicit taxation caused by marketing board policy (Franzel et al. 1989). In countries with wide consumer-producer price margins, whether the household is a net consumer will also affect output prices (Table 13.4).

What factors have influenced nutrient-to-grain price ratios in African countries, and which factors will influence those ratios in the future? At the world level, real fertilizer prices and the real prices of major cereal crops have both fallen for much of the twentieth century, driven by technical change (Tomich, Kilby, and Johnston 1995). From the early 1970s to the early 1990s, there was an insignificant, slight downward trend in the ratio of world urea prices (free on board [f.o.b.] Rotterdam) to world maize prices (f.o.b. U.S. Gulf).[5] Nitrogen prices may rise somewhat in the late 1990s as capacity lags behind demand, but cereal prices may not

Table 13.4 Lilongwe, Malawi: Effect of Price Assumptions on the Profitability of Alternative Maize Technologies in 110 On-Farm Demonstrations, 1990 and 1991

	Local Maize with Fertilizer	Hybrid Maize with Fertilizer
Fertilizer applied (kg nutrient/ha)	55	145
Yield increase observed over unfertilized local maize (kg/ha)	750	2,400
	Marginal rate of return (%)[a]	
Subsidized input prices[b]		
Maize-deficit households[c]	133	237
Maize-surplus households[c]	64	136
Unsubsidized input prices		
Maize-deficit households[c]	79	145
Maize-surplus households[c]	27	72

Source: Authors' calculations based on data provided by the Ministry of Agriculture/UNDP/FAO Fertilizer Demonstration Program.

Notes: a. Marginal rate of return on input expenditures. A return above 100% is usually assumed to be necessary for widespread adoption by farmers.

b. Subsidy of 25% on fertilizer and about 40% on hybrid seed.

c. The price of maize in households that purchase maize is about 40% above the farm gate selling price.

continue to fall, making future trends in world price ratios hard to predict. For most African countries, real border fertilizer-to-maize price ratios have not increased over time and are unlikely to increase significantly in the immediate future.

Instead, costs of fertilizer in Africa are inherently high because of infrastructural and marketing constraints (discussed later). As subsidies are removed and exchange rates liberalized, sharp increases in the fertilizer-to-maize price ratios, occurring over a relatively short time, almost inevitably follow (Table 13.3). In some countries, such as Ethiopia (Franzel et al. 1989), liberalization of maize markets can improve maize prices and more than compensate for increased input prices. In others, liberalization does not appear to halt declines in the real price of maize. Thus the overall effect of "liberalization" policies on price ratios should be determined empirically.

Risk Aversion and Credit Constraints

Although risk aversion is commonly assumed to be an important influence on farmers' choice of technology, many observers have concluded that for farmers who have already adopted fertilizer, risk aversion can account for a reduction in fertilizer use of 20% or less below "optimal" application rates (Binswanger and Sillers 1983). Ahmed, Falcon, and Timmer (1989), however, have argued that these studies usually focus only on production risk, not on price risk in a general equilibrium context. Certainly, output price instability poses an important risk for fertilizer users in western Africa (Vlek 1990). Furthermore, production risk is likely to be considerably more important in marginal crop production areas, such as drought-prone areas of Kenya (McCown et al. 1992).

Constraints on cash or credit availability can often cause farmer behavior that resembles risk aversion (Binswanger and Sillers 1983). For smallholders, fertilizer expenditures represent a considerable proportion of total cash expense in crop production. In Malawi, a liquidity crisis in the smallholder credit system, not a change in the fertilizer-maize price ratio per se, caused a dramatic reduction in smallholders' fertilizer use between the 1992–1993 and 1993–1994 maize seasons (HIID/EPD 1994).

Availability

Despite differences of opinion on other issues, many analysts of fertilizer use and policy in Africa and the rest of the developing world agree that basic problems of availability—"getting the right fertilizer to the right place at the right time"—are at least as important as, if not more important than, price-response interactions in determining fertilizer use (Fontaine 1991). These factors are often referred to as nonprice factors, but they can be accommodated within a pricing framework, because they raise the real price of fertilizers to farmers.

Although the features of the African fertilizer economy that lead to high prices are often intertwined with those that constrain availability, policymakers have often focused solely on the one effect (high prices) rather than on availability and have ignored the underlying causes completely.

Supply Constraints on Fertilizer Use and Policy Responses

Procurement and Distribution

Because of the small size of present and potential markets for fertilizer in most individual African countries, it is generally more economical to import fertilizer than to produce it locally (Vlek 1990).[6] Nonetheless, the differences between world f.o.b. fertilizer prices and landed cost tend to be at least twice as high for many African countries as they are in Asia (Shepherd and Coster 1987). Bumb (1988) states that this large difference results from the small volumes of fertilizer most African countries import, which increases their transportation costs and weakens their bargaining position in negotiating for lower prices.

In 1990, almost one-third of fertilizer imports in Africa were financed by aid. In fact, for 21 countries with small markets, all fertilizer was financed through donor programs. Donors of fertilizer impose conditions (such as limitations on the origin, transporters, and type of fertilizer) that may lead to excessive marketing costs and margins that ultimately translate into higher fertilizer prices (Gerner and Harris 1993).

Another reason for the relatively high fertilizer prices in Africa is the high cost of distribution. First, landlocked countries such as Malawi face additional high transport costs from the seaport to port of entry. Second, internal distribution costs are usually considerably higher in Africa than in other regions of the developing world. High transportation costs, in turn, are the result of poor physical infrastructure. Africa lags far behind India and China by the measure of kilometers of road per 100 km^2 (Vlek 1990). Furthermore, attention to road maintenance and finding the appropriate balance between rural-feeder and trunk roads are important infrastructural issues (Mwangi 1995). Finally, lack of competition in fertilizer distribution systems, often as a result of public-sector operation, can also contribute to inefficiencies and higher marketing margins (Pinstrup-Andersen 1993).

Indirect Policy Effects

Numerous indirect pricing effects can be caused by timing, financing, and managing fertilizer procurement and distribution. Losses from public-sector organizations trading fertilizer are often made good by the government or are offset by profits in commodity trading. Wastage, costs of capital, or costs of extending credit to farmers are often underestimated (Shepherd 1989). Perhaps the most universal indirect effects in Africa, however, are the practice of pan-territorial pricing[7] and exchange rate overvaluation.

Pan-territorial pricing combined with public-sector fertilizer distribution means marketing organizations are often unaware of the true costs of marketing fertilizer. This practice in effect subsidizes farmers at a higher rate in regions with higher transport costs. Even when other aspects of the market are liberalized and subsidies are reduced or eliminated, as in Ghana, pan-territorial pricing may persist (Bumb et al. 1994). One suggestion for an intermediate step in removing pan-territorial pricing is to control prices up to regional distribution centers within a country but not beyond those points (Shepherd 1989).

Exchange rate overvaluation has been a common feature of macroeconomic management in nearly all sub-Saharan countries (Ghura and Grennes 1991), although it is much less evident today. As is well-known, in the long run exchange rate overvaluation tends to discriminate against the agricultural sector and will therefore work against increased fertilizer use. In the short run, however, imported fertilizers as a tradable input will be implicitly subsidized by overvaluation.

Fertilizer Policy: Issues and Options

In this section, we consider specific policy options for the fertilizer sector. We first look at the demand side, moving from the relatively short-run question of fertilizer subsidies to the longer-run issues of determining and targeting high-potential areas and developing appropriate agricultural research programs. We then discuss the supply side. Again, shorter-term concerns such as donor aid, credit, and distribution costs in general precede intermediate options such as privatization and long-run considerations such as local production capacity and infrastructure development.

An excellent approach to fertilizer policy is provided by Ahmed, Falcon, and Timmer (1989), who state that "some combination of market forces and government interventions is needed, with the market providing allocative signals, the government stabilizing them around a market trend [based in some way on world prices], and a competitive private sector delivering the goods at low cost." In most countries in Asia, the Near East, and Latin America, however, such a fertilizer strategy has not been attained.[8]

Fertilizer Subsidies: Theory and Practice

Many countries in Africa have attempted to promote the use of fertilizer through subsidies on fertilizer prices, credit, or both, despite the fact that many other factors besides explicit subsidies are important in determining the growth of fertilizer consumption. Much of the literature on fertilizer consumption has focused on the subsidy issue, perhaps because subsidies and their effects on government agricultural budgets are relatively easy to observe.

Although credit subsidies are almost universally regarded as policy doomed to fail, fertilizer subsidies have received some support from policy analysts. In a world of market and information failures, where policymakers often choose nonefficiency objectives, a subsidy on inputs might be justified (Shalit and Binswanger 1985). If the government's goal is to achieve food self-sufficiency, in many cases a subsidy on fertilizer is relatively more efficient than a subsidy on output (Barker and Hayami 1976). Given broader policy objectives, such as food security or growth with equity, however, a fertilizer subsidy may no longer necessarily be the preferred policy instrument (Quizon 1985). For countries pursuing the goal of local fertilizer production capacity, some form of subsidization might also be part of the policy package (Ahmed, Falcon, and Timmer 1989).

The justification for subsidies that might receive the most consideration in relation to maize in Africa is that subsidies might promote adoption of fertilizer where farmer learning costs and other bottlenecks in the system slow or halt movement toward a socially optimal level of fertilizer use (Shalit and Binswanger 1989). Miller and Tolley (1989) show that the social benefits from an optimal subsidy policy are expected, theoretically, to be relatively small, although the parameters in their model were derived primarily from the Asian, not the African, experience. In practice, Dalrymple (1975) and Ndayisenga and Schuh (1995) have concluded that subsidies have not been particularly efficient means of encouraging fertilizer adoption in Africa.[9] Furthermore, rent seeking on the part of public- and private-sector actors means subsidies, once instituted, can be difficult to remove.

In sum, the question of whether the social return to fertilizer subsidies is greater than the return to alternate forms of investment—such as agricultural research, extension, or infrastructure development—is an empirical one. A complete analysis is likely to be complicated by the fact that the payoff period for these investments tends to be longer than the payoff period for fertilizer subsidies. Also, some complementarities may exist between fertilizer subsidies and investments with longer horizons.

Certainly, the period of heavy subsidization in many African countries (lasting roughly until the mid-1980s) was not associated with particularly rapid growth in fertilizer consumption. The record from individual countries such as Nigeria is mixed. Smith et al. (1994a) have argued that despite problems of fertilizer supply in Nigeria, the heavy subsidy played a secondary role in the adoption and spread of maize seed-fertilizer technology. Nonetheless, Daramola (1989) concluded that chaotic and untimely supply of fertilizer was one of the most important reasons for nonadoption. It appears that continuing the fertilizer subsidy cannot be justified on either efficiency or equity grounds (Nwosu 1995).

In many countries, subsidy removal, exchange rate liberalization, or both have led to reductions—sometimes rather sharp—in fertilizer consumption. Ghana and Zambia are countries where policy changes have been largely responsible for the decline in fertilizer consumption in recent years. Changes in prices and exchange rates can be expected to induce similar reductions in fertilizer consumption in other maize-producing countries, such as Tanzania and Malawi.

Targeting High-Potential Areas

There are many strong arguments for a country to give priority to high-potential areas in increasing food production. From the point of view of fertilizer use, there are two major determinants of what constitutes a high-potential area. The first is agronomic response; the second is economic potential, which, in turn, is related to location and infrastructure (Byerlee et al. 1994).

Several considerations modify the definition of agronomic response. First, absolute potential yield level is as important as marginal response.[10] Also, responsiveness is likely to be related to the availability of improved, more responsive maize varieties.[11] In any case, enough attention has been given to the issue of focusing on high-potential areas that it is time for proponents of such a strategy to begin defining exactly where those areas are and to spell out as many of the important efficiency and equity issues as possible.

Developing, Collating, and Distributing Information Through Research and Extension Systems

An emphasis on focused agricultural research is crucial for the development of the fertilizer sector. For "the right fertilizer to be available at the right time in the right place," the right fertilizer must be known. What are the major nutrient limitations for current maize production systems? Have they simply been mismeasured, or has the limiting factor been misidentified?[12]

Two avenues of investigation must be pursued. The first is the collection and analysis of existing information on soil fertility research with the express goal of making that information relevant to fertilizer policy development. This effort must go far beyond the repetitive calculation of value-cost ratios that are given little policy interpretation. Crop modeling and geographic information systems may assist in this enterprise. Good data from on-farm research will be much more valuable than results from experiment stations. The second avenue is the development of new knowledge by extending both the spatial and the temporal dimensions of soil fertility research. Even in a country like Kenya, which has a relatively long, strong history of such research, too little information may be available to develop a comprehensive fertilizer policy (P. L. G. Vlek, personal communication). There is evidence that researchers are giving greater attention to such issues (Waddington and Ransom 1995), but the declining support for agricultural research is a major underlying problem. In some countries, such as Zimbabwe, private-sector initiatives may fill part of the research gap, but the long-term research strategy will still require substantial public-sector involvement—which, in turn, will require higher, not lower, investments in agricultural research.

Similarly, the process of developing research recommendations, making them consistent with policy, and turning them into more effective (but more complicated) extension advice is far from satisfactory in most African countries. This, however, is another crucial factor in transforming agronomic potential into effective fertilizer demand (Gandhi and Desai 1992).

Short-Run Supply Issues

Major maize-producing and fertilizer-consuming countries in which significant amounts of donor-financed fertilizer have been available include

Tanzania, Kenya, Malawi, Ethiopia, and Ghana (Gerner and Harris 1993). Donors should move quickly toward untying fertilizer aid from the purchase of particular types of fertilizer from specific sources and also toward providing fertilizer aid in cash rather than in-kind (Ndayisenga and Schuh 1995), since tied fertilizer aid often results in higher costs and the provision of inappropriate types of fertilizer.

Other measures to reduce fertilizer distribution costs include the consolidation of orders within a country (and possibly pooling orders among small neighboring countries), easing the process of obtaining foreign exchange for fertilizer imports, ending public-sector favoritism within the marketing channel, and reducing bureaucratic obstacles in general (Ndayisenga and Schuh 1995). Countries that have not moved toward the use of high-analysis fertilizers (such as urea, diammonium phosphate, and triple super phosphate) can also bring down fertilizer prices by doing so.[13]

Credit problems are rife throughout the fertilizer marketing channel. In Ethiopia, for example, international shippers with a load of fertilizer at the Eritrean port of Assob require a guarantee of U.S.$6 million before the fertilizer can even be unloaded for transshipment to Addis Ababa. Government-guaranteed loans to fertilizer importers, wholesalers, and large traders selected by banks on strict commercial criteria would be one way to overcome this problem (Ndayisenga and Schuh 1995). Government-sponsored credit schemes featuring group lending, credit extended by traders, and effective rural financial intermediation based on small community savings and credit schemes have all been proposed as solutions to small farmers' liquidity problems. Experience with government schemes has been disappointing, and even credit programs meeting all of the standard criteria for success, such the Malawian program, have collapsed (HIID/EPD 1994). Nonetheless, both experience with and analysis of other means of providing credit to smallholders have been limited to date.

Privatization of Supply

In the intermediate run, the efficiency of the market mechanism will be enhanced if governments create a policy environment that aids the development of privately operated businesses in the fertilizer sector and provide basic institutions and infrastructure (Ndayisenga and Schuh 1995; Ahmed, Falcon, and Timmer 1989). Actual experience with privatization in Africa, however, has been mixed. If private-sector firms find privatization unprofitable because of larger infrastructural constraints or other factors (such as fixed marketing margins, uneven application of subsidies to different actors in the system, or uneven risk sharing in the case of large stock buildup), they are not going to enter the system. Inviting the private sector in at the time a market is shrinking is hardly a prescription for success, as the Ghanaian case shows (Bumb et al. 1994). Experience from Cameroon, on the other hand, shows that once a market is developed, the private sector can import and deliver inputs at a lower cost—provided there is appropriate support from the public sector, such as the provision of market information (Truong and Walker 1990).

Caution in the Development of Local Production Capacity

In Africa, only Nigeria and Zimbabwe produce nitrogenous fertilizers in any large amounts. Although 29 African countries possess phosphate reserves, only 3 currently use local reserves in the production of phosphatic fertilizers (Gerner and Harris 1993). Nonetheless, local production is unlikely to make a substantial contribution to large-scale increases in African fertilizer consumption in the foreseeable future. Forty years of experience in the development of local production capacity in the developing world suggests, for nitrogen at least, that most countries—even some with substantial feedstocks—would have been better off importing fertilizer rather than manufacturing it locally. Economic evaluation of potential plant construction has been consistently plagued by excessive optimism regarding finished product–feedstock price ratios and capacity utilization. Utilization of local rock phosphate deposits might prove to be an exception, but potential projects should be subject to more careful ex ante economic analysis than has been the case in the construction of nitrogenous fertilizer plants (Tomich, Kilby, and Johnston 1995).

Infrastructual Development

For a bulky input such as fertilizer, reductions in transportation and storage costs are indispensable if long-run consumption is ever to approach the social optimum. Since infrastructure affects far more than the fertilizer sector alone, we will not consider it in detail here. Construction and maintenance of rural-feeder roads, as well as more general attention to maintenance within the transport sector, are, however, likely to play a key role in reducing fertilizer distribution costs. Furthermore, public provision of legal and social infrastructure may also help to reduce risks of fertilizer distribution (Ndayisenga and Schuh 1995).

Conclusions

Increased fertilizer use will be an essential component of a strategy to increase per capita maize production in Africa. Until recently, policy debates about the fertilizer sector in African countries focused particularly on the questions of subsidies and macroeconomic management, giving little attention to investments in agricultural research or infrastructural development.

Without doubt, fertilizer subsidies in many countries have been considerably higher than could be justified by any economic rationale.[14] We suggest that the long-run goal in any country is the complete removal of fertilizer subsidies. Countries with limited fertilizer consumption at present could lift subsidies in relatively short order. For countries with higher current consumption (say, 25,000 t nutrient per annum on a relatively sustained basis), the withdrawal of subsidies should be conditional on the development of an agricultural-sector strategy (Lele 1992). Opinion is divided between advocates of "short, sharp, shock" shifts in policies and attempts to move more slowly toward a more optimal policy regime. Our considered opinion is that gradual reforms with careful planning are likely to give better long-run results than shock treatment.

At present, policymakers often appear to be driven more by influence of donor or lending agencies or by crisis management than by longer-run strategic considerations. In some cases, procurement, distribution, and pricing arrangements are maintained for years and then changed drastically at short notice. The result can be an environment in which signals to both public agencies and private-sector participants in the fertilizer sector, right down to the farm level, are highly variable and even conflicting. Fertilizer policy must move from the current state of crisis management to one of crisis avoidance. Whether because of subsidy withdrawal or exchange rate liberalization, mismanagement of a donor-assisted fertilizer grant (von Braun and Puetz 1987), or collapse of a supporting institution like the credit system, sharp fluctuations in fertilizer use are a common feature in Africa.

For the fertilizer sector to be effective, the government, in consultation with the private sector and donors, will need to develop a detailed national fertilizer-sector policy and plan within the framework of a comprehensive agricultural strategy. Most countries in Africa lack such policies and plans. These policies must be broadly consistent with one another, and present and potential actors in the system must understand policy objectives and the means of attaining them. This, in turn, means governments must develop strong internal policy capacity and the ability to communicate forcefully with donors and the private sector (Martin and Lele 1992). Strengthening policy capacity through providing better incentives within public service is actually complementary to, not competitive with, greater private-sector participation in the fertilizer sector (Ndayisenga and Schuh 1995).

Two issues emerge from our analysis of strategies for increasing fertilizer use on maize. First, over time, governments should withdraw from fertilizer procurement, distribution, and pricing and concentrate increasingly on the provision of information, legal institutions, and infrastructure. In the process of privatization, a clear definition of the roles of government and the private sector in the development and operation of the fertilizer system is necessary. Short- and long-term roles and how these roles change as the sector develops will need to be clearly spelled out so mutual trust and confidence can exist. Initially, training of fertilizer distributors may be needed. In countries with relatively large markets, a fertilizer industry association can aid the process of public-sector–private-sector dialogue.

Some government functions will remain important after the privatization of fertilizer marketing. These include setting and enforcing standards and quality control; estimating demand in consultation with the private sector; monitoring and evaluating sector performance; setting up mechanisms for private-sector–government consultations; in general, creating a conducive environment for private-sector participation; and giving strong support to research, extension, and the development of infrastructure, all of which will be important over the longer term (Sodhi 1993). Appropriate policymakers in these areas should be involved in decisions about fertilizer policy.

Second, future work on fertilizer policy should give greater attention to institutional issues. Government commitment to agricultural development, active collaboration with the private sector, and more thoughtful, coordinated donor action will all be necessary. In the real world, agricultural strategies, strong government policy capacity, and financially viable private-sector fertilizer distributors are unlikely to come into existence all at once, so an improved understanding of institutions should contribute substantially to the design of second-best solutions.

Notes

1. This would be equivalent to 30% of Africa's *total* nitrogen consumption in 1993 if all nitrogen consumed were obtained from urea.

2. Tomich, Kilby, and Johnston (1995, pp. 247–248) have contended, however, that "for twenty-seven of the past thirty years, the *maximum point of [world] price variation* for nitrogenous fertilizer fell *below* the [economic] cost of domestic manufacture" in developing countries.

3. When marginal response is calculated from an estimated response function, different functional forms can also result in widely varying results for the same data.

4. Fertilizer is subsidized in many of these countries as well, but differentials in nutrient-to-grain price ratios would be likely to remain even in a completely unsubsidized world.

5. This downward trend is even less impressive when the speculative nitrogen price rise at the time of the world food and oil crises in the early 1970s is taken into account.

6. Exploitation of local rock phosphate deposits might be an exception.

7. Pan-territorial pricing for an input like fertilizer means the price to farmers throughout a country is the same despite actual differences in transportation, storage, and other costs related to marketing.

8. See, for example, chapter 7 in Tomich, Kilby, and Johnston (1995).

9. Desai (1991) has argued that the effects of fertilizer subsidies on fertilizer adoption in India were negligible; the observable effects of India's subsidy policies were far greater in the development of the domestic fertilizer industry.

10. A high marginal response to initial application of fertilizer in an area with naturally poor soils, where organic matter levels may never become sufficiently high for long-term sustainable yield increases, is not nearly as favorable an indicator as a high marginal response to fertilizer on relatively good soils where nutrients have become depleted because they are continuously extracted by crops (P. L. G. Vlek, personal communication).

11. This is a different issue from that of improving nutrient-use efficiency in high-yielding varieties that are already widely used by farmers.

12. Apart from nitrogen and phosphorous, sulfur, zinc, and potassium might be particularly important for large areas where maize is grown in Africa (G. Edmeades, personal communication). Furthermore, either custom blending of fertilizers or the addition of micronutrients tends to increase the price of fertilizer by U.S.$15–$20/t, so discovering the "right fertilizer" has economic implications that extend beyond the economics of supplying the major nutrients.

13. Malawi has made a partial transition to high-analysis fertilizers. Pessimism has been expressed about some of the other short-term measures (HIID/EPD 1994).

14. Those who argue that Africa is somehow "different" should be challenged to develop formal policy models that clearly specify policy objectives. Good examples are provided by Barker and Hayami (1976), Quizon (1985), and Miller and Tolley (1989). Model parameters could then be changed to reflect empirical conditions in Africa.

14

Maize Marketing and Pricing Policy in Eastern and Southern Africa

Thomas S. Jayne, Stephen Jones,
Mulinge Mukumbu & Share Jiriyengwa

Since the early 1980s, donors and international lending agencies have promoted the reform of agricultural marketing as a central component of economywide structural adjustment programs in Africa. Between 1980 and 1995, agricultural policy reforms were initiated in 35 African countries (World Bank 1994a); although these reforms have been implemented slowly and unevenly, staple food marketing has been transformed. In almost all African countries, the magnitude of state marketing board operations has diminished. Reform measures have gone beyond the relatively limited agenda for agricultural marketing defined by the watershed Berg Report in 1981 (World Bank 1981). Although advocating market liberalization, the Berg Report nevertheless acknowledged that the social and political sensitivity of food prices made it inevitable that the state would intervene to some degree in staple food markets.

Donor advocacy of food market reform reflected both optimism about the private sector's capacity to organize markets efficiently and increasing pessimism about the performance of state involvement in markets (Lipton 1991; Jones 1994). These views were reinforced by positive models of the agricultural economy, which contended that state interventions in agricultural markets—ostensibly designed for rural development or to correct for specific market failures—were often fundamentally designed to serve the interests of a dominant elite of bureaucrats, urban consumers, and industry (e.g., Bates 1981). Most international donors and lenders had withdrawn their earlier support of state food marketing agencies by the mid-1980s. By the early 1990s, donors had increasingly gained leverage over domestic agricultural policies through aid conditionality, which led to partial marketing reforms in Zimbabwe, Kenya, Malawi, and South Africa and almost total reliance on the private sector in Zambia and Tanzania.[1] The prevailing wisdom is that by reducing marketing costs, reform will reduce food prices for consumers, raise producer prices, and generally stimulate the adoption of farm technology and growth in agricultural productivity.[2]

This chapter surveys the record of maize marketing reform in the anglophone eastern and southern African countries of Zimbabwe, Zambia, Kenya, Malawi, Tanzania, and South Africa. The maize marketing systems of these countries share a common ideological and institutional heritage stemming from the model of controlled, single-channel marketing boards in the UK and dominions. This colonial heritage has shaped policy far into the postcolonial period. These countries' experiences also provide a particularly important test of the effects of maize market liberalization, since state agencies have played a much more prominent role in staple food markets than in many other African countries. A historical perspective on maize market reform in these countries is thus useful both to understand the forces that have shaped the structure and performance of the maize sectors in the region and to assess the prospects that recent policy reforms can be sustained.

In the sections that follow, we address several issues that have major implications for food policy throughout Africa.

- Why the anticipated supply response to market liberalization has not yet occurred
- Why the Batesian model of discrimination against farmers is not appropriate in most of these countries
- Why the successes of the state-led approach to stimulating smallholder maize production were unsustainable
- Why the elimination of government food subsidies associated with market reform has not adversely affected urban or rural consumers in most countries
- Why marketing board deficits have risen rather than declined in the liberalized marketing environment in most countries
- What additional changes in government policy may be necessary to support the fledgling food policy reforms already implemented and to reduce the likelihood that such reforms will be reversed

Finally, we argue that two critical issues remain unresolved, both of which are likely to determine the ultimate political and economic viability of the food marketing policy model that is emerging in these countries. These issues are the capacity of liberalized marketing systems to contain the effects of price instability and to support productivity improvements in smallholder agriculture within the constraints of the highly dualistic agricultural systems inherited from the colonial period.

The Evolution of Maize Marketing Policy, 1900–1995

Two principal factors have shaped the evolution of food marketing systems in eastern and southern Africa. First, from their inception in the 1930s and

1940s, controlled marketing systems were designed to mitigate the effects of unstable food production in the region, where droughts are frequent and cereal yields are among the most variable in the world.[3] The staple crop, white maize, is thinly traded on world markets. Most of the population of southern and eastern Africa lives in landlocked cities and remote rural areas that face high transport costs to coastal ports. Maize yields in the major production areas of southern Africa are highly correlated (Jayne, Takavarasha, and van Zyl 1994), which limits the potential for regional trade to offset national production deficits. These structural features suggest that reliance on private trade alone to offset production fluctuations would involve large price fluctuations between export and import parity levels unless countries maintained substantial maize stocks from year to year (Muir and Takavarasha 1989). The social and economic disruptions caused by instability in staple food prices have given rise to the region's historical commitment to food price stabilization and the associated regulation of food markets.

The second dominant influence over the evolution of food marketing systems in the anglophone countries of eastern and southern Africa was the significant presence of European settlers in agriculture, which created a rural economy with an intensely dualistic structure. In general, the greater the importance of European agriculture during the early colonial period, the greater the degree of state intervention in food marketing. At one end of the spectrum were South Africa and Zimbabwe, where controlled maize marketing was designed to prevent African farmers from eroding the viability of the less efficient European maize producers—a prospect viewed as antithetical to homesteader development and rural colonization. At the other extreme were Malawi and Tanzania, which experienced limited European settlement or industrialization.[4]

Grain marketing policy in the region has passed through three clearly identifiable phases. The first phase, under colonial regimes, was shaped by the interests of white settler agriculture and industry. The second phase, following the end of colonial rule, was marked by attempts to overcome the inherited dualism in the provision of rural services. The newly independent states sought to improve smallholders' access to markets and end nonindigenous minorities' domination of marketing activity by expanding the existing state marketing system while discouraging private trade. The attempts prompted by donors to liberalize and privatize food marketing systems under structural adjustment represent the third phase of grain marketing policy. Some countries are now moving to a fourth phase, in which the state has withdrawn almost completely from maize trading and pricing.

Phase 1: Colonial Marketing Policies

The regulatory framework governing staple food marketing in eastern and southern Africa was established between the 1930s and the 1940s. Only in the last five years have reforms fundamentally altered this framework.

Prior to 1920, Africans supplied the bulk of the food needs in Zimbabwe, Zambia, Malawi, and Kenya. European-owned land in Zimbabwe and Zambia was used primarily for speculation, mining, or both. The fledgling development of mines, livestock interests, urban employment, and the British starch market (which offered a premium for white maize) greatly expanded the demand for grain, which fueled the rapid growth in maize production by both Europeans and Africans.[5] Grain trade was virtually devoid of government regulation during this period.

As the number of Europeans engaged in farming rose greatly during the 1920s, African farmers were increasingly perceived as a threat. Substantial evidence from Kenya, Zambia, and Zimbabwe indicates that African maize surpluses were capable of being generated at prices below the cost of production on most European farms and that European farm organizations lobbied successfully in the colonial legislatures for protection on the grounds that they could not compete without it.[6] The depression of the early 1930s brought this problem to a head, and under pressure from an organized European farm lobby, the colonial governments responded with the Maize Control Acts of the 1930s in Zimbabwe and Zambia and the Native Produce Ordinance in Kenya in 1935.

These acts shared several common features. They established restrictions on grain movement from African areas to towns, mines, and other demand centers where African production could otherwise undercut European maize production; they created monopoly state crop buying stations in European farming areas without similar investments in African farming areas; and, for European farmers, they offered prices (typically above export parity) that were financially sustained through "rake-off" taxes on maize sales by Africans to licensed private traders operating as agents of the boards (Table 14.1). These taxes effectively maintained a two-tiered pricing structure, with European farmers receiving prices that were 30 to 60% higher than those received by African farmers. The cross-subsidies involved enabled farm prices to be supported for European farmers without increasing the price of wage goods for industry. Other colonial regulations not related specifically to maize were passed to reinforce Europeans' dominance of the market, including the continued forced removal of Africans from most high-potential farming land and various taxes levied on African households to induce them to move off their farms and work as wage laborers.

The combination of maize legislation, land evictions, and fiscal policies eroded Africans' dominance over food marketing and simultaneously contributed to the growth of European agriculture in Kenya, Zambia, and Zimbabwe after 1935. Figure 14.1, p. 219, shows that African per capita grain production in Zimbabwe's communal lands declined steadily from a high of about 300 kg in 1925 to about 200 kg at independence in 1980.[7]

Pricing policy was based on the setting of official purchase and selling prices, which were usually fixed over the season and at all locations where

Table 14.1 Phase 1 of Maize Marketing Policy in Eastern and Southern Africa: Colonial Regime (service provision to settler farmers)

Country and Period	State Marketing Agency	Market Regulation	Pricing Policy
Kenya, 1922–1963	Kenya Farmers Association statutory board established in 1923 to buy grain from European producers. Monopoly marketing boards established in 1930s. Provincial boards purchased African production, Maize Board settler production.	Native Foodstuffs Ordinance (1922) restricts African maize sales across district boundaries. Marketing of Native Produce Ordinance (1935) tightens controls at all stages of grain marketing. Monopoly by marketing board on sales to registered millers established, 1942.	Prices for European delivery to Maize Board 36% higher on average than official price for African maize offered by provincial boards (1941–1962).
Malawi, 1943–1971	Maize Control Board established, 1947. Succeeded by Farmer's Marketing Board with limited price stabilization mandate.	Colonial policy bans maize production by British South Africa Company in Nyasaland to restrict competition against Rhodesian and Kenyan producers (1933). Native Produce Ordinance (1943) introduces state trading in response to drought and price instability.	Limited price stabilization intervention.
South Africa, 1935–1987[a]	Maize Board established in 1935. Cooperatives function as agents of Maize Board in marketing and storage.	Maize Board monopoly on marketing in main production areas. Regulations restrict African farmers' access to main urban markets. Maize Board monopoly on sales to registered millers.	Maize Board buying and selling prices set by government. Fixed and pan-territorial.
Tanzania, 1942–1955	Grain Storage Department operated from 1949 to 1955. No government intervention from 1955 to 1962.	Rationing and price controls introduced during World War II. Cereal pool operated jointly with Kenya and Uganda.	Grain Storage Department set prices with limited stabilization objectives.

(continues)

Table 14.1 continued

Country and Period	State Marketing Agency	Market Regulation	Pricing Policy
Zambia, 1936–1964	Maize Control Board established in 1936, serving European producers.	Restrictions on interdistrict maize movements.	Single price to commercial farmers close to the railway line; 41% higher than for Africans, on average (1936–1958).
Zimbabwe, 1931–1980	Maize Control Board established, 1931; replaced by Grain Marketing Board, 1950. Buying stations built almost exclusively in European farming areas. African surplus production channeled mainly to licensed agents of Grain Marketing Board.	Grain Marketing Board monopoly over grain marketing in commercial farming and urban areas. African producers barred from selling grain outside of their "reserve" areas, except to Marketing Board or licensed agents. Marketing Board monopoly on sales to registered millers.	Prices for European delivery to Grain Marketing Board 56% higher than for African smallholders, on average (1936–1960).

Source: Jayne and Jones (1996).
Note: a. Although South Africa was not a colony during this period, its food marketing policies were similar in design and intent to those of the former British colonies examined here.

Figure 14.1 Zimbabwe: Per Capita Grain Production in African Communal Lands, 1914–1994

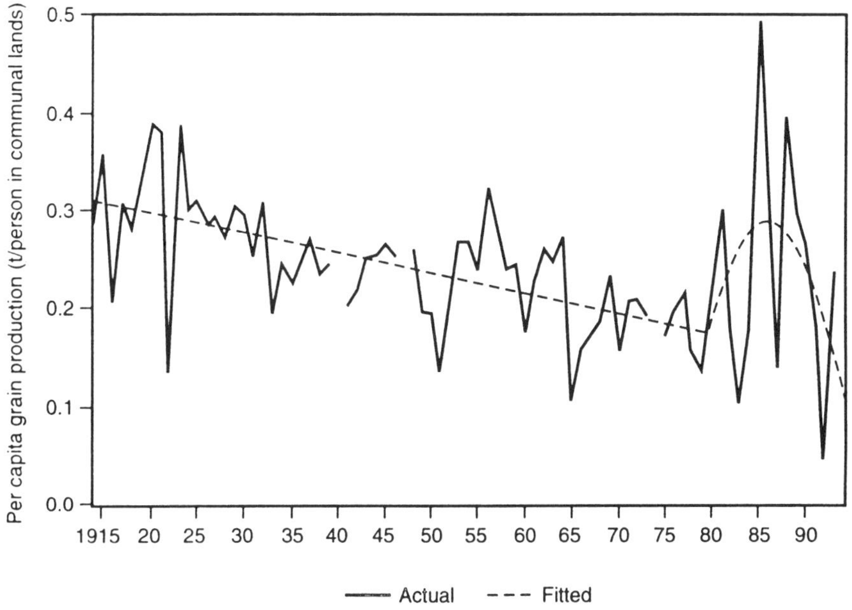

Source: Data from the Annual Reports of the chief native commissioner, presented in Mosley (1983).

the marketing agency traded. Consequently, returns to private storage and marketing were reduced or eliminated. Price fluctuations between seasons were partially stabilized, and the price setting process rapidly became the focus of organized lobbying by white farmers and industrial grain purchasers (primarily millers and feeders), who also controlled management of the marketing boards.

The centralization and concentration of maize sales through single-channel grain boards shaped the development of downstream marketing activities, most notably processing. Since the early 1900s, small-scale hammer mills had been used increasingly to mill maize into *posho* (whole maize meal), but the single-channel systems developed regulations to sell maize to only a few licensed millers, which encouraged the rise of large-scale, concentrated grain milling industries. The roller mill technology used in these industries was first employed on a large scale in Kenya and Zimbabwe in the 1950s. By controlling the movement of maize by private means into urban and grain-deficit rural areas, government policy encouraged the distribution of relatively costly industrial meal in urban areas at the expense of whole meal and the development of the small-scale milling industry. This policy also made rural households in grain-deficit areas dependent on the large-scale milling firms once their own production was exhausted. Over the period 1985–1993, the regulated processing margins awarded to the registered roller mills were from two to nine times higher than margins observed for informal hammer mills.[8] These controls inflated the price consumers paid for maize meal, but a concentrated large-scale milling sector reduced per-unit transaction costs for the boards (compared with selling small amounts to numerous buyers) and, more important, facilitated the implementation and monitoring of price controls on maize meal. Therefore, the rise of a few large industrial maize processors to link downstream distribution activities into the official maize marketing system created a manageable system of supplying the urban population with staple food at prices easily controlled by the state.[9] Within a span of three decades (1955–1985), urban consumers had switched almost entirely from whole maize meal to more expensive refined meal produced by roller mills (Jayne et al. 1995).

This system of regulation, described in Table 14.1, was highly effective in achieving its principal objectives. European grain production expanded and benefited from prices generally exceeding export parity. All countries except Zambia (and Zimbabwe during World War II) became self-sufficient in maize production. The cost of supporting maize production by European farmers was paid for largely by African farmers (through the two-tiered pricing format) and consumers rather than taxpayers, making the system fiscally sustainable. Opposition to agricultural price supports—for instance, from mining and livestock interests in South Africa and Zimbabwe—was accommodated by selective consumer subsidies. The stability of the policy and pricing environment (and the limited competition faced by incumbents) contributed to downstream investment and rapid growth of commercial agriculture, including commercial farmers' adoption of hybrid maize varieties developed by national research systems (Eicher 1995).

Phase 2: Independence

An important policy objective at independence in Kenya, Malawi, Tanzania, Zambia, and Zimbabwe was to expand smallholder grain production. The newly independent governments of each country responded to this objective in fundamentally similar ways. The key features of their strategy were the expansion of state crop buying stations in smallholder areas that had been excluded from these benefits under the colonial regime, continuation of direct state control over grain supplies and pricing, efforts to stabilize and often subsidize urban consumer prices without reliance on imports, and elimination of the dominant role of nonindigenous minorities (Table 14.2). Although control of the grain boards was wrested from white

Table 14.2 Phase 2 of Maize Marketing Policy in Eastern and Southern Africa: Independence (expansion of marketing services to smallholders)

Country and Period	State Marketing Agency	Market Regulation	Pricing Policy
Kenya, 1963–1988	Maize Board and provincial boards combined (1963). Combined with Wheat Board to form National Cereals and Produce Board (1980). Five hundred new Produce Board buying centers created, 1980–1982. Threefold increase in number of Produce Board employees, 1980–1987.	Licensing system for interdistrict trade creates rents for local officials. Licensing system for interdistrict maize meal trade creates rents for licensed millers.	Increase in National Cereals and Produce Board purchase prices (1977). Produce Board margin squeezed as imported maize sold at subsidized prices by the board (1984).
Malawi, 1971–1987	Agricultural Development and Marketing Corporation (ADMARC) established in part to develop smallholder maize production (1971). ADMARC established up to 1,400 marketing points by the early 1980s.	No legal restrictions on trade in smallholder produce by Malawian Africans, but such trade is officially disapproved. Expansion of ADMARC crowds out private trading. Malawian Asians restricted from trade in smallholder produce.	Pan-territorial pricing to encourage production in northern districts. ADMARC subsidizes maize producer prices and fertilizer through taxation of smallholder cash crops.
Tanzania, 1963–1984	National Agricultural Products Board established to procure from cooperatives; board replaced by National Milling Corporation. Cooperatives abolished in 1975 and replaced by village procurement.	Statutory single-channel, three-tiered system (primary cooperatives, regional cooperative unions, marketing boards). Replaced by two-tiered system, 1975. Crackdown on "economic saboteurs," 1983, attempted to suppress parallel markets.	Maize producer prices determined as residual after deduction of cooperative and National Agricultural Products Board costs. Pan-territorial and pan-seasonal price setting. National Milling Corporation margins squeezed. Regional pricing introduced, 1981.

(continues)

Table 14.2 continued

Country and Period	State Marketing Agency	Market Regulation	Pricing Policy
Zambia, 1964–1985	Agricultural and Rural Marketing Board established to serve "nonviable" producers (1964). Merged with Grain Marketing Board to form National Agricultural Marketing Board (Namboard) in 1969. Nationalization of grain mills (1986).	No changes.	Increases in consumer subsidies. Implicit taxation of agriculture through currency overvaluation increases from 1975. Expansion of pan-territorial pricing to smallholder farmers.
Zimbabwe, 1980–1992	Development of limited Grain Marketing Board depot network in smallholder areas from 1970s; network rapidly expanded after 1980. Marketing Board staff doubled in numbers.	No changes.	Consumer and producer subsidies increased in late 1970s. Expansion of pan-territorial pricing, taxing commercial maize producers. Consumer subsidies phased out by 1985; reintroduced from 1991 to 1993.

Source: Jayne and Jones (1996).

commercial farming interests, attempts to accommodate wider interests occurred without fundamental changes to marketing institutions or the policy framework. In general, the expansion of state market infrastructure in smallholder areas facilitated the disbursement of credit and subsidized inputs to smallholders through allied state organizations designed to recoup loans through farmers' grain sales to the marketing boards (Rohrbach 1989; Howard 1994; Jones 1994).

A key objective of pricing policy during this period was to avoid maize imports. The thinness of the world market for white maize had often made it necessary to import yellow maize, and this less-preferred substitute had come to be associated with failed agricultural policies (Pinckney 1988). Government stockholding of grain, primarily maize, thus expanded as a result of national food security concerns (particularly exposure to drought) and as an unplanned result of pricing policy. An important consequence of the expansion of the depot network under pan-territorial pricing was greater implicit taxation of those grain producers (mainly white commercial farmers) located near consumer markets. In most cases, however, taxpayers bore most of the cost of expanding marketing services to smallholders. Political independence was also associated in some cases with increased consumer subsidies and the squeezing of marketing board margins (especially in Zambia and, at a later stage, in Kenya). Indirect subsidies were conferred through the operating deficits of the marketing boards. In Kenya, Zambia, and Zimbabwe, the grain boards' actual operating costs normally exceeded the margin between the controlled producer and selling price by 30% or more (Argwings-Kodhek, Mukumbu, and Monke 1993; McKenzie and Chenoweth 1992; GMB, various years).

Although the postindependence model of providing services to smallholders appears to have had important successes in boosting grain production and incomes in some rural areas, by the mid-1980s several major problems had emerged in all the countries and militated for reform.

1. The costs of marketing boards escalated, as the scale and complexity of their activities increased and governments failed to reimburse the losses incurred from implementing government policies. In some cases, especially in Zambia during the 1980s, the treasury costs of state marketing operations became so large as to have major macroeconomic effects on rates of inflation, interest, and currency exchange.

2. The state systems for input delivery and crop payment became increasingly unreliable and tardy, especially in Zambia, Tanzania, Kenya, and Malawi (Westlake 1994; Howard 1994). Smallholders' repayment of credit also became problematic in many cases. In Zimbabwe, almost 80% of smallholder recipients of government credit were in arrears in 1990 (Chimedza 1994).

3. In Zimbabwe and Zambia, the implicit taxation of large-scale commercial farmers that had helped to finance the expansion of the official

marketing system became more difficult to sustain, as producers switched to other, uncontrolled crops. Both the weakened level of service and implicit taxation increased incentives for parallel marketing, especially in Kenya (Bates 1989).

4. An increasing proportion of the maize sold to marketing boards came from smallholder farmers who generally grew maize on poorer land in areas where rainfall was less reliable. This increased the instability of marketing board purchases and sales and hence that of the fiscal demands made by the marketing system.

5. A growing body of empirical evidence found the controlled marketing systems to be responsible for several problems. These systems inflated prices paid by rural consumers for grain because of their controls on private grain movement (Jayne and Chisvo 1991; Odhiambo and Wilcock 1989). They suppressed the development of relatively low-cost, small-scale grain processors, thereby imposing unnecessary costs on urban consumers (Mukumbu 1992; Jayne and Rubey 1993). Finally, particularly in Zimbabwe and Kenya, controlled marketing systems inflated the costs to consumers, producers, and/or taxpayers associated with holding large stockpiles of white maize, largely an outgrowth of postindependence pricing policy oriented toward self-sufficiency (Buccola and Sukume 1988; Pinckney 1988). Many politicians and bureaucrats in the region put only limited trust in such analyses, at least initially, especially when they were conducted and disseminated primarily by expatriate researchers and donor organizations.

A major difference in the implementation of maize policy before and after independence was the method of supporting food production growth. The colonial governments drew the resources to support European farming largely from African farmers and consumers (through the cross-subsidies implicit in market regulations). By contrast, the postindependence governments drew the resources to promote African smallholders primarily from the treasury. The resulting large budget deficits since independence have made the resource transfers to support maize production highly visible, yet the impressive growth in maize production seen during the colonial period involved major resource transfers as well.

Phase 3: Market Liberalization and Privatization

Along with problems in the performance of controlled marketing systems, shifts in the political balance of power pushed the grain marketing systems of eastern and southern Africa toward liberalization in the mid-1980s. After first trying to strengthen the performance of state marketing boards in the 1960s and 1970s, donors and international lending agencies began promoting the reform of food marketing and pricing in Africa in the 1980s as a central component of economywide structural adjustment programs. The framework of policy-based lending within which market reforms have occurred in each country (except South Africa) has strongly influenced the path of reforms and has expanded external leverage over domestic agricultural policy through aid conditionality.

As noted earlier, these measures have exceeded the relatively limited agenda for food market reform outlined in the Berg Report (World Bank 1981). Several factors shaped this change. Especially in the 1990s, donors lost patience with phased and partial reform programs that were increasingly seen as propping up costly and otherwise unsustainable pricing and marketing policies rather than facilitating reforms. In addition, models of the political economy of agricultural policy (such as Bates 1989) suggested that interventions by the state in agricultural markets, although ostensibly designed for rural development or to correct for market failures, were in fact driven by the objectives of eliminating rural opposition, generating rents, and nurturing political patronage. As a result, donor-driven policy reforms have embraced both "liberalization" (the removal of regulatory restrictions over the private sector) and "privatization" (the withdrawal of the state from performing direct marketing functions).

The main maize marketing reform measures implemented over the past decade (Table 14.3) have had several key features.

1. Reform generally began with attempts to reduce the level of subsidies provided to the marketing system by widening the margins between producer and selling prices, reducing the number of state crop buying stations (in Kenya, Malawi, and Zimbabwe), and reducing maize stocks (especially in Zimbabwe and Kenya). This phase has continued over several years without major changes to the structure of the marketing system. Attempts to enact reforms in Kenya and Zambia have broken down and subsequently been reintroduced.

2. Major regulatory reforms to liberalize the domestic market have occurred only in the last three years in South Africa, Zambia, Zimbabwe, and Kenya.

3. In every country except Zambia and Tanzania, the main state maize marketing agency continues to exist and remains a major player in the market. This contrasts sharply with cereal market reforms elsewhere in western and eastern Africa, where state grain marketing agencies are now largely defunct.

4. Key elements of official pricing policy have remained unchanged through much of the reform process in Zimbabwe, Malawi, Kenya, and South Africa. Regional and seasonal variations in official buying and selling prices for grain have (with some minor exceptions) not been introduced. Official price setting procedures have not become appreciably more sensitive to market conditions. Controls over the price of maize meal,

Table 14.3 Phase 3 of Maize Marketing Policy in Eastern and Southern Africa: Structural Adjustment (marketing liberalization and privatization)

Country and Period		State Marketing Agency		Market Regulation		Pricing Policy
Kenya, from 1988	1988	Financial restructuring of National Cereals and Produce Board. Phased closure of Produce Board depots, but depot construction continues. Produce Board debts written off; crop purchase fund established but not replenished.	1988–1989	Phased increase in permitted purchases from traders by millers. Relaxation of unlicensed maize movement and milling.	1988	Cereal Sector Reform Programme envisages widening of Produce Board price margin. In fact, margin narrows.
			1991	Further relaxation of district trade limits.	1992	Produce Board unable to defend ceiling prices.
	1994	Produce Board to be restricted to role of limited buyer and seller of last resort.	1992	Movement restrictions tightened.	1993	Limits set on Produce Board purchases. Unable to defend floor price.
			1993	Abolition of mill quotas.		
			1994	Liberalization of internal and external trade.		
Malawi, from 1987	1987	ADMARC market closure program initiated. Strategic Grain Reserve established, managed by ADMARC.	1987	Private maize trade officially promoted.	1987	Premiums set for maize delivered to main depots. Target of zero loss on ADMARC maize account.
			1992	Provisions of 1987 legislation (governing prices, location of trade) abolished.		
South Africa, from 1987	1987	Subsidies to Maize Board trading account ended.	1993	Direct farmer-to-miller trade permitted, subject to levy payment of approval.	1987	Fixed producer price replaced with advance price and supplementary payment (*agterskot*) to ensure maize account breaks even.
			1995	Removal of restrictions on private maize purchases and sales, subject to stabilization levy. Imports liberalized (with zero tariff).	1995	Maize price controls removed.

(continues)

Table 14.3 continued

Country and Period		State Marketing Agency		Market Regulation		Pricing Policy
Tanzania, 1984–1990	1986	Reestablishment of cooperatives.	1984	Relaxation of controls on grain movement. De facto toleration of private trade.	1984	Maize flour price decontrolled.
	1990	National Milling Corporation access to crop finance ended. Strategic Grain Reserve transferred to Ministry of Agriculture.	1987	Movement restrictions abolished.	1985	Maize grain price decontrolled.
					1986	Maize flour subsidies removed.
	1991	New Cooperative Act recognizes cooperatives as private institutions.	1988	Private traders allowed to purchase from cooperatives.	1987	Official producer prices to be regarded as minimum cooperatives.
			1990	All restrictions on purchase of grain by traders eliminated.		
Zambia, 1985–1994	1989	Namboard abolished and marketing functions transferred to cooperatives.	1986	Liberalization of interdistrict trade.	1986	Consumer subsidies removed, then reintroduced.
			1992	Deregulation of small-scale milling. Restrictions on external trade relaxed.	1989	Maize meal coupon program introduced (until 1992).
	1993	State provides crop purchase finance to approved buyers.			1993–1994	Consumer subsidies, official floor and into mill prices abolished.
	1995	Food Reserve Agency formed.	1994	Completion of liberalization of import and export trade.		
Zimbabwe, from 1986	1986	Number of maize collection points reduced.	1991–1992	Phased elimination of controls on trade between smallholder areas.	1987	Producer prices allowed to decline in real terms, then rise after 1992 drought.
			1993	Grain Marketing Board monopsony-seller status restricted to large mills, then eliminated (1994). External trade still controlled by Marketing Board.	1993	Maize meal subsidies abolished; consumer prices decontrolled.

Source: Jayne and Jones (1996).

however, had been abolished in all countries by 1994.[10] Marketing agencies have either not been granted, or have been unwilling to exercise, autonomy in setting domestic prices.

5. The role of the marketing agency in an increasingly liberalized market has not been clarified. Although this role has usually been conceived in terms of stabilizing prices, providing marketing services to smallholders, and providing a buyer and seller of last resort (BSLR), the reforms enacted have usually been inconsistent with this role. This discrepancy has occurred mainly because of the unresolved conflict between fiscal objectives (implying a withdrawal from unprofitable activities) and the BSLR objective (implying a withdrawal from profitable activities so they can be assumed by the private sector and a focus on activities the private sector will not undertake). Considerable confusion also remains about the links between price stabilization and reserve policy.

6. Initiatives to provide positive support to private trading, and to develop the public institutions required for a privatized marketing system to function effectively, have been limited (Jones 1994).

Assessment of Market Liberalization Experiments

The effects of food market reform over the past decade are difficult to assess for three main reasons. First, the effects of the reforms are difficult to isolate from other processes affecting the broader economy, especially broader macroeconomic adjustments and extreme weather conditions. Second, the reforms described in the previous section have been partial and have been subject to reversal in some cases, and (in almost all cases) decisive reform measures have been implemented only very recently. Third, because only weak and partial data on factor productivity are available, the welfare implications of these reforms are unclear. Notwithstanding these caveats, several consistent trends appear to be emerging out of the market reform experiments in Zimbabwe, Zambia, Kenya, Tanzania, and Malawi: reduced food marketing costs from surplus to deficit areas, improved access to food by low-income urban consumers, limited supply response to the partial food market reforms that have been implemented, a gradual movement of the region to structural food deficits, and mixed impact on marketing board financial deficits.

Reduced Food Marketing Costs to Grain-Deficit Rural Areas

Before reforms were enacted, the controlled marketing systems featured a flow of grain that extended from farmers to the marketing boards, moved onward to large-scale industrial processors in urban areas, and went from there to consumers in both rural and urban areas (Bryceson 1993; Kirsten and Sartoius von Bach 1992; Jayne and Chisvo 1991; Mukumbu 1992).[11] Regulatory barriers prevented informal traders or consumers from purchasing grain once it was in the hands of the marketing board. This system was based on the implicit assumption of rural self-sufficiency in maize grain. On the surface, this assumption seemed plausible, because grain sales normally rose rapidly in most smallholder areas where marketing board infrastructure was developed (Bryceson 1993; Rohrbach 1989). This provided some evidence of a "surplus" in excess of a particular area's consumption requirements. Microlevel research, however, showed that grain deliveries to the marketing boards should not be mistaken for a "surplus" from a given region over and above consumption requirements. Marketed output from a small segment of well-equipped farmers often masks considerable grain deficits among a large proportion of households. Official restrictions on private trade and weak market infrastructure have often made it easier for farmers producing a surplus to sell it to the marketing boards than to their maize-deficit neighbors a few kilometers away (Jayne and Chisvo 1991).

This marketing structure—characteristic in Zimbabwe, Kenya, Zambia, and South Africa prior to the reforms—created a circuitous and expensive movement of grain from rural to urban areas, where it was milled by high-cost urban millers only to be transported back to rural areas for consumption by grain-deficit households.[12] After locally produced supplies were depleted, rural households had no legal means of acquiring grain from outside their "zone." The controls on interzone grain movement provided the industrial urban millers with a de facto monopoly on maize distribution into grain-deficit areas (Mukumbu 1992; Kirsten and Sartoius von Bach 1992; Jayne and Chisvo 1991). The milling costs charged by urban milling firms and passed along to consumers, taxpayers, or both, however, were substantially higher than those charged by informal, small-scale operators of hammer mills in rural Zimbabwe, Zambia, Kenya, and Malawi. In Zimbabwe, the suppression of private trade (to assure the dominance of the controlled marketing system) inhibited the development of direct trade from surplus to deficit rural areas. This situation made households largely dependent on relatively expensive refined maize meal distributed through the official market channel, which reduced cash incomes of the poorest rural households by as much as 30% (Jayne and Chisvo 1991).

Increased recognition of the food purchasing status of many rural households heightened decisionmakers' awareness of the need for more decentralized and efficient food distribution networks serving the semiarid areas. Although few quantitative assessments of the reforms are available, the emerging picture is that the legalization of interdistrict grain movement in Zimbabwe, Zambia, and Kenya has reduced the difference between prices realized by producers and those paid by consumers for maize meal

(GRZ 1995; Rubey 1995; Argwings-Kodhek and Jayne 1996). This reduction in marketing costs has been achieved primarily through the expanded role of small-scale trading and milling networks in fulfilling the residual grain needs of rural households.

*Changes in Urban Consumption Patterns
and Improved Access to Food by Low-Income Consumers*

Prior to the reforms, urban maize milling was dominated by several large private firms that purchased maize from the state marketing boards—often at subsidized prices—processed it using roller mill technology,[13] and distributed meal to retailers at government-controlled margins that were substantially higher than margins observed by the informal, small-scale mills for whole meal.[14] By the early 1990s, private trading had been liberalized, and subsidies on roller-milled meal were eliminated in Zambia, Kenya, and Zimbabwe. In each case, the large-scale millers swiftly lost a major part of their market share to small urban processors. Urban consumption of hammer-milled meal rose to 45 to 55% of total meal consumption in Zimbabwe and to about 40% in Kenya and Zambia (Jayne et al. 1995).[15] During the 1993–1994 and 1994–1995 seasons, the cost of whole meal was about 60 to 75% of the retail price of roller-milled meal in most urban areas of Zimbabwe, Zambia, and Kenya (Table 14.4, columns C and E). Household surveys indicate that low-income consumers in particular shifted quickly to hammer-milled meal (Rubey 1995; Jensen and Luckett 1993; Mukumbu and Jayne 1994). The reforms have clearly promoted urban household food security in much of the region by providing a lower-cost alternative to roller-milled meal, especially after the withdrawal of direct and indirect subsidies conferred through the official marketing channel, which caused prices of roller-milled meal to rise (Table 14.4, column C).

It is important to note that the rapid shift from roller-milled to hammer-milled meal has likely been accentuated by a decline in urban consumers' real incomes following the introduction of macroeconomic adjustment policies in much of the region since the late 1980s and early 1990s. This effect is suggested by survey evidence of an inverse relationship between consumption of hammer-milled whole meal and household income (Rubey 1995; Jensen and Luckett 1993; Mukumbu and Jayne 1994).

Limited Supply Response to Food Market Liberalization

Although it is commonly asserted that state marketing boards in Africa depressed food production by keeping producer prices below border price equivalents,[16] this view is invalid for most of the countries examined here. Throughout the 1980s and up to the time market reforms were first initiated, official producer prices usually exceeded export parity prices in the major production regions of Zimbabwe, South Africa, and Kenya (all typically

Table 14.4 Eastern and Southern Africa: Real Maize Grain and Maize Meal Price Trends in Selected Countries, 1980–1994

	Maize Grain Price[a]		Index of Real Maize Meal Prices		
Country and Period	As Percentage of Import Parity[b] (A)	In Constant 1994 Local Currency Units (B)	Roller-Milled Meal, Retail (C)	Roller-Milled Meal, Including Subsidies[c] (D)	Hammer-Milled Meal[d] (E)
Zimbabwe					
1980–1989[e]	.78	100	126	158	n.a.
1990	.65	83	132	132	n.a.
1991	.46	76	132	141	n.a.
1992	.92	154	164	245	167
1993	.90	129	215	215	144
1994	.86	120	178	178	147
Zambia					
1980–1989	.71	100	149	n.a.	n.a.
1990	.60	101	162	186	n.a.
1991	.70	111	136	152	n.a.
1992	.56	73	198	208	n.a.
1993	.56	121	214	214	160
1994	.77	132	228	228	154
Kenya					
1980–1989	.91	100	117	n.a.	n.a.
1990	.93	89	110	122	95
1991	.97	93	111	123	98
1992	1.11	119	167	181	129
1993	.69	86	160	n.a.	98
1994	1.14	82	121	143	94
Malawi					
1980–1989	.78	100	180	n.a.	n.a.
1990	.87	108	163	n.a.	n.a.
1991	.90	103	151	n.a.	n.a.
1992	1.19	140	179	n.a.	n.a.
1993	1.14	141	271	n.a.	n.a.
1994	1.21	190	311	n.a.	n.a.
South Africa					
1980–1989	.80	100	232	232	n.a.
1990	.85	84	223	223	n.a.
1991	.92	86	228	228	n.a.
1992	.89	86	224	224	n.a.
1993	.90	80	238	238	163
1994	.74	66	212	212	149

Sources: White maize prices, f.o.b. U.S. Gulf: Fisher (various years). Freight rates: IWC (1993). Maize grain prices in Zimbabwe: GMB (various years); Zambia: Howard (1994) until 1992, LACE (1995) thereafter; Kenya: CBS (various years) until 1990, agricultural commodity prices, *Daily Nation,* various issues; Malawi: ADMARC, courtesy of R. Goldman and C. Pinckney; South Africa: Maize Board (various years). Consumer price index and exchange rate data for all countries: IMF (1995). Roller-milled meal prices in Zimbabwe: Central Statistics Office unpublished data, various years; Zambia: Howard (1994) until 1992, LACE (1995) thereafter; Kenya: Tegemeo (1996); South Africa: Maize Board files. Consumer subsidies in Zimbabwe: Central Statistics Office unpublished data, various years; Zambia: Sipula (1993); Kenya: CBS (various years).

(continues)

Table 14.4 continued

Notes: n.a. = data not available.

a. Marketing board selling prices in Zimbabwe and South Africa; marketing board selling prices in Malawi (until 1985), Kenya (until 1990), and Zambia (until 1993); then average retail maize grain prices in the Blantyre, Nairobi, and Lusaka markets.

b. Import parity derived as f.o.b. white maize, U.S. Gulf, plus international shipping, demurrage, and rail transport to Harare, Lusaka, Nairobi, Blantyre, and Johannesburg.

c. Refers to direct consumer subsidies on roller-milled meal.

d. Calculated as the retail price of maize grain in informal market channels (Harare, Lusaka, Nairobi, and Johannesburg) plus custom-milling fee; does not include cost of transporting meal from mill to home or opportunity cost of labor time.

e. Years are marketing years (e.g., 1990 = 1990–1991).

exporters during this period).[17] Maize price supports were continued in part because of the farm lobbies' sustained influence in the formation of food prices, a process established during the colonial era, but a more important force was the high priority postindependence African governments gave to achieving maize self-sufficiency and avoiding the political risks associated with importing yellow maize. Only in Tanzania and Zambia, which typically imported small quantities of maize, did depressed producer prices (relative to import parity) apply. Although currency overvaluation did introduce an often substantial indirect tax on food producers (Schiff and Váldes 1992), this tax was significantly offset by a package of state investments designed to reduce farm production and marketing costs (primarily input subsidies, concessional credit, and investments in state crop buying stations).[18]

The period in which these public investments, transfers, or both reached their peak coincided closely with the brief, dramatic rise in smallholder grain production experienced in many of these countries. Most important were the expansion of marketing board buying stations in smallholder areas (Zimbabwe, 1980–1986; Zambia, 1983–1989; Kenya, 1980–1982; Malawi, 1974–1985); the expansion of state credit disbursed to smallholders (Zimbabwe, 1980–1986; Zambia, 1983–1988; Kenya, 1975–1983); and explicit or implicit subsidies on inputs (Zimbabwe, 1980–1991; Zambia, 1971–1991; Malawi, 1980–1993), albeit with the degree of subsidization varying widely among countries and years. These pricing and market support policies clearly encouraged farmers to adopt newly available hybrid maize seed and stimulated growth in smallholder grain area and yields during the 1980s in Kenya, Zimbabwe, and Zambia (Table 14.5).[19] Smallholder grain production in Zimbabwe increased by 51%, from 117 kg per person during the period 1975–1979 to 177 kg during the period 1985–1989 (Table 14.5, column A), whereas per capita maize production in Zambia rose 47% over the same period. In Kenya, per capita grain production rose 30% between 1970–1974 and 1980–1984. The major

Table 14.5 Eastern and Southern Africa: Trends in Maize Production Per Capita, Area, Yield, Net Exports, and Fertilizer Use in Selected Countries

		Three-year Centered Moving Average			
Country and Period	Maize Production Per Capita (kg) (A)	Maize Area (ha) (B)	Maize Yield (t/ha) (C)	Net Maize Exports (000 t) (D)	Fertilizer Use (000 t) (E)
Zimbabwe, smallholder sector					
1970–1974	70	685	0.66	n.a.	8.6
1975–1979	75	721	0.66	418	27.1
1980–1984	101	1,083	0.69	201	97.2
1985–1989	143	1,071	1.18	311	119.0
1990–1994	78 (95)[a]	1,002	0.73	49	86.6
Zambia[b]					
1970–1974	224	577	1.51	−78	47.9
1975–1979	160	626	1.22	−94	65.3
1980–1984	188	989	1.03	−181	74.3
1985–1989	235	848	1.56	−161	80.4
1990–1994	173 (193)[a]	836	1.46	−239	68.2
Malawi					
1970–1974	317	1,026	1.14	14	14.1
1975–1979	277	1,019	1.15	−5	21.8
1980–1984	260	1,113	1.16	30	33.4
1985–1989	225	1,176	1.13	−24	43.9
1990–1994	179 (190)[a]	1,294	1.03	−215	58.0
Kenya					
1970–1974	87	1,108	.93	77	144.2
1975–1979	110	1,252	1.26	71	130.2
1980–1984	121	1,238	1.76	59	155.7
1985–1989	117	1,404	1.89	120	235.1
1990–1994	85 (92)[a]	1,366	1.70	−102	241.5
South Africa					
1970–1974	327	4,250	1.77	2,435	n.a.
1975–1979	332	4,393	1.97	2,909	n.a.
1980–1984	311	4,235	2.19	3,069	n.a.
1985–1989	206	3,947	1.81	1,428	n.a.
1990–1994	204 (216)[a]	3,437	2.27	1,090	n.a.

Sources: Population data: Urban and Nightingale (1993). Grain data: Ministry of Agriculture (Zimbabwe); Jones (1994) (Zambia); Ministry of Agriculture, compliments of J. Rusike (Malawi); Egerton University, Kenya Market Development Program/Policy Analysis Matrix database (Kenya); Maize Board (various years), and RSA (1994) (South Africa).

Notes: n.a. = data not available.
a. Figures in parentheses exclude 1992, a drought year.
b. Figures for Zambia are for coarse grains. The share of maize in total coarse grain production during the period 1980–1989 is estimated at 91% for Zimbabwe, 98% for Malawi, 95% for Zambia, 92% for Kenya, and 94% for South Africa (USDA 1992).

spurts of growth in food grain production in the 1980s in Zimbabwe, Kenya, and Zambia were fueled by large state subventions to the maize sector (Jayne et al. 1994; Mosley 1994; Howard 1994) and by viable agricultural technologies available through decades of agricultural research.

This state-led model of providing services to foster growth in smallholder maize production, however, has proven politically and economically unsustainable. Evidence in some areas indicates that growth in production was achieved at a cost greater than the value of the increased maize output, especially in Zambia, Zimbabwe, and South Africa (Howard 1994; Jayne and Rukuni 1993; Wright and Nieuwoudt 1993).[20] As fiscal deficits increasingly came under attack in the prevailing atmosphere of structural adjustment, state marketing services and associated deficits were gradually or sharply cut back.[21]

In Zimbabwe, even though 17 additional permanent buying stations were established between 1985 and 1992, the number of seasonal rural buying stations declined from 135 in 1985 to 42 in 1989 and 9 in 1991. Disbursement of government credit to smallholders declined steadily from a peak of Z$195 million in 1987 to under Z$40 million in 1994 (in constant 1994 Z$). In the three cropping seasons since 1993, when major maize policy reforms went into effect, fertilizer use on all crops has declined to less than 75% of the 1985–1989 level. In Zambia, grain area, fertilizer use, hybrid seed purchases, and maize production have also declined since the late 1980s as a result of lower real producer prices, higher real fertilizer prices, deteriorating state marketing services, and a reduction in the availability of credit (associated with the collapse of the state credit agency and a sudden rise in the level of real interest rates in 1993–1994). Fertilizer nutrient use, which peaked in 1986–1987 at 88,000 t, diminished to less than 60,000 t in 1994–1995. Hybrid maize seed purchases declined from 15,000 t in 1989–1990 to 4,799 t in 1994–1995. Both total cropped area and maize area have contracted by about 15% since state support to agriculture peaked in the late 1980s (Howard, Nakaponda, and Ferris 1995). In Kenya, fertilizer use and maize yields have continued to rise. Yet population growth in all three countries has outstripped smallholder grain production since the mid-1980s (Table 14.5). The stagnation in yields and per capita production is especially noteworthy in Zimbabwe, Kenya, and Zambia, where the use of improved maize hybrids and fertilizer use per ha are the highest in Africa.[22] Some of the decline in food production between 1990 and 1994 can be attributed to the 1992 drought, the worst in decades. But when the effects of the drought are removed (see Table 14.5, note a), it becomes clear that the decline is not simply a transitory phenomenon brought about by drought.

Welfare effects cannot be inferred from the decline in per capita production and yield stagnation associated with food market reform. Information on factor productivity is largely unavailable. Furthermore, grain market reform has created complex distributional effects, expanding opportunities for farmers and consumers in some regions while withdrawing benefits in others. The withdrawal of benefits has occurred in Tanzania and Zambia (and will occur in Zimbabwe and Malawi), particularly in the more remote smallholder areas where pan-territorial pricing has encouraged production of grain surpluses that were not economically viable given existing market and infrastructural conditions. The extent to which these state pricing policies have skewed the allocation of productive resources away from that of comparative advantage is unclear.

The major conclusion about food market reform is that, contrary to most donor expectations, the removal of government controls on private grain trading generally did not raise production incentives or expand market opportunities for smallholder farmers. This conclusion is corroborated by the data presented in Table 14.5. The primary reason for the weak production response has been the withdrawal of substantial state transfers to the maize sector associated with market liberalization. In contrast to other cases in Africa, where liberalization facilitated trade in illegal but relatively well-functioning parallel markets, the transition from the controlled to market-oriented systems in the countries studied here has had an ambiguous or even a negative effect on farm production, at least so far. There is evidence, however, that the number of private traders serving smallholders has increased, especially in high-potential areas, within several years after the initiation of partial food market reform (Amani and Maro 1992; Kaluwa 1992). Given sufficient public-sector support and time, private systems of input delivery, finance, and commodity marketing will likely develop and partially or fully offset the effects of the withdrawal of public marketing services.

Gradual Movement of the Region to Structural Food Deficits

The policy-related decline in the growth of cereal production in the region has been associated with a general movement toward a structural food deficit. Five of the six countries examined have experienced a marked decline in net maize exports, especially South Africa and Zimbabwe, the two reliable surplus producers in the 1970s and 1980s (Table 14.5). Malawi and Kenya have moved from net surplus to net deficit status during the 1990s,[23] and average net maize imports to Zambia have doubled since the late 1970s.

The decline in net grain surplus may be attributed to the same sources that caused the decline in per capita grain production. The removal of state transfers to the maize sector as part of structural adjustment has reduced production growth in Zimbabwe (since 1985), Zambia (since 1990), Malawi (since 1987), and South Africa (since 1987). Second, population is growing at 3% per year, and urban demand for maize is growing even more rapidly. Neither yields nor area has expanded quickly enough since 1980 to keep grain production even with population growth in any country in the region.

This macrolevel picture of declining national food surpluses is paralleled by microlevel information on household grain marketing behavior. During the 1970s and early 1980s, it was commonly believed throughout the region that most smallholder areas, and most households within those areas, were surplus grain producers. Regardless of whether this perception was erroneous or whether household food marketing patterns have changed over the past several decades because of declining farm size, population pressure, and soil degradation in some areas, abundant household survey data from the late 1980s and the 1990s reveal that a large proportion of rural farm households are actually net buyers of grain, even in a normal year (see, for example, Cousins, Weiner, and Amin 1992; Kandoole and Msukwa 1992; Kirsten and Sartoius von Bach 1992; Odhiambo and Wilcock 1990; Weber et al. 1988).[24] In Zimbabwe, which was a typical food exporter during the 1980s, the proportion of rural farm households that are net grain buyers has reached 70% or more in the drier areas, where over 60% of the smallholder population lives. Of these households, about half purchase over 50% of their annual grain requirements (Jayne and Chisvo 1991). In Malawi, more than 65% of the rural population were net buyers of maize in 1991, a year of normal weather (Kandoole and Msukwa 1992). These findings indicate that, in much of the region, the food security implications of food market reform depend upon how reform has affected the capacity of consumers to acquire food, especially smallholders in the food-deficit and generally poorer regions.

Impact of Liberalization on Marketing Board Deficits

Although the fiscal burden of the controlled marketing systems has been the principal factor driving reform, marketing board deficits have actually increased in Zimbabwe, Malawi, Zambia, and Kenya since the late 1980s. Senior policymakers in the region have been generally reluctant to relinquish control over setting marketing board prices in a liberalized marketing environment and to allow prices to move more consistently with prevailing market conditions. As a result, when official prices have diverged substantially from market conditions, the private sector has increasingly bypassed or undercut the official marketing system, often with a disastrous impact on the boards' trading accounts.

The most dramatic example of this situation occurred in Zimbabwe. After the drought of 1992–1993, the Grain Marketing Board announced substantially higher official producer prices fixed well in advance of planting. The 1993 harvest was a bumper crop, and in eight months the board accumulated unsalable stocks and a trading deficit equal to 2.8% of gross national product. Subsequently, the board has set a producer price that is too low relative to market prices (in 1995–1996) and has lost market share to private traders and millers. In Tanzania and Zambia, the state marketing boards continued to fix prices without due reference to prevailing market conditions until their access to credit was cut. The resulting financial crises caused these boards to be abolished. In Kenya, the board's deficit increased over the period of the Cereal Sector Reform Programme, as its price margins were squeezed. Only South Africa has managed to reform its maize pricing system without imposing large trading deficits on the marketing board.[25]

Although the need for more flexible price setting in a market environment has been highlighted by both analysts and the management of the boards, little progress has been made in devolving price setting authority to the boards themselves, which is a first step in allowing the boards to conduct their operations in a more commercial manner. Senior politicians continue to exercise control over price setting by marketing boards in Zimbabwe, Malawi, and Kenya. The main concerns with devolving price setting authority are that the more autonomous and commercially oriented boards might increase price volatility by frequently altering prices as market conditions change and that they might also pay less attention to the social objectives historically pursued through food marketing policy. As has been shown in Zambia, the transition from pan-territorial to market pricing has benefited large-scale European farmers close to urban areas and has reduced grain prices received by small-scale African farmers in the more remote grain-surplus areas. A transition to market-oriented pricing is anticipated to have similar effects in Zimbabwe (GMB 1991; Masters and Nuppenau 1993), which could have major political ramifications and be portrayed as another step backward in attempts to promote smallholder maize production and erase the dualism inherited from the colonial marketing system.

For these reasons, little progress has been made toward establishing more flexible pricing strategies that allow the marketing boards to respond to, or influence at the margin, prevailing prices in private trading channels (Jones 1994). Where reforms have increased the marketing agencies' formal autonomy over prices, these agencies have often been unwilling to exercise this power (for instance, in Malawi), apparently in part from fear of the political consequences of being involved in implementing unpopular policies and in part for technical and managerial reasons (e.g., an inability to determine on what basis regional differentiation of prices should be introduced).

Looking to the Future: Strategies to Strengthen the Performance of Grain Marketing Systems in the Region

The empirical record of food marketing reforms in eastern and southern Africa highlights two major issues with generalizable implications for food policy elsewhere in Africa: first, the importance of moving beyond a reform agenda of liberalization toward a financially sustainable growth path involving the coordination and strengthening of food, financial, and input marketing tasks; and second, the need to devise market-oriented mechanisms to reduce vulnerability to price and supply instability and thus reduce the stabilization burden that has been borne by the state.

Beyond Food Market Liberalization

The history of food marketing policy reviewed in this chapter has demonstrated that policy's political centrality in the deeply established conflicts between urban and rural interests and between the large- and small-farm sectors. Structural adjustment has not removed these conflicts, although it has (probably temporarily) enhanced the influence of external donors over policy. At one level, structural adjustment has been a defeat for the aspiration of independent governments to overcome the dualism and poverty of the rural economy. On the other hand, there is evidence that some features of the reformed market systems promote growth and do indeed reduce some elements of dualism (e.g., by improving direct access for smallholder produce to urban markets). Market liberalization, however, is certainly not an end in itself. Schultz's "efficient but poor" observation of low-resource farmers also describes the functioning of marketing systems in many developing areas (Shaffer et al. 1985). Marketing margins may approximate costs, but these costs may be too high to encourage the rapid private investment necessary to promote productivity growth. Although private food trade in eastern and southern Africa has grown and has brought important tangible benefits, especially to urban consumers, the evidence so far suggests that the anticipated stimulus to technology adoption and growth in food production has been weak.

Food marketing and food security policy strategies will need to alter their emphasis from liberalizing food markets to promoting growth in productivity throughout the entire food system through the development and coordination of markets—most notably for commodities, inputs, and finance—in a financially sustainable way. The experience of Zimbabwe, Zambia, Malawi, and Kenya demonstrates that we know how to promote farm-level productivity growth for smallholder farms temporarily. The former state-controlled system addressed the coordination problem successfully (from the standpoint of many farmers) by offering credit, supplying needed inputs, and tying repayment to the sale of the crop upon harvest. These schemes, however, usually involved subsidies on inputs and credit, low repayment rates, and losses on marketing board trading operations, which basically amounted to shifting the costs and risks of the food system from one group to another rather than reducing the total costs for society as a whole. Eventually, production gains, achieved disproportionately by well-equipped farmers in high-potential areas, became unsustainable as the budgetary transfers provoked decisive internal and external opposition. Moreover, the subsidized controlled systems inhibited the development of potentially better coordinated and sustainable private input-credit-output commodity systems.

The challenge for the future is to design integrated and financially sustainable systems of input delivery, farm finance, and reliable output markets to provide both the incentives and the ability to increase farmers' use of purchased inputs (fertilizer and hybrid seed) and other productivity-enhancing investments. So far, liberalization and privatization appear to have replaced often unreliable, high-cost, and centralized forms of state marketing with private markets that are competitive but often lacking in information and infrastructure and poorly integrated with other key activities. Market transactions in the region mainly involve sale by private negotiation in a context of price uncertainty and poorly functioning credit markets. Farmers do not have reliable access to spot markets in which a high level of trade occurs on standardized quality, quantity, and contract terms. Out of such spot markets, more sophisticated market forms may emerge that allow hedging of price risks, reductions in transactions costs, and explicit coordination of credit, input delivery, and output price between farmers and trading firms through more complex transaction mechanisms.

Stabilizing Food Prices in an Era of Fiscal Austerity

In most countries in the region, maize meal makes up a large part of the diet. Demand is generally inelastic. Declines in supplies and related price surges have a disproportionate impact on the poor. To avoid pricing the poor out of food markets, private traders or the government must be able to release food onto the market from either stocks or imports during poor harvests. Most studies of private trade indicate an underprovision of interyear storage because of high risks and market failures.[26] In some cases, these risks have been exacerbated by government behavior (Sahn and Delgado 1987; Steffen 1995). Some evidence indicates that private traders question whether governments would really allow them to profit from high prices in drought years. Especially in landlocked countries where the gap between import and export parity is large, the potential magnitude of price volatility may be unacceptably high to many African governments.

In spite of the strong rationale for moderating extreme price fluctuation, the marketing board "buyer and seller of last resort" approach has not emerged as a successful model in the current liberalized market environment for two reasons. First, as mentioned earlier, the costs of such a system may be enormous, especially if floor and ceiling prices are set without reference to market conditions (Pinckney 1993; Buccola and Sukume 1988). Second, stabilization schemes have impeded private investment in the marketing system by dampening spatial and temporal price variation and by the unpredictable and uneven implementation of these schemes.

An alternative approach to the instability problem would be for states to make market-facilitating investments so as to reduce marketing costs, shrink the wedge between import and export parity prices, and reduce the cost of stabilizing food consumption. Such a strategy would include several features: improving road and rail infrastructure, introducing seasonality in marketing board prices, and strengthening alternative marketing channels that distribute low-cost "self-targeted" staple commodities.

1. Improving road and rail infrastructure is essential. A considerable part of the food price instability in eastern and southern Africa results from the high cost of transportation, which makes import parity prices two to four times higher than export parity prices in much of the region (Koester 1986). National rail transport rates are generally very high compared to rates elsewhere. For example, the cost of railing a tonne of white maize from the Western Transvaal in South Africa to the Copper Belt of Zambia is about U.S.$90, roughly the amount South African farmers are paid to grow it (Scott 1995).

2. Introducing seasonality in marketing board prices would increase investment in private storage within years and thereby expand the scope for a wider range of private marketing activities.

3. Alternative marketing channels distributing low-cost "self-targeted" staple commodities should be strengthened. Self-targeted commodities (those that are more important in the consumption patterns of low-income groups) offer a safety net for the poor in the event of price hikes on the more expensive refined meals. Examples of self-targeted commodities are whole maize meal and yellow maize.[27] Policy constraints in several countries, however, still hamper the development of potentially cost-effective, market-oriented approaches to mitigating food price spikes during years when production is low.

Can the Fledgling Market Reforms Be Sustained?

The experimental nature of food market liberalization in eastern and southern Africa is apparent from the fact that, prior to the recent reforms, almost no one in the region had experienced a market-oriented food system in his or her lifetime. The market reform process initiated in the late 1980s in eastern and southern Africa has not been driven by, and has not yet created, a strong domestic political base for maintaining a liberalized, privatized, and unsubsidized food marketing system. Smallholder farmer groups have generally opposed the state's partial withdrawal from food marketing activities and generally remain suspicious of private marketing. Meanwhile, something of a schism has developed among large-scale farmers. Large-scale farmers of European descent have lobbied increasingly for liberalization, as their influence over the official marketing system has waned since the transition to majority rule, whereas a new class of black commercial farmers—many of whom are politicians or otherwise have close ties to government—supports the continuation of state marketing. Within governments, increasing fiscal difficulties and foreign exchange shortages have strengthened the hand of finance ministries arguing for budgetary restraint, both for the sake of macroeconomic stability and to maintain access to foreign borrowing. These considerations may have overcome the patronage attractions of preserving the former controlled marketing system.

It is difficult to envisage a wholesale reintroduction of trade controls, but pressures for subsidies and price supports are likely to reemerge, especially if the new systems fail to buffer producers and low-income consumers sufficiently against severe production and price shocks. But without significant public-sector investments that reduce the magnitude of food price instability and provide the conditions for more efficient private trade, the sustainability of the reforms may be jeopardized by calls to reimpose state food purchasing and price controls. The current situation in the region is therefore one in which marketing policy has moved only recently to a fundamentally new stance, whose longer-run implications remain to be tested. At the heart of the problem is the level of price variability that can be expected under the new system, how the problems this variability poses (especially for smallholders) can be accommodated, and what will happen to the marketing boards—which are now envisaged as playing a limited price supporting role but in which the old processes of government food price setting have remained largely intact.

Notes

1. For simplicity, the former colonial names Southern Rhodesia, Northern Rhodesia, and Nyasaland are referred to in this chapter by their current names—Zimbabwe, Zambia, and Malawi, respectively.

2. For a theoretical critique of the microeconomic foundations of food market liberalization, see Barrett and Carter (1994).

3. Zimbabwe, Zambia, and other countries of southern Africa have the highest variability in cereal yields in Africa (cereal yield variability was defined as the detrended coefficient of variation). Yield variability in these countries was double or triple that of major food-producing countries in Asia and Latin America from 1961 to 1993 (see Chapter 2).

4. The place of Malawi within colonial policy is illustrated by legislation established in 1933 prohibiting the British South Africa Company from growing maize there to prevent competition with European maize producers in Zambia and Zimbabwe (Deininger and Binswanger 1995).

5. Over the period 1920–1929, European maize cultivation increased from 13,129 ha to 95,499 ha in Kenya, from 10,056 ha to 17,903 ha in Zambia, and from 70,803 ha to 132,787 ha in Zimbabwe.

6. See, for example, Keyter (1975), Mosley (1975), and Jansen (1988). Without protection, according to the secretary of agriculture of Rhodesia in 1934, "the

extinction of the European farmer through native competition must be merely a question of time" (National Archives of Zimbabwe: S1542/M2, Darwin to CNC, July 1934). European farmers' opposition to competition from Africans was not universal, however, and some of the more efficient farmers supported free trade. Yield records of European farms in Southern Rhodesia indicate a wide dispersion, with the majority falling into lower-yield categories, and it was from this group that the most vociferous demands for protection against African competition were heard in the colonial legislature (Mosley 1983:172–183). Perhaps not surprisingly, the protection of European maize producers was strongly opposed at first by white consumer interest groups in Kenya and Zimbabwe, primarily animal feeders and plantation farms, on the grounds that protection would substantially raise the cost of maize.

7. Per capita maize output by Africans rose after the 1950s, reflecting a major substitution in production from millet and sorghum to maize, especially during the latter half of the century.

8. See Jayne et al. (1995) for a historical cross-country analysis. The problem of inflated marketing costs through the official marketing system appeared to be a long-standing phenomenon. As reported to the 1944 Native Production and Trade Commission in Zimbabwe, "The Africans' complaint is not so much what they can sell maize at, but the great differences when they sell and buy. . . . If they got 6/- and it cost only 6/6 to buy a bag of meal it would be all right, but they do not understand why they have to pay 22/- . . . for a bag of mealie meal" (National Archives of Zimbabwe, ZBJ 1/1/1, reported in Mosley 1987).

9. Examples in which state maize procurement and milling were vertically integrated in a single agency included Tanzania and Malawi; examples where the marketing boards served as de facto procurement agents for large-scale private milling firms included Zimbabwe, Zambia, Kenya, and South Africa.

10. Implicit subsidies are still conferred to varying degrees through financial losses on the marketing boards' trading accounts.

11. The boards' role as procurement agent for the large processors was most evident in Zimbabwe, South Africa, and Kenya, where board depots and mills were often located side by side and were occasionally linked by conveyer belts.

12. The situation in Malawi was somewhat different, since historically ADMARC had been organized to sell grain to consumers through its rural networks, although stockouts appear to have been common (Scarborough 1990).

13. In Zimbabwe prior to reform, the largest miller handled 65% of all industrial meal sales, whereas the two largest millers handled 85%. In Kenya and South Africa, prohibitions on interdistrict movement of maize meal resulted in regional oligopolies, with the largest miller accounting for 30 to 80% of all sales within particular regions.

14. This does not count the opportunity cost of time for waiting in the queue to custom mill grain or the travel costs. The advantages of using hammer mills must be considered against the fact that roller mill technology produces important by-products for the stock-feed and oil processing industries that hammer-milled whole meal does not.

15. Although the increased demand for whole meal was handled to some extent by unutilized capacity, investment in small-scale milling has been rapid since the reforms were implemented. In Zambia, the number of hammer mills increased from an estimated 4,156 to around 6,000 between 1992 and 1994. The number of hammer mills operating in the capital cities of Nairobi and Harare has risen by 80% in the past six years and by 57% in the past two years, respectively (Jayne et al. 1995).

16. See, for example, World Bank (1981); Cleaver (1985); and Schiff and Váldes (1992).

17. For example, in Zimbabwe during the period 1980–1989, the net protection coefficient for maize was 31% above export parity at official exchange rates and 15% below export parity at imputed real exchange rates (Jansen and Muir 1994). Wright and Nieuwoudt (1993) have argued that maize pricing policy in South Africa prior to 1993 involved major income transfers from consumers to producers relative to free-market pricing.

18. For example, Jabara (1985) demonstrated that despite falling real food prices in Kenya in the 1970s, the profitability of grain production actually increased because of growth in farm productivity, which was achieved in part through government investments in agriculture.

19. For detailed analyses of the effects of these state interventions on maize technology adoption, see Rohrbach (1989) and Howard (1994).

20. Howard (1994) derived a negative rate of return to the package of maize technology investments over the period 1978–1992 when the costs of associated marketing investments were included.

21. The decline in maize production was sometimes temporarily deliberate, as in Zimbabwe and South Africa during the late 1980s, because of the unintended accumulation of massive maize stockpiles.

22. Use of hybrid seed continues to rise in Kenya, Malawi, and Zambia, and hybrids have been almost universally adopted by Zimbabwean smallholders. As argued by Rohrbach (1989) for Zimbabwe and HIID/EPD (1994) for Malawi, however, without increased nutrient use the productivity gains from these hybrids are virtually exhausted.

23. The deficit position of Kenya is understated, because the data do not capture the estimated 100,000 to 250,000 t of maize informally imported from Uganda and Tanzania each year.

24. Mellor and others made this point decades ago for Asia, but the perception of rural food self-sufficiency in Africa has been modified more slowly, because lower population densities in most of Africa were often equated with an abundance of land and a relatively egalitarian distribution of productive resources.

25. This was achieved by paying farmers the residual revenue collected from the sale of maize by the Maize Board after deducting the costs of its domestic and export operations.

26. See Lele (1971); Southworth, Jones, and Pearson (1979); and Sahn and Delgado (1987). Failures in financial markets also contribute to an underprovision and a concentration of interannual storage by the private sector, since without access to credit, long-term storage can be borne only by relatively large-scale traders who can finance inventories with their own capital and bear substantial risk.

27. Recent evidence indicates that, when given a choice between white and yellow maize, some proportion of consumers chose to buy yellow maize at a price discount (Rubey 1995), revealing its potential role in stabilizing expenditures by consumers on food during poor harvests. Yellow maize benefits from a well-functioning international commodity exchange, typically costs 10 to 20% less than the limited supplies of white maize on international markets, and is subject to much smaller price fluctuations. This provides countries in the region with a wider range of buying and risk management options to procure needed supplies in the event of domestic production shortfalls.

Part 4

Conclusions

15

Accelerating Maize Production: Synthesis

Carl K. Eicher & Derek Byerlee

Africa's Food Crisis: The Urgency of the Task

Africa's food crisis is an enduring problem. With population growing at 2.7% and food demand at 3.0 to 3.5% per year, Africa faces the same kind of long-term food challenge India faced in the early 1960s (Schultz 1964). The urgency of the food problem is underlined by the fact that many countries in eastern and southern Africa have already moved, or will soon move, from being largely self-sufficient in (or exporters of) maize to being increasingly dependent on maize imports most years. There is little doubt that rapid technical change in food staples—such as maize, cassava, sorghum, and millet—is the key to solving the food crisis. Increased productivity in staple food production is also critical to raising rural incomes and stimulating broad-based economic growth.

Whereas broad-based technical change in food staples is central to renewed economic growth and poverty alleviation in Africa, the chapters in this book underscore the need to develop country-specific strategies for realizing these gains. The role of maize in production systems and consumer diets varies markedly, from its dominant role in eastern and southern Africa to a secondary but rapidly expanding role in western Africa. Within a region, considerable variation also exists in the degree of land scarcity, even between neighboring countries such as Malawi and Zambia. This variation has important implications for whether strategies to increase food production should emphasize intensification through increased yields or continued expansion of cropped area. But the situation is evolving quickly because of rapid population growth, as well as increasing integration of farm households into the market economy. The role of maize can change dramatically in a surprisingly short period of time, as shown by the transformation of maize production systems in the Nigerian savanna over the past 20 years, stimulated by investments in technology and infrastructure.

Since rural development is a learning process, the challenge African nations face is how to learn from each other and from other regions ways to develop productive and sustainable food production systems that will address the long-term food crisis (Sanders, Shapiro, and Ramaswamy 1996). Because maize is Africa's most important food staple, Africa's emerging maize revolution represents a ray of hope that the crisis can be addressed. At least a dozen African countries have developed maize production, input, and marketing systems that have been instrumental in transforming maize into a major food and cash crop. The findings of this book sound a note of cautious optimism that maize can play an important role in increasing food production to enhance food security and rural income growth. Maize is generally grown in relatively favorable production environments, so it can be used as a catalyst for accelerating food production. Nonetheless, maize yields in Africa are still low relative to average yields in other continents, and many countries, after initial successes, have found it difficult to finance and sustain maize research and production programs.

This book was written with the objectives of analyzing Africa's experience in increasing smallholder maize production and identifying the crucial decisions and priorities for expanding maize production throughout Africa. Since these complex issues are so strongly intertwined, specialists in various disciplines were brought together to provide the first Africa-wide study of the maize industry in 30 years.[1] This chapter synthesizes the study's main findings and offers guidelines for accelerating maize production in Africa over the next 10 to 20 years.

Africa's Emerging Maize Revolution: A Qualified Success

The country case studies in this book have shown that maize is a politically important crop that has been a major focus of food production efforts in Africa over the past 30 years. These campaigns have promoted the adoption of seed-fertilizer technology in maize, supported by extensive government programs to supply inputs, support producer prices, provide credit, and invest in marketing and transport infrastructure.

In eastern and southern Africa, where most maize is produced in mid- to high-altitude environments, the technological package that was the basis for these campaigns was developed from research conducted initially for commercial farmers in Kenya and Zimbabwe. Subsequently, varieties and hybrids were adapted to meet the special production, consumption, farm storage, and processing requirements of small-scale farms producing maize for family food needs and for the market. In western Africa, where most maize is produced in tropical environments, international centers provided many of the improved open-pollinated varieties (OPVs) that spearheaded the rapid expansion of maize over the past two decades. Nonetheless, nearly all countries have devoted considerable research resources to adapting these technologies to local smallholder conditions. In areas where this research capacity was lacking (at least for significant periods of time), the adoption of the new technologies has been slow.

All of the countries studied have complemented these research efforts through aggressive extension and input supply and marketing policies to provide incentives for the adoption of the packages. In particular, public seed companies have been established in nearly all of the countries; fertilizer has been subsidized in all of the countries at some point; and in some cases, cheap credit, usually with inputs provided in-kind, has been available. Several of the countries in eastern and southern Africa also invested extensively in parastatal marketing companies to assure markets for the increased output and to "protect" producers and consumers from price instability. Some of the countries have also invested heavily in road and marketing infrastructure that has often been decisive in adopting the technology, or even in initiating maize production itself, in more isolated regions. In some cases—especially in Zimbabwe—these were homegrown efforts, but in most others a substantial amount of donor assistance was critical, although often this support has been "stop and go" as one donor project has been replaced by another with a different flavor.

At first sight, these national production campaigns appear to have been successful in stimulating increased maize production. Improved varieties and hybrids have been the lead input in these campaigns. Today, Africa's maize seed research and delivery system is more advanced than that for any other food crop. Improved varieties of maize are now grown on at least 40% of the maize area. In several countries of eastern and southern Africa, hybrids cover the majority of the maize area, whereas in western and central Africa, improved OPVs are widely grown. Adoption of fertilizer has also accompanied the adoption of improved varieties; however, fertilizer use on maize has lagged and is still low in many areas where improved seed has been widely adopted. In western Africa, maize area has expanded rapidly in several countries, providing the bulk of the production increases.

Although progress has been uneven, the following points reveal the Africa-wide nature of the emerging maize revolution.

- In Kenya, the average national maize yield doubled and maize production grew fivefold from 1965 to 1980.
- Zimbabwean smallholders doubled national maize production between 1980 and 1986.
- In Zambia, maize production rose nearly fourfold from the early 1960s to the late 1980s.
- In Malawi, the area planted to hybrids expanded from 3% of the maize area in 1987 to 30% in 1995.

- Maize production campaigns in the savanna areas of Ghana and Nigeria have more than doubled maize production over the past two decades.

Nonetheless, the performance of the maize subsector as measured by production increases has been only a qualified success, as demonstrated by the following statistics.

- Maize yields in Africa have increased by less than 1.0% per year over the past 20 years to reach the current average of 1.3 t/ha. This is only half the rate experienced in Latin America and Asia despite similar levels of adoption of improved varieties.
- In Zimbabwe, where nearly all smallholders use hybrid seed, smallholder yields average less than 1 t/ha and have risen very slowly over the past 10 years.
- The bulk of the maize production increase in Africa has been achieved by area expansion.
- Maize production per capita has declined in most major producers in eastern and southern Africa, and most countries are increasingly dependent on imports.

Several factors help explain the inability of maize production to keep up with population growth and food demand despite the substantial investment made in research and delivery systems over the past two decades. First is the lack of technical progress in some large maize producers such as Tanzania, Ethiopia, and Zaire. Second, in some countries (e.g., Zimbabwe and Zambia) maize production has shifted from high-yielding commercial farms to lower-yielding small-scale farms in more marginal areas. Third, the impressive adoption of improved maize seed has been coupled with limited adoption of complementary inputs, especially fertilizer and other soil fertility–related practices.

The country case studies in this book also reflect considerable concern about the sustainability of the gains that have been achieved. Yield gains in many systems are threatened by soil fertility depletion, increasing pest problems (especially the parasitic weed *Striga*), and overdependence on maize monoculture. In addition, increased maize production has often been achieved at a high fiscal cost that has not been sustainable. Recent structural adjustment programs have removed subsidies and eliminated pricing and marketing supports in many countries, resulting in sharp declines in maize area and input use in several countries.

We conclude that the emerging maize revolution in Africa, based on widespread adoption of improved maize varieties and hybrids, has not yet stimulated a sustained general increase in food production sufficient to result in a broad-based increase in rural incomes. Only in the savanna areas of western Africa (e.g., Ghana and Nigeria) has an agricultural transformation occurred that is of the same magnitude as the green revolution in Asia; even in these areas, evidence shows that this transformation has slowed considerably in the 1990s. In eastern and southern Africa, where increasing the productivity of the dominant food crop—maize—is the key to agricultural transformation, the impacts of maize production strategies have been less than expected and have often been achieved at high fiscal costs. Clearly, then, widespread adoption of modern varieties of maize is a necessary but insufficient condition to reverse the downward trend in per capita production.

Strategic Issues in Increasing Maize Production

One inescapable conclusion flows from Africa's food production experience over the three decades since independence: There are no shortcuts. Many African countries have announced bold green revolution projects to increase food production, but these projects have invariably failed because they were often donor driven, lacked high-level political support, and typically focused on only one component of the delivery system—such as seed, fertilizer, or extension. The experience in maize over the past 30 years reveals that African countries must resolve four strategic decisions in designing a national food production strategy: the appropriate role for the public and private sectors at different stages of development; the political decision to focus and concentrate scarce scientific, managerial, and financial resources on a few basic food crops in a few areas; the role of smallholders versus commercial farmers in addressing the food crisis; and the extent to which technological strategies should emphasize intensification and external inputs. Addressing these strategic decisions is a cumulative process, leading to the creation of a capacity within African governments to manage food economies during times of both scarcity and abundance.

The first strategic issue is the politically sensitive one of defining the roles of the state and the private sector. The country studies and technical chapters in this book reveal that the government has played a critical role in conducting maize research and providing farmer support and marketing services, but these programs incurred large and unsustainable government subsidies in the process. For example, in Zambia expenditures in support of maize production and marketing consumed 17% of the government's budget in the late 1980s, which explains why market liberalization programs were necessary to move to fiscally sustainable roles for the government in the maize sector.

Both the state and the private sector have legitimate, critical roles to play in both production and marketing. Historically, African governments have dominated the early stages of development of smallholder maize economies by undertaking research, providing inputs, setting up seed inspection services, and providing guaranteed maize prices and market outlets. Gradually, private investments were made in seed, fertilizer distribution, marketing, and maize milling. Today, the private sector has gained ascendancy in input distribution and wholesale and retail marketing in most countries. African governments and donors are reaching agreement on a reduced but vital role for the state in providing strong support to research and extension, setting up rules and regulations to guide market forces, and maintaining a strategic grain reserve—especially in southern African countries exposed to extremely variable weather. No consensus exists, however, on the appropriate role of the state in areas of marginal production potential, poor infrastructure, or very small farms—where limited capital and purchasing power, high transport costs, and small market size will inevitably limit investment by the private sector but concerns of equity and rural poverty alleviation demand state intervention.

The second crucial decision in designing a national maize strategy is whether to focus scarce financial and managerial resources on one or a few key crops, such as maize, and a few areas or to sprinkle those resources over a wide array of crops across the entire country. Most African countries cannot afford the luxury of carrying out research on 20 to 30 commodities and building a comprehensive research system even for major commodities, such as maize. We recommend concentrating high-level political leadership and financial and managerial resources on one or two "political" crops and staying the course for the coming 10 to 20 years. Likewise, the country studies lend support to a strategy of developing national commodity campaigns for maize (and other key crops) in natural resource areas that are favorable for agriculture. The decision to focus initially on favorable areas will be politically difficult, but Asia's experience is encouraging on this point.[2]

The third key issue is the relative emphasis on the smallholder sector. The evidence in this book has powerfully demonstrated that small farms are competitive with large farms if they receive technology tailored to their special circumstances and are given economic incentives and efficient farmer support services. The key to sustainable development in Africa is consistent and significant growth in smallholder agricultural productivity. This strategy should give particular attention to the critical role of women in African smallholder maize production.

The final major set of issues concerns the technological strategy that should be employed. First, it is important to decide on the relative emphasis to place on intensification through yield-increasing technologies versus expansion in crop area. In the 1960s, Africa was a land-surplus continent where labor and capital limited production. Today, with the exception of some land-abundant countries such as Zambia, Zaire, and Mozambique, yield intensification is becoming the primary source of increasing food production. Malawi's smallholder sector is a textbook case of an Asian-type country that has a high population density and a small, declining average farm size.

In areas where intensification is the way forward (i.e., most of Africa), the next decision relates to the relative emphasis to place on using external inputs versus low-input systems to increase maize yields. Over the past decade, some donors and nongovernmental organizations (NGOs) have promoted low-input strategies, which require little or no chemical fertilizer or other external inputs. The low-input strategy, however, is not a new model of farming in Africa. Under low population growth rates (1.0% per annum), shifting cultivation was a sustainable food production system. But with annual population growth of 2.7%, it is unrealistic to expect that low-input farming systems can meet Africa's future demand for food. In addition, many low-input systems require increased labor inputs, a scarce factor in African smallholder farming systems (Low 1993). Nonetheless, it is equally clear that sole dependence on external inputs—especially chemical fertilizer, as emphasized in the recent past—cannot arrest the growing problem of soil degradation. For the future, an approach that aggressively promotes the use of external inputs, especially chemical fertilizer, and at the same time steps up efforts to integrate the use of organic sources of nutrients is required.

Seed technology also offers the choice between an external input, hybrid seed, where seed must be purchased every year, and OPVs whose seed can be recycled and used for a few years. In eastern and southern Africa, researchers have concentrated on developing hybrids, and seed systems have been established that have successfully delivered hybrid seed to millions of smallholders. This program has been reinforced by evidence that indicates that, contrary to conventional wisdom, hybrids often perform well under low-input conditions. In western Africa, the main research emphasis has been on developing improved OPVs, which have also spread rapidly but have usually been based on a one-time adoption of new seed. Because the private sector lacks interest in producing the seed of OPVs, the trend in research throughout Africa is toward maize hybrid development as the appropriate strategy for smallholders, especially in countries with better road networks.

Integrating Technology and Policy to Accelerate Maize Production

Even when a strategy is formulated that establishes an appropriate role for the public and private sectors, selects crops and areas for concentration on smallholder production, and balances low-input and external input systems, a number of technological, institutional, and policy measures must be integrated in ways that recognize the interactions among these efforts. All of these efforts must be coordinated with ongoing macroeconomic reforms to level the playing field for smallholders.

The country studies and technical chapters in this volume have shown that many African countries have gained considerable experience in increasing maize productivity through maize production campaigns, but many of these efforts have been fragmented, fiscally unsustainable, and unable to address problems of soil degradation. Many governments have launched food production campaigns with unrealistic time frames (usually five years). Meanwhile, donors and NGOs have often helped strengthen one factor in the system, such as extension, which they have regarded as the key to generating a green revolution. The end result is that Africa has served as a laboratory for highly diffuse, fragmented, and poorly conceived food production campaigns operating with unrealistic time frames.

National policy analysts rather than donors must take the lead in pragmatically piecing together a coherent program to increase maize production, starting with the problems of farmers and moving on to research priorities, input delivery, processing, marketing, and consumer preferences. To be sure, such integrated commodity campaigns are complex and are difficult to design, implement, evaluate, modify, and sustain for a decade or two, especially if the basic agricultural support institutions are weak. Given the nature of maize—a crop amenable to hybrid seed development that is grown in relatively favored circumstances and is widely consumed in urban areas—it is likely that production strategies for other food staples in Africa (except rice) will need to be tailored to fit the specific production and marketing circumstances of that crop.

The final section synthesizes the empirical experience based on the best available information on technological and policy reforms for accelerating maize production throughout Africa.

Technology-Generating Strategies

The release and widespread adoption of improved maize seed has undoubtedly been the major success story of agricultural research in Africa over the past three decades. Most countries now have the capacity to screen for local adaptation of genetic materials obtained from several sources, especially the international centers. Only a few countries possess the scientific capacity and assured funding to develop newer generations of improved varieties and hybrids on a continuous basis. In addition, maize breeders are giving more attention to developing varieties that meet the needs of small-scale producers in terms of environmental niche, consumer preferences, and home processing requirements, as well as addressing major constraints on maize production such as drought and *Striga* infestations.

Africa's maize experience over the past two decades, however, strongly vindicates the assertion that improved varieties alone are not enough to increase and sustain a high rate of growth in maize production (Blackie 1994a). Varietal improvement research must be complemented with strong research programs on crop and resource management. This capacity is still generally weak throughout Africa, despite considerable allocation of research resources. National programs have difficulty matching researchers' skills and interests with farmer priority problems, and international centers have yet to demonstrate their effectiveness in creating a technology pipeline in which several management techniques must be integrated and applied in location-, system-, and even season-specific conditions.

The major crop and resource management problem for many production systems is declining soil fertility, brought about by increasing population density and the demise of traditional fallow methods for maintaining soil fertility. In areas where population density is high, yields are likely declining in the absence of fertilizer use. Although about half of all fertilizer used in Africa is applied to maize, this amount has been insufficient to reverse the decline in soil fertility. The priority for developing and disseminating new technologies over the next decade or more should therefore be intensified research efforts to improve soil fertility management. This research will involve a long-term effort to adapt an array of methods within a farming-systems framework to the heterogeneous agroecological and socioeconomic landscape of African maize production. Given the extent of nutrient mining and land degradation, both organic and inorganic supplies of nutrients must be rapidly increased. Fertilizer recommendations must be developed for specific locations and seasonal conditions to improve efficiency of use. The promotion of legumes in various forms (grain legumes, green manure, and agroforestry) will also be central to this effort, but considerable research is still required to develop appropriate strategies for selecting and incorporating legumes into Africa's heterogeneous farming systems. This research must also consider the labor, cash, and food calendars of small-scale farm households, as well as rural wage rates. Even where organic sources of nutrients can be developed to provide some of the nitrogen requirements of maize production, inorganic nutrients—especially phosphorus and sometimes micronutrients—will be needed to jump-start the system.

The productivity of many maize production systems in land-extensive areas is also limited by high labor inputs and poor weed control, because hand-hoe agriculture is the norm throughout Africa, and few countries have a class of landless laborers. Even in land-scarce countries, seasonal labor availability is usually a major constraint on the adoption of new technologies for intensification of maize production. Therefore, enhanced efforts to provide labor-saving solutions are an important part of any maize production strategy. In many cases, this will involve some form of minimum or reduced tillage, which often depends on the use of herbicides to substitute for tillage and weeding operations. Such systems have been widely adopted by smallholders in Mexico and Central America, and recent experiences in southern Africa are promising, although much more on-farm verification is needed.

Strengthening Research and Extension Delivery Systems

Effective research and extension systems to develop and disseminate these new technologies are a vital component of any long-term effort to accelerate maize production. Many African countries are at an earlier stage of scientific development and have weaker research and extension systems than was the case with India on the eve of its green revolution in the mid-1960s.[3] With one or two exceptions, organized research on food crops for smallholders in Africa did not really begin until after World War II. At independence in the 1960s, 90% of agricultural researchers were still expatriates. Nonetheless, steady progress has been made in developing maize research capacity over the past two decades. The returns to this investment have been as high in Africa as in other regions of the developing world, although returns have been somewhat more variable because of the small size of many African countries and the diversity of production environments.

The recent decline in government support for agricultural research in many African countries is especially worrisome in light of the critical role new technologies must play in intensifying Africa's maize revolution. The lack of domestic political support for research and extension systems is reflected in declining budgets for these essential public goods. Donors, including the development banks, currently fund more than half of all agricultural research in Africa, but this pattern is not sustainable over the long term. Nor is it a healthy trend in terms of the likely influences of donors on research priorities.

Effective national research systems in the twenty-first century will be pluralistic, including public research institutes, universities, the private sector, and NGOs. Major changes are also required in the way public-sector research and extension systems are organized and how they interact with farmers, farmers' organizations, the private sector, and NGOs (Rukuni 1996). The crisis in research funding in part reflects the new era of fiscal austerity but also mirrors many governments' lack of confidence in the management and performance of public research systems, which have grown dramatically (in terms of the number of scientists) over the past three decades. Reductions in the size of many national research systems in terms of the number of scientists and support staff, as well as the number of research programs and stations, will be part of the solution to developing sustainable funding by national governments. Given the small size of many African countries (relative to those in Asia) and their environmental diversity, most national research programs will have to develop the capacity to become intelligent "borrowers" of technology from the global (public and private) agricultural research system. Regional research collaboration is another strategy that is gaining popularity as a way of increasing market size for research products and using scarce research resources more efficiently.

Africa's agricultural extension systems are also in disarray and must move beyond the Training and Visit model to systems that more effectively address financial sustainability, institutional pluralism, and farmer participation. The common extension practice of providing one standard recommendation, or "message," for farmers in a region or an entire country should be replaced with problem-solving approaches that address the complex, situation-specific crop and soil management technologies farmers require. Increasingly, extension will have to focus on improving farmers' skills to adapt technologies to their own needs. In addition, many countries will be faced with the difficult task of reducing the number of extension workers while at the same time improving the quality of extension staff members and their coverage of farmers, especially women farmers who have usually been bypassed in extension programs. The involvement of private input dealers, NGOs, and farmers' associations in technology diffusion provides hope that a more diversified system of technology and information supply will reduce the heavy dependence on state-financed extension services.

Strengthening Seed and Fertilizer Delivery Systems

Successful dissemination of improved maize technologies involves the coordination of seed supply, fertilizer importation and distribution, credit, and marketing. Although the public sector has sometimes successfully provided these services and promoted widespread adoption of new maize technologies, this success has often been achieved through financially unsustainable input subsidies (see Chapters 13 and 14). The obvious solution to this problem is to devolve as far as possible the functions of input and product marketing and credit supply to the private sector, and this process is now well underway in most countries. The public sector's abrupt withdrawal from many of these functions, however, carries an element of risk if the private sector lacks the skills, capital, or perceived market size to successfully manage input supply and output markets. The challenge is how to sequence the transition from the public to the private sector in both input and output markets in a way that allows the private sector to play an increasing role in input delivery and marketing over time.

This challenge is illustrated by the case of the seed and fertilizer industries. The maize seed industry is undergoing rapid change, especially in eastern and southern Africa. In the early stages, the public sector has played the dominant role by developing appropriate varieties and hybrids and a seed certification system that signals assured seed quality to farmers. Seed production and marketing should be passed quickly to the private sector, although in the case of OPVs Africa still lacks good examples of sustainable seed production systems managed by either the public or the private sector. Over time, the private sector can also gradually take on some of the applied maize breeding work in response to a growing market share for hybrids and improved varieties, although legal safeguards and liberalization of seed regulations and seed trade must be in place before private companies will be willing to invest in maize research and development. Although the number of public, public-private, and private seed companies in five countries—Kenya, Malawi, Tanzania, Zambia, and Zimbabwe—tripled from 5 in 1980 to 15 in 1996, only the most scientifically advanced countries such as Zimbabwe, Kenya, and South Africa are at a stage where the private sector can take on much of the applied maize breeding work.

In most countries, further liberalization of seed regulations and seed trade is needed to induce private companies to invest in research and development and release public resources now invested in maize breeding. But even under a liberalized policy regime, seed companies have been slow to invest in countries whose seed markets are perceived to be thin and whose legal environments are regarded as uncertain. In fact, a major multinational seed company recently withdrew from the seed markets of several African countries because of unfavorable policy regimes, poor infrastructure, inadequate legal safeguards, and the small market size. This situation underlines the need for African governments to support the development of local private seed companies through reliable access to maize varieties and hybrids from the public sector and the international centers.

The private and public sectors' roles in fertilizer distribution are also complex and changing over time. Many issues surrounding the appropriate roles of the public and private sectors in marketing fertilizer in Africa remain to be resolved. Given the limited success and high cost of fertilizer importation and distribution by the public sector, most countries are now committed to having the private sector manage fertilizer imports and distribution. In several countries, however, the private sector has been slow to take over management from the government, and fertilizer distribution has virtually collapsed. Because of the unique characteristics of fertilizer (seasonal demand, multiple products, and high bulk-to-value ratio), inexperienced private dealers, thin markets, and economies of size in purchasing and importation, the transition to private-sector distribution has been difficult.

Even when the private sector is actively involved in importing and marketing fertilizer, fertilizer prices are generally higher in Africa relative to Asia and Latin America because of the small size of imported shipments and high internal transport costs. Concerted efforts are required to strengthen efficiency and competitiveness in fertilizer importation and distribution to reduce farm-level prices for fertilizer, as well as to promote fertilizer use and efficiency at the farm level. The public sector has an important role to play in providing technical information, training, and farm-level adaptive research and extension to develop site-specific fertilizer recommendations. But initially, the public sector may have to provide loan guarantees to facilitate importation, as well as facilitate bulk purchasing of fertilizer by several private dealers to realize efficiencies in buying and importing fertilizer. Also, the public sector has a role in catalyzing the development of rural financial markets and small-scale savings groups to help overcome the cash constraint on fertilizer purchases.

It is now generally agreed that fertilizer subsidies have not been cost-effective in promoting fertilizer use in most countries of Africa. It must be recognized, however, that for several reasons (including the serious soil fertility problem, the high cost of fertilizer as a result of poor infrastructure, and the poverty of smallholders) economic incentives or other programs may be required to help farmers make long-run investments in soil fertility improvements featuring both organic and inorganic sources of nutrients. Fresh thinking is needed on how these incentives can be most efficiently provided, targeted effectively to smallholders, and then phased out after a fixed period.

Public and Private Roles in Maize Marketing

The liberalization of maize marketing, now well underway in Africa, aims eventually to move away from the monopoly of government grain boards and restrictions on private trading.[4] Zambia, for example, privatized all wholesale and retail maize marketing over a period of three years, from 1993 to 1996. Today, the private sector handles all wholesale and retail maize marketing, and the government of Zambia is responsible for maintaining a national maize food reserve. An analysis of marketing reforms reveals that the main beneficiaries of these reforms to date have been maize consumers rather than farmers. For example, the relaxation of restrictions on grain movement across district and provincial boundaries in Zimbabwe has improved the cost-effectiveness of marketing, promoted small-scale processing, and reduced maize prices to consumers—especially to food-deficit rural households. In addition, the relaxation of restrictions on maize markets and grain imports has provided poor consumers with a wider range of maize products. Although white maize is almost universally preferred, consumers in eastern and southern Africa have shown a willingness to consume yellow maize at a price discount. Food security at the national level and for poor households can be enhanced if countries are willing to import yellow maize rather than depend on a very thin world market for white maize.

The transition to the private sector has not been easy, and most countries are still struggling to find an appropriate balance between the state and the private sector. Maize producers and consumers currently face considerable uncertainty with respect to food prices and the future policy environment. Some state intervention is justified, given the array of market failures in staple food marketing and the intrinsic instability of prices of food staples (Smith 1995:561). The difficult issue now being debated in many countries is the appropriate role of governments in stabilizing maize markets, especially in landlocked countries where sharp fluctuations in rainfall and a wide gap between import and export parity prices generate highly unstable prices for both farmers and consumers.

Progress in developing stable and efficient markets in the private sector is closely related to investments in rural roads and transport, much of which must be financed by the public sector. Rural roads, for example, are less developed in Africa today than they were in India at the beginning of the green revolution (Spencer and Badiane 1995). There is little doubt that even with renewed investments in rural roads, weak rural infrastructure will slow efforts to develop efficient markets for many years. An unresolved issue is the role of the state in the short to medium term in addressing the needs of farmers in isolated areas with poor infrastructure.

Concluding Note

Africa faces a long-term food crisis that cannot be resolved by crash food production programs or low external input production models. It is evident that no single input, such as improved varieties or chemical fertilizer, or intervention, such as extension or price policy reform, will provide a "magic bullet" that can transform agricultural productivity and restore food self-sufficiency.

As the research in this volume indicates, success in increasing maize production offers hope for ameliorating Africa's long-term food crisis. The six country studies and the technical chapters have provided evidence of the beginning of a maize revolution in Africa that is being fueled by national maize production strategies integrating research, extension, input delivery, and marketing.

The country studies and technical chapters have highlighted the strategic decisions to resolve in the process of designing sustainable national strategies to increase maize production. Given the time lag before the payoffs to investments—such as plant breeding research, soil fertility improvements, and human resource development—become apparent, it is important that maize production campaigns have an implementation horizon of at least a decade or longer. The Ghana case study demonstrated the high payoff to sustained support by a single donor for the Ghanaian maize and cowpea research program over 15 years, complemented by an NGO-supported extension program for more than 10 years.

Africa's maize experience underlines the failure of prepackaged institutional models and the need for countries to invest time and resources in crafting demand-driven research, extension, seed, fertilizer, and marketing systems (Rukuni 1996). The country studies point to the need for a mix of private- and public-sector efforts to provide the requisite input supply and marketing support, with the specific mix depending on the stage of development.

Public investments are needed in rural roads and in marketing and policy reforms, which are essential for input and output markets to function efficiently. Public investments are also needed in developing local analytical capacity in food policy analysis, because excellence in research and in managing a food economy comes through experience. The recent maize marketing reforms in Zambia, Kenya, Malawi, and Zimbabwe have been greatly facilitated by local scholars' studies of food security and marketing. A major gap still exists, however, in our knowledge about the relative payoff to competing priority public investments—such as research, extension, infrastructure, and market price stabilization—as a guide to the allocation of scarce government revenues and administrative skills. The answer to this critical question is likely to be fairly specific to each country depending on its institutional development, infrastructure, agroclimatic situation, and population pressure, and the issue merits in-depth study by social scientists.

Finally, donors and international development banks must be prepared to provide consistent and coordinated long-term support for food production for the next 10 to 20 years. Although donors have switched resources to natural resource management, the critical choice ahead is how to integrate agricultural productivity and sustainable resource management into the design of long-term focused national food production strategies. Much of this effort must emphasize human capital development, institutional strengthening, and the capacity to develop and articulate homegrown strategies. Success will require an unprecedented degree of active collaboration among donor agencies, senior African policymakers, experienced and committed field workers, and last but not least, farmers and farming communities.

Notes

1. See Miracle (1966) and Johnston (1958).
2. In the mid-1960s, the government of India took a high-level political decision to launch its national wheat production campaign in 17 favorably endowed districts (those with water, roads) and to target seed, fertilizer, credit, and extension

to farmers in those districts (Mellor 1976). Today, 3 of India's 25 states form the country's breadbasket by producing 36% of its national food grain production and 80% of its public food procurement (Lele and Bumb 1995).

3. For example, food crop research began in India in the early part of the twentieth century. By independence in 1947, India had accumulated 40 years of research experience on its main food staples, and Indian scientists held most key research positions.

4. In western Africa, governments have played a limited role in maize marketing.

Acronyms and Abbreviations

ADMARC	Agricultural Development and Marketing Corporation (Malawi)
BSA	British South Africa Company
BSLR	buyer and seller of last resort
CIMMYT	Centro Internacional de Mejoramiento de Maíz y Trigo (International Maize and Wheat Improvement Center)
CRI	Crops Research Institute (Ghana)
CUSA	Credit Union and Savings Association (Zambia)
DR&SS	Department of Research and Specialist Services (Zimbabwe)
FAO	Food and Agriculture Organization (United Nations)
f.o.b.	free on board
FSR/E	farming systems research/extension
GGDP	Ghana Grains Development Project
GNP	gross national product
IAR	Institute for Agricultural Research (Zaria, Nigeria)
IARC	international agricultural research center
IAR&T	Institute for Agricultural Research and Training (Moor Plantation, Ibadan, Nigeria)
IITA	International Institute of Tropical Agriculture (Nigeria)
IMF	International Monetary Fund
KARI	Kenya Agricultural Research Institute
MAFF	Ministry of Agriculture, Food, and Fisheries (Zambia)
masl	meters above sea level
MDI	market-driven systems, intensification phase
NARSs	national agricultural research systems
NCRI	National Cereals Research Institute (Badeggi, Nigeria)
NCZ	Nitrogen Chemicals of Zambia
NGOs	nongovernmental organizations
NSCM	National Seed and Cotton Milling Company (Malawi)

OPVs	open-pollinated varieties
PDE	population-driven systems, expansion phase
PDI	population-driven systems, intensification phase
R&D	research and development
SAFGRAD	Semi-Arid Food Grain Research and Development Project (Ouagadougou, Burkina Faso)
SG 2000	Sasakawa-Global 2000 (Ghana)
SIDA	Swedish International Development Authority
SR52	Southern Rhodesia 52
T&V	Training and Visit extension system
UNIP	United National Independence Party (Zambia)
USAID	United States Agency for International Development
ZCF	Zambian Cooperative Federation
ZNFU	Zambian National Farmers Union
ZSPA	Zambian Seed Producers' Association

References

Ahmed, J. K.; Falcon, W. P.; and Timmer, C. P. 1989. *Fertilizer Policy for the 1990s.* Development Discussion Paper No. 293 AFP. Cambridge: Harvard Institute for International Development (HIID).

Akposoe, M. K., and Edmeades, G. 1981. *Ghana Grains Development Project Second Annual Report 1980. Part Two: Research Results.* Accra: Ghana Grains Development Project (GGDP).

———. 1982. *Ghana Grains Development Project Third Annual Report 1981. Part Two: Research Results.* Accra: Ghana Grains Development Project (GGDP).

Alderman, H. 1991. *Downturn and Economic Recovery in Ghana: Impacts on the Poor.* Cornell Food and Nutrition Policy Monograph No. 10. Ithaca: Cornell Food and Nutrition Policy Program (CFNPP).

———. 1992a. *Incomes and Food Security in Ghana.* Cornell Food and Nutrition Policy Paper No. 26. Ithaca: Cornell Food and Nutrition Policy Program (CFNPP).

———. 1992b. *Food Security and Grain Trade in Ghana.* Cornell Food and Nutrition Policy Paper No. 28. Ithaca: Cornell Food and Nutrition Policy Program (CFNPP).

Alderman, H., and Shively, G. 1991. *Prices and Markets in Ghana.* Cornell Food and Nutrition Policy Paper No. 10. Ithaca: Cornell Food and Nutrition Policy Program (CFNPP).

Alderman, H., and Higgins, P. 1992. *Food and Nutritional Adequacy in Ghana.* Cornell Food and Nutrition Policy Paper No. 27. Ithaca: Cornell Food and Nutrition Policy Program (CFNPP).

Alexandratos, N. 1995. *World Agriculture: Towards 2010. An FAO Study.* Chichester: John Wiley.

Allan, A. Y. 1971. "The Influence of Agronomic Factors on Maize Yields in Western Kenya with Specific Reference to the Time of Planting." Ph. D. thesis, University of East Africa, Nairobi.

Amani, H. K. R., and Maro, W. 1992. "Policy to Promote an Effective Private Trading System in Farm Products and Farm Inputs in Tanzania," in J. B. Wyckoff and M. Rukuni (eds.), *Food Security Research in Southern Africa: Policy Implications.* Harare: University of Zimbabwe.

Amanor, K. S. 1991. "Managing the Fallow: Weeding Technology and Environmental Knowledge in the Krobo District of Ghana." *Agriculture and Human Values* 8(1,2): 5–13.

———. 1994. *The New Frontier. Farmer Responses to Land Degradation: A West African Study*. Geneva: United Nations Research Institute for Social Development (UNRISD).

Andrews, D. J., and Kassam, A. H. 1976. *The Importance of Multiple Cropping in Increasing World Food Supplies*. Multiple Cropping Special Publication No. 27. Madison, Wisconsin: American Society of Agronomy.

Araki, S. 1993. "Effect on Soil Organic Matter and Soil Fertility of the *Chitemene* Slash-and-Burn Practice Used in Northern Zambia," in K. Mulongoy and R. Merckx (eds.), *Proceedings: Soil Organic Matter Dynamics and Sustainability of Tropical Agriculture*. Chichester: Wiley-Sayce.

Argwings-Kodhek, G.; Mukumbu, M.; and Monke, E. 1993. "The Impacts of Maize Market Liberalization in Kenya." *Food Research Institute Studies* 12(3): 331–348.

Argwings-Kodhek, G., and Jayne, T. S. 1996. *Relief Through Development: Maize Market Reform in Urban Kenya*. Tegemeo Working Paper No. 2. Njoro: Egerton University.

Asafo-Adjei, B., and Soza, R. 1993. *Ghana Grains Development Project Fourteenth Annual Report*. Kumasi: Ghana Grains Development Project (GGDP).

Atsu, S. Y. 1974. *The Focus and Concentrate Programme in the Kpandu and Ho Districts: Evaluation of an Agricultural Extension Programme*. Technical Publications Series No. 34. Legon: Institute of Statistical, Social, and Economic Research (ISSER).

Balasubramanian, V. 1984. "Maize Research by MIDAS," in *Ghana Grains Development Project, Ghana National Maize Workshop: Proceedings 1981–1984*. Kumasi: Ghana Grains Development Project (GGDP).

Balcet, J. C., and Candler, W. 1981. *Farm Technology Adoption in Nigeria*. Washington, D. C.: World Bank.

Bangwe, L. 1995. *Effects of Market Liberalization on Maize Production and Marketing for Smallholders*. Adaptive Research Planning Team Report. Lusaka: Ministry of Agriculture, Food, and Fisheries.

Barker, R., and Hayami, Y. 1976. "Price Support vs. Input Subsidy for Food Self-Sufficiency in Developing Countries." *American Journal of Agricultural Economics* 58(4): 617–628.

Barrett, C., and Carter, M. 1994. "Microeconomically Coherent Agricultural Policy Reform in Africa." Prepared for the World Bank research program, "Reforms in Socialist Economies in Africa." Madison: University of Wisconsin Press.

Bates, R. H. 1981. *Markets and States in Tropical Africa: The Political Basis of Agricultural Policies*. Berkeley: University of California Press.

———. 1989. *Beyond the Miracle of the Market: The Political Economy of Agrarian Development in Kenya*. Cambridge: Cambridge University Press.

———. 1993. "'Urban bias': A Fresh Look." *Journal of Development Studies* 29: 219–228.

Bindlish, V., and Evenson, R. 1993. *Evaluation of the Performance of T&V Extension in Kenya*. Agricultural and Rural Development Series No. 7. Washington, D.C.: World Bank.

Binswanger, H. P., and Sillers, D. A. 1983. "Risk Aversion and Credit Constraints in Farmers' Decision Making: A Reinterpretation." *Journal of Development Studies* 20: 5–21.

Binswanger, H. P., and Pingali, P. 1988. "Technological Priorities for Farming in Sub-Saharan Africa." *World Bank Research Observer* 3(1): 81–98.

Blackie, M. J. 1982. "A Time to Listen: A Perspective on Agricultural Policy in Zimbabwe." *Zimbabwe Agricultural Journal* 79(5): 151–156.

———. 1987. "The Elusive Peasant: Zimbabwe's Agricultural Policy, 1965–1986," in M. Rukuni and C. K. Eicher (eds.), *Food Security for Southern Africa*. Harare: UZ/MSU Food Security Project, University of Zimbabwe.

———. 1990. "Maize, Food Self-Sufficiency, and Policy in East and Southern Africa." *Food Policy* 15: 383–394.

———. 1994a. "Maize Productivity for the 21st Century: The African Challenge." *Outlook on Agriculture* 23: 189–195.

———. 1994b. "Realizing Smallholder Agricultural Potential," in M. Rukuni and C. K. Eicher (eds.), *Zimbabwe's Agricultural Revolution*. Harare: University of Zimbabwe.

Blackie, M. J., and Jones, R. B. 1993. "Agronomy and Increased Maize Productivity in Southern Africa." *Biological Agriculture and Horticulture* 9: 147–160.

Bockari-Kugbei, S. 1994. "The Role of Small-Scale Enterprises in African Seed Industries." Ph. D. thesis, University of Reading, Reading, UK.

Bonnen, J. 1990. "Agricultural Development: Transforming Human Capital, Technology, and Institutions," in C. K. Eicher and J. M. Staatz (eds.), *Agricultural Development in the Third World*. Baltimore: Johns Hopkins.

Borlaug, N. E. 1996. "Mobilizing Science and Technology for a Green Revolution in African Agriculture," in S. A. Breth (ed.), *Achieving Greater Impacts from Research Investments in Africa*. Mexico City: Sasakawa Africa Association.

Borlaug, N. E., and Dowswell, C. R. 1995. "Mobilising Science and Technology to Get Agriculture Moving in Africa." *Development Policy Review* 13(2): 115–129.

Bosque-Perez, N. A., and Mareck, J. H. 1990. "Distribution and Species Composition of Lepidopterous Maize Borers in Southern Nigeria." *Bulletin of Entomological Research* 80: 363–368.

Boughton, D., and de Frahan, B. H. 1994. *Agricultural Research Impact Assessment: The Case of Maize Technology Adoption in Southern Mali*. MSU International Development Working Paper No. 41. East Lansing: Michigan State University.

Boxall, R. (ed.). 1994. *Ghana Larger Grain Borer Project. Seventh Quarterly Report*. Accra: Ministry of Agriculture/Overseas Development Agency.

Bratton, M. 1994. "Micro Democracy? The Merger of Farmer Unions in Zimbabwe." *African Studies Review* 37: 9–37.

Bryceson, D. F. 1993. *Liberalizing Tanzania's Food Trade: Public and Private Faces of Urban Marketing Policy*. Geneva and London: United Nations Research Institute for Social Development (UNRISD) in association with James Currey.

Buccola, S., and Sukume, C. 1988. "Optimal Grain Pricing and Storage Policy in Controlled Agricultural Economies: Application to Zimbabwe." *World Development* 16(3): 361–371.

Buddenhagen, I. 1992. "Prospects and Challenges for African Agricultural Systems: An Evolutionary Approach." Carter Lecture Series on Sustainability in Africa: Integrating Concepts. Gainesville: University of Florida Press.

Bumb, B. L. 1988. "Fertilizer Supply in Sub-Saharan Africa: An Analysis," in T. B. Tshibaka and C. A. Baanante (eds.), *Fertilizer Policy in Tropical Africa*. Lomé: International Fertilizer Development Center (IFDC) and International Food Policy Research Institute (IFPRI).

Bumb, B. L.; Teboh, J.; Atta, S.; and Asenso-Okyere, W. 1994. *Ghana: Policy Environment and Fertilizer Sector Development*. Muscle Shoals: International Fertilizer Development Center (IFDC).

Bumb, B. L., and Baanante, C. A. 1996. "The Role of Fertilizer in Sustaining Food Security and Protecting the Environment to 2020." Food and Agriculture Discussion Paper 17. Washington, D.C.: International Food Policy Research Institute (IFPRI).

Bunderson, W. T. 1994. "Comparative Maize Yields Under Alley Cropping with Different Hedge Species in the Lilongwe Plains." Mimeo.

Byerlee, D., and Heisey, P. W. 1993. "Performance of Hybrids Under Low-Input Conditions in Eastern and Southern Africa." Unpublished paper, International Maize and Wheat Improvement Center (CIMMYT), Mexico City.

Byerlee, D., and López-Pereira, M. A. 1993. "Technical Change in Maize: A Global Perspective." First International Maize Symposium, 16–19 March, Guadalajara, Mexico.

Byerlee, D.; with Anandajayasekeram, P.; Diallo, A.; Gelaw, B.; Heisey, P. W.; López-Pereira, M. A.; Mwangi, W.; Smale, M.; Tripp, R.; and Waddington, S. 1994. *Maize Research in Sub-Saharan Africa: An Overview of Past Impacts and Future Prospects.* CIMMYT Economics Working Paper 94-03. Mexico City: International Maize and Wheat Improvement Center (CIMMYT).

Byerlee, D., and Heisey, P. W. 1996. "Past and Potential Impacts of Maize Research in Sub-Saharan Africa: A Critical Assessment." *Food Policy* 21(3): 255–277.

Carr, S. 1994. "The Unique Challenge of Malawi's Smallholder Agricultural Sector." Mimeo.

Carter, S. E. (ed.). 1993. *Soil Fertility Management in Mutoko Communal Area, Zimbabwe: Report of a Field Exercise, August 12–September 3, 1992.* Nairobi: Tropical Soil Biology and Fertility Program (TSBF).

CBS (Central Bureau of Statistics, Kenya). Various years. *National Agricultural Statistical Sampling Frame II.* Nairobi: Ministry of Planning and National Development.

Chang, J. F., and Shibles, R. M. 1985. "An Analysis of Competition Between Intercropped Cowpea and Maize. II. The Effect of Fertilizer and Plant Population." *Field Crops Research* 12: 145–152.

Chimedza, R. 1994. "Rural Financial Markets," in M. Rukuni and C. K. Eicher (eds.), *Zimbabwe's Agricultural Revolution.* Harare: University of Zimbabwe.

CIMMYT (International Maize and Wheat Improvement Center). 1981. *World Maize Facts and Trends, Report One: An Analysis of Changes in Production, Consumption, Trade, and Prices over the Last Two Decades.* Mexico City: CIMMYT.

———. 1988. *From Agronomic Data to Farmer Recommendations: An Economics Training Manual.* Mexico City: CIMMYT.

———. 1990. *1989/90 CIMMYT World Maize Facts and Trends: Realizing the Potential of Maize in Sub-Saharan Africa.* Mexico City: CIMMYT.

———. 1992a. *Enduring Designs for Change: An Account of CIMMYT's Research, Its Impact, and Its Future Direction.* Mexico City: CIMMYT.

———. 1992b. *1991/92 CIMMYT World Maize Facts and Trends: Maize Research Investment and Impacts in Developing Countries.* Mexico City: CIMMYT.

———. 1993. "CIMMYT Survey of Maize Seed Production, Sales and Prices." Mexico City: CIMMYT. Unpublished.

———. 1994. *CIMMYT 1993/94 World Maize Facts and Trends. Maize Seed Industries, Revisited: Emerging Roles of the Public and Private Sectors.* Mexico City: CIMMYT.

Cleaver, K. 1985. *The Impact of Price and Exchange Rate Policies on Agriculture in Sub-Saharan Africa.* World Bank Staff Working Paper No. 728. Washington, D.C.: World Bank.

———. 1993a. "Making Agricultural Extension Work in Africa," in N. C. Russell and C. R. Dowswell (eds.), *Policy Options for Agricultural Development in Sub-Saharan Africa*. Mexico City: CASIN/SAA/Global 2000.

———. 1993b. *A Strategy to Develop Agriculture in Sub-Saharan Africa and a Focus for the World Bank*. World Bank Technical Paper No. 203. Washington, D.C.: World Bank.

Cleaver, K. M., and Donovan, W. G. 1995. *Agriculture, Poverty, and Policy Reform in Sub-Saharan Africa*. World Bank Discussion Paper No. 280. Washington, D.C.: World Bank.

Collinson, M. P. 1987. "Farming Systems Research: Procedures for Technology Development." *Experimental Agriculture* 33: 365–386.

Conroy, A. C. 1990. "Fertilizer Use and Distribution in Zimbabwe." World Bank Background Paper, Zimbabwe Agriculture Sector Memorandum. Harare: World Bank.

———. 1993. "The Economics of Smallholder Maize Production in Malawi with Reference to the Market for Seed and Fertiliser." Ph. D. thesis, University of Manchester, Manchester, UK.

Conroy, A. C., and Kumwenda, J. D. T. 1995. "Risks Associated with the Adoption of Hybrid Seed and Fertilizer by Smallholder Farmers in Malawi," in D. C. Jewell, S. R. Waddington, J. K. Ransom, and K. V. Pixley (eds.), *Maize Research for Stress Environments: Proceedings of the Fourth Eastern and Southern Africa Regional Maize Conference, Harare, Zimbabwe, 28 March–1 April 1994*. Harare: International Maize and Wheat Improvement Center (CIMMYT).

Corbett, D., and Coulter, J. 1995. "Report on Reorienting the South African Agricultural Research System to Meet the Needs of Small Farmers." Pretoria: Agricultural Research Council.

Cousins, B.; Weiner, D.; and Amin, N. 1992. "Social Differentiation in the Communal Lands of Zimbabwe." *Review of African Political Economy* 53: 5–24.

Cromwell, E. 1996. *Government, Farmers, and Seeds in a Changing Africa*. New York: Oxford University Press.

Cromwell, E., and Zambezi, B. T. 1993. *The Performance of the Seed Sector in Malawi: An Analysis of the Influence of Organizational Structure*. London: Overseas Development Institute.

CSO/MAFF (Central Statistical Office and Ministry of Agriculture, Food, and Fisheries). 1991, 1992, 1993, 1994, and 1995. *Final Crop Forecast*. Lusaka: CSO.

Dakurah, A. H., and Arias, F. R. 1987. "Cropping Systems in the Ejura District of Ashanti Region, Ghana: Results and Implications of a Formal Survey." Mimeo. Kumasi: Ghana Grains Development Project (GGDP).

Dalal, R. C. 1974. "Effects of Intercropping Maize with Pigeonpea on Grain Yield and Nutrient Uptake." *Experimental Agriculture* 10: 219–224.

Dalrymple, D. G. 1975. *Evaluating Fertilizer Subsidies in Developing Countries*. AID Discussion Paper No. 30. Washington, D.C.: United States Agency for International Development (USAID).

Danagro. 1987. *Regional Seed Production and Supply Project: A Report Prepared for the Southern African Development Coordination Conference*. Gaborone: SADCC.

Dapaah, S. K., and Otinkorang, E. S. 1988. "The Place of Fertilizer in Ghana's Quest for Increased Agricultural Productivity," in T. Tshibaka and C. A. Baanante (eds.), *Fertilizer Policy in Tropical Africa*. Lomé: International Fertilizer Development Center (IFDC) and International Food Policy Research Institute (IFPRI).

Daramola, B. 1989. "The Study of Socioeconomic Factors Influencing Fertilizer Adoption Decisions in Nigeria: A Survey of Oyo State Farmers." *Fertilizer Research* 20: 143–151.

De Capitani, A., and North, D. C. 1994. *Institutional Development in Third World Countries: The Role of the World Bank*. Washington, D.C.: World Bank.

de Janvry, A., and Sadoulet, E. 1994. "Structural Adjustment Under Transaction Costs," in F. Heidhues and B. Knerr (eds.), *Food and Agricultural Policies Under Structural Adjustment*. Frankfurt: Peter Lang.

De Woronin, B. 1993. "The Zimbabwe Seed Industry." Proceedings of the Food and Agriculture Organization/University of Zimbabwe Regional Workshop on Marketing of Agricultural Inputs in Eastern and Southern Africa, Harare, Zimbabwe, 5–9 October 1992.

Deininger, K., and Binswanger, H. P. 1995. "Rent Seeking and the Development of Large-Scale Agriculture in Kenya, South Africa, and Zimbabwe." *Economic Development and Cultural Change* 43(3): 493–522.

Dennison, E. B. 1961. "The Value of Farmyard Manure in Maintaining Fertility in Northern Nigeria." *Empire Journal of Experimental Agriculture* 29: 330–336.

Dent, J. B., and Thornton, P. K. 1988. "The Role of Biological Simulation Models in Farming Systems Research." *Agricultural Administration and Extension* 29: 111–122.

Desai, G. M. 1991. "Fertilizer Policies: A Perspective on Price and Subsidy Issues." National Consultation on Fertilizer Pricing: An Aspect of Public Policy, Jaipur, India, 30 September–2 October.

Dickie, M.; Fisher, A.; and Gerking, S. 1987. "Market Transactions and Hypothetical Demand Data: A Comparative Study." *Journal of the American Statistical Association* 82(397): 69–75.

DOA (Department of Agriculture, Gold Coast). 1959. *Annual Report for 1956–57*. Accra: DOA.

Donovan, W. G. 1995. *Agriculture and Economic Reform in Sub-Saharan Africa*. Washington, D.C.: World Bank, Africa Technical Department.

Doran, J. W.; Fraser, D. G.; Culik, M. N.; and Liebhardt, W. C. 1987. "Influence of Alternative and Conventional Agricultural Management on Soil Microbial Processes and Nitrogen Availability." *American Journal of Alternative Agriculture* 2: 99–106.

Douglas, J. E. 1980. *Successful Seed Programs: A Planning and Management Guide*. Boulder, Colorado: Westview.

Dudal, R., and Byrnes, B. H. 1993. "The Effects of Fertilizer Use on the Environment," in H. van Reuler and W. H. Prins (eds.), *The Role of Plant Nutrients for Sustainable Food Crop Production in Sub-Saharan Africa*. Leidschendam: Dutch Association of Fertilizer Producers (VKP).

Eberhart, S. A., and Sprague, G. F. 1973. "A Major Cereals Project to Improve Maize, Sorghum, and Millet Production in Africa." *Agronomy Journal* 65: 365–373.

Edmeades, G. 1990. "Trip Report." Mexico City: International Maize and Wheat Improvement Center (CIMMYT). Internal document.

Edmeades, G.; Dankyi, A. A.; Marfo, K.; and Tripp, R. 1991. "On-Farm Maize Research in the Transition Zone of Ghana," in R. Tripp (ed.), *Planned Change in Farming Systems: Progress in On-Farm Research*. Chichester: John Wiley.

Eicher, C. K. 1982. "Facing Up to Africa's Food Crisis." *Foreign Affairs* 61(1): 151–174.

———. 1988. "An Economic Perspective on the Sasakawa-Global 2000 Initiative to Increase Food Production in Sub-Saharan Africa." Presented at the workshop

"Reviewing the African Agricultural Projects," Sasakawa-Global 2000 Africa Initiative, 18 March, Nairobi, Kenya [Carter Center, Atlanta].
———. 1989. *Sustainable Institutions for African Agricultural Development.* Working Paper No. 19. The Hague: International Service for National Agricultural Research (ISNAR).
———. 1990. "Building African Scientific Capacity for Agricultural Development." *Agricultural Economics* 4(2): 117–143.
———. 1995. "Zimbabwe's Maize-Based Green Revolution: Preconditions for Replication." *World Development* 23(5): 805–818.
Eicher, C. K., and Rukuni, M. 1996. *Reflections on Agrarian Reform and Capacity Building in South Africa.* Staff Paper No. 96-3. East Lansing: Michigan State University, Department of Agricultural Economics.
Ellis, R. T. 1959. "The Food Properties of Flint and Dent Maize." *East African Journal* 24: 251–253.
Experience Incorporated. 1969. *A Time Phased, Seed Multiplication, Distribution and Implementation Plan for the Republic of Tanzania: Project Report Prepared for the Republic of Tanzania and the United States Agency for International Development.* Minneapolis: Experience Incorporated.
Eyzaguirre, P. 1996. *Agricultural and Environmental Research in Small Countries: Innovative Approaches to Strategic Planning.* New York: John Wiley.
Fafchamps, M.; de Janvry, A.; and Sadoulet, E. 1995. "Transaction Costs, Market Failures, Competitiveness and the State," in G. H. Peters and D. Hedley (eds.), *Agricultural Competitiveness: Market Forces and Policy Choices.* Aldershot: Darthmouth.
Fakorede, M. A. B.; Adenola, O. A.; Kim, S. K.; Nweke, F. I.; Iken, J. E.; Akinwumi, J. A.; and Alofe, C. O. 1995. "Research Priorities for the Improvement of Maize Production in Nigeria: Farmers' Views." West and Central African Regional Maize and Cassava Workshop, Cotonou, Benin Republic, 28 May–2 June.
FAO (Food and Agricultural Organization). 1971, 1979, and 1993. *Fertilizer Yearbook.* Rome: FAO.
———. Various years. *Agrostat PC.* Rome: FAO.
———. 1994a. *Agrostat.* Rome: FAO.
———. 1994b. *The State of Food and Agriculture, 1994.* Rome: FAO.
Farrington, J. 1995. "Mobilising Science and Technology or Fostering Organizational Change? A Response to Borlaug and Dowswell." *Development Policy Review* 13: 131–133.
Ferguson, A. E.; Millard, A. V.; and Khaila, S. W. 1990. "Crop Improvement Programmes and Nutrition in Malawi: Exploring the Links." *Food and Nutrition Bulletin* 12: 273–278.
Fisher, O. Various years. "White Maize Prices, U. S. Gulf." Data files, Orville Fisher Agribusiness Consultant, Topeka, Kansas.
Fontaine, J.-M., with Sindzingre, A. 1991. *Macro-Economic Linkages: Structural Adjustment and Fertilizer Policy in Sub-Saharan Africa.* OECD Technical Paper No. 49. Paris: Organisation for Economic Cooperation and Development (OECD).
Franzel, S.; Colburn, F.; and Getahun, D. 1989. "Grain Marketing Regulations: Impact on Peasant Production in Ethiopia." *Food Policy* 14(4): 347–358.
Freeman, H. A.; Roe, T. L.; and Smith, J. Forthcoming. *The Effect of Agricultural Policies on Farming Households' Crop Choice: Economic Evidence and Environmental Implications in Northern Nigeria.* Environment and Development Economics.
Freeman, H. A., and Smith, J. 1996. "Intensification of Land Use and the Evolution of Agricultural Systems in the West African Northern Guinea Savanna." *Quarterly Journal of International Agriculture* 35: 109–124.

Fresco, L. O., and Kroonenberg, S. B. 1992. "Time and Spatial Scales in Ecological Sustainability," in *Land Use Policy* (London) 9: 155–168.
Friis-Hansen, E. 1992. *The Performance of the Seed Sector in Zimbabwe: An Analysis of the Influence of Organizational Structure.* London: Overseas Development Institute.
Gandhi, V. P., and Desai, G. M. 1992. "Converting Potential into Effective Demand for and Use of Fertilizers: A Study of Small Farmers in Gazaland District of Zimbabwe," in S. Wanmali and J. M. Zamchiya (eds.), *Service Provision and Its Impact on Agricultural and Rural Development in Zimbabwe.* Washington, D.C.: International Food Policy Research Institute (IFPRI).
Gausi, R. K. 1970. *Seed Industry Development in Malawi.* Lilongwe: Chitedze Research Station.
Gerhart, J. 1975. *The Diffusion of Hybrid Maize in Western Kenya.* Mexico City: International Maize and Wheat Improvement Center (CIMMYT).
Gerner, H., and Harris, G. 1993. "The Use and Supply of Fertilizers in Sub-Saharan Africa," in H. van Reuler and W. H. Prins (eds.), *The Role of Plant Nutrients for Sustainable Food Crop Production in Sub-Saharan Africa.* Leidschendam: Dutch Association of Fertilizer Producers (VKP).
Gevers, H. O. 1988. "Highlights of Maize Breeding in South Africa," in *Proceedings of the Eighth South African Maize Breeding Symposium.* Pretoria: Department of Agriculture and Water Supply.
GGDP (Ghana Grains Development Project). 1991. *A Study of Maize Technology Adoption in Ghana.* Mexico City: GGDP.
Ghura, D., and Grennes, T. J. 1991. *The Impact of Real Exchange Rate Misalignment and Instability on Macroeconomic Performance in Sub-Saharan Africa.* International Agricultural Trade Research Consortium Working Paper No. 91-4. Raleigh: North Carolina State University, Department of Economics and Business.
Giller, K. E.; McDonagh, J. F.; and Cadisch, G. 1994. "Can Biological Nitrogen Fixation Sustain Agriculture in the Tropics?" in J. K. Syers and D. L. Rimmer (eds.), *Soil Science and Sustainable Land Management in the Tropics.* Wallingford: Commonwealth Agricultural Bureau (CAB) International.
Giller, K. E.; Itimu, O.; and Masamba, C. 1996. "Nutrient Sourcing by Agroforestry Species and Soil Fertility Maintenance in Maize Associations," in S. R. Waddington (ed.), *Research Results and Network Outputs in 1994 and 1995: Proceedings of the Second Meeting of the Soil Fertility Working Group for Maize-Based Farming Systems.* Harare: International Maize and Wheat Improvement Center (CIMMYT).
Giri, G., and De, R. 1980. "Effects of Preceding Grain Legumes on Growth and Nitrogen Uptake of Dryland Pearl Millet." *Plant and Soil* 56: 459–464.
GMB (Grain Marketing Board, Zimbabwe). Various years. *GMB Report and Accounts.* Harare: Grain Marketing Board.
GOM/NSO (Government of Malawi/National Statistics Office). 1984. *National Sample Survey of Agriculture 1980/81.* Zomba: Government of Malawi, 1984.
Gordon, H., and Spooner, N. 1992. "Kenyan Grain Marketing Reform in Principle," in *Proceedings of the Conference on Maize Supply and Marketing Under Market Liberalization.* Nairobi: Egerton University, Policy Analysis Matrix Project.
Grant, P. M. 1981. "The Fertilization of Sandy Soils in Peasant Agriculture." *Zimbabwe Agricultural Journal* 78: 169–175.
Grant, R. M. 1991. *Contemporary Strategy Analysis.* Cambridge: Blackwell.
Gryseels, G., and Anderson, J. R. 1991. "International Agricultural Research," in P. G. Pardey, J. Roseboom, and J. R. Anderson (eds.), *Agricultural Research*

Policy: International Quantitative Perspectives. New York: Cambridge University Press.
GRZ (Government of the Republic of Zambia). 1989. *Report on Fertilizer Use.* Lusaka: Ministry of Agriculture and Water Development.
———. 1991. *An Evaluation of the Agricultural Credit System in Zambia.* Lusaka: Agricultural Credit Study Team.
———. 1995. *A Review of Maize Marketing Liberalisation During 1994: The Transition Programme 1995/96.* Lusaka: Ministry of Agriculture, Food, and Fisheries (MAFF), Food Security Division.
Harrison, M. 1970. "Maize Improvement in East Africa," in C. L. A. Leakey (ed.), *Crop Improvement in East Africa.* Farnham Royal: Commonwealth Agricultural Bureaux.
Hassan, R. M.; Ngure, M.; and Njoroge, K. 1994. "Adoption Patterns and Performance of Maize Seed and Fertilizer Technologies in Kenya." Nairobi: Kenyan Agricultural Research Institute (KARI). Unpublished paper.
Hassan, R. M. (ed.). Forthcoming. *New Tools and Emerging Challenges for Agricultural Research: Methods and Results from a GIS Application to Maize in Kenya.*
Hassan, R. M.; Karanja, D. D.; and Mulamula, H. A. Forthcoming. "Availability and Effectiveness of Agricultural Extension Services for Maize Farmers in Kenya," in R. M. Hassan (ed.), *New Tools and Emerging Challenges for Agricultural Research: Methods and Results from a GIS Application to Maize in Kenya.*
Haugerud, A., and Collinson, M. P. 1990. "Plants, Genes and People: Improving the Relevance of Plant Breeding in Africa." *Experimental Agriculture* 26: 341–362.
Hayenga, M. L. 1979. "Market Information and Price Reporting in Thinly Traded or Imperfect Markets," in *Market Information and Price Reporting in the Food and Agricultural Sector, Proceedings of a Conference Sponsored by the North Central Regional Committee 117.* Madison: University of Wisconsin Press.
Heisey, P. W. 1990. "Comment: Maize Research in Malawi." *Journal of International Development* 2(2): 243–253.
———. 1994. "Are Green Revolution Varieties Riskier? The Case of Hybrid Maize in Malawi." Production Economics/Management Workshop, Department of Agricultural and Applied Economics, University of Minnesota, St. Paul, May.
Heisey, P. W., and Smale, M. 1995. *Maize Technology in Malawi: A Green Revolution in the Making?* CIMMYT Research Report No. 4. Mexico City: International Maize and Wheat Improvement Center (CIMMYT).
Heisey, P. W., and Mwangi, W. 1996. *Fertilizer Use and Maize Production in Sub-Saharan Africa.* CIMMYT Economics Program Working Paper 96-01. Mexico City: International Maize and Wheat Improvement Center (CIMMYT).
Heisey, P. W., and Waddington, S. R., (eds.). 1993. *Impacts of On-Farm Research: Proceedings of a Workshop on Impacts of On-Farm Research in Eastern and Southern Africa.* CIMMYT Eastern and Southern Africa On-Farm Research Network Report No. 24. Lilongwe and Harare: International Maize and Wheat Improvement Center (CIMMYT).
Herbst, J. 1990a. *Economic Reform in Africa: The Lessons of Ghana.* Field Staff Report No. 15. Indianapolis: Universities Field Staff International.
———. 1990b. *State Politics in Zimbabwe.* Berkeley: University of California Press.
Heyer, J.; Maitha, J. K.; and Senga, W. M. 1976. *Agricultural Development in Kenya: An Economic Assessment.* Nairobi: Oxford University Press.

HIID/EPD (Harvard Institute for International Development/Economic Planning and Development Department, Government of Malawi). 1994. *Fertilizer Policy Study: Market Structure, Prices, and Fertilizer Use by Smallholder Maize Farmers*. Lilongwe: Department of Economic Planning and Development.

Hikwa, D., and Mukurumbira, L. 1995. "Highlights of Previous, Current, and Proposed Soil Fertility Research by the Department of Research and Specialist Services (DRSS) in Zimbabwe," in S. R. Waddington (ed.), *Report on the First Meeting of the Network Working Group. Soil Fertility Research Network for Maize-Based Farming Systems in Selected Countries of Southern Africa*. Lilongwe and Harare: Rockefeller Foundation Southern Africa Agricultural Sciences Program and the International Maize and Wheat Improvement Center Maize Program.

House, W. J., and Zimalirana, G. 1992. "Rapid Population Growth and Poverty Generation in Malawi." *Journal of Modern African Studies* 30: 141–161.

Howard, J. A. 1994. "The Economic Impact of Improved Maize Varieties in Zambia." Ph. D. thesis, Michigan State University, Department of Agricultural Economics, East Lansing, Michigan.

Howard, J. A.; Nakaponda, B.; and Ferris, J. 1995. "Factors Affecting Maize Supply, Demand, and Prices in Zambia." Unpublished paper. East Lansing: Michigan State University.

Huchu, P., and Sithole, P. N. 1994. "Rates of Adoption of New Technology and Climatic Risk in the Communal Areas of Zimbabwe," in E. T. Craswell and J. Simpson (eds.), *Soil Fertility and Climatic Constraints in Dryland Agriculture*. ACIAR Proceedings No. 54. Canberra: Australian Centre for International Agricultural Research (ACIAR).

IFA/IFDC/FAO (International Fertilizer Industry Association/International Fertilizer Development Center/Food and Agriculture Organization). 1992. *Fertilizer Use by Crop*. Rome: IFA, IFDC, FAO.

IMF (International Monetary Fund). 1995. *IMF Financial Statistics*. Washington, D.C.: IMF.

ISSER (Institute of Statistical, Social, and Economic Research). 1994. *The State of the Ghanaian Economy in 1993*. Legon: University of Ghana.

IWC (International Wheat Council). 1993. World Grain Statistics. London: IWC.

Jabara, C. 1985. "Agriculture Pricing Policy in Kenya." *World Development* 13(5): 611–626.

Jansen, D. 1988. *Trade, Exchange Rate, and Agricultural Pricing Policies in Zambia*. World Bank Comparative Studies. The Political Economy of Agricultural Pricing Policy Series. Washington, D.C.: World Bank.

Jansen, D., and Muir, K. 1994. "Trade, Exchange Rate, and Agriculture in the 1980s," in M. Rukuni and C. K. Eicher (eds.), *Zimbabwe's Agricultural Revolution*. Harare: University of Zimbabwe Publications.

Janssen, B. H. 1993. "Integrated Nutrient Management: The Use of Organic and Mineral Fertilizers," in H. van Reuler and W. H. Prins (eds.), *The Role of Plant Nutrients for Sustainable Food Crop Production in Sub-Saharan Africa*. Leidschendam: Dutch Association of Fertilizer Producers (VKP).

Jayne, T. S., and Chisvo, M. 1991. "Unravelling Zimbabwe's Food Insecurity Paradox." *Food Policy* 16(5): 319–329.

Jayne, T. S., and Rubey, L. 1993. "Maize Milling, Market Reform, and Urban Food Security: The Case of Zimbabwe." *World Development* 21(6): 975–988.

Jayne, T. S., and Rukuni, M. 1993. "Distributional Effects of Maize Self-Sufficiency in Zimbabwe: Implications for Pricing and Trade Policy." *Food Policy* 18(4): 334–341.

Jayne, T. S.; Takavarasha, T.; Attwood, E. A.; and Kupfuma, B. 1993. *Postscript to Zimbabwe's Maize Success Story: Policy Lessons for Eastern and Southern Africa*. Staff Paper No. 93-68. East Lansing: Michigan State University, Department of Agricultural Economics.

Jayne, T. S.; Khatri, Y.; Thirtle, C.; and Reardon, T. 1994. "Determinants of Productivity Change Using a Profit Function Approach: The Case of Zimbabwe." *American Journal of Agricultural Economics* 76(3): 613–616.

Jayne, T. S.; Takavarasha, T.; and van Zyl, J. 1994. "Interactions Between Food Market Reform and Regional Trade in Zimbabwe and South Africa: Implications for Food Security." *Agrekon* 33(4): 184–201.

Jayne, T. S.; Rubey, L.; Tschirley, D.; Mukumbu, M.; Chisvo, M.; Santos, M.; Weber, M.; and Diskin, P. 1995. *Effects of Food Market Reform on Access to Food in Four Countries in Eastern and Southern Africa*. MSU International Development Paper 19. East Lansing: Michigan State University.

Jayne, T. S., and Jones, S. 1996. *Grain Marketing and Pricing Policy in Eastern and Southern Africa: A Survey*. MSU International Development Paper. East Lansing: Michigan State University.

Jensen, H., and Luckett, B. 1993. *A Profile of Poverty in Zambia Based on the 1991 Household Expenditure and Incomes Survey*. CARD Staff Report 93–SR61. Ames: Iowa State University.

Jiggins, J., Reijntjes, C., and Lightfoot, C. 1996. "Mobilising Science and Technology to Get Agriculture Moving in Africa: A Response to Borlaug and Dowswell." *Development Policy Review* 14: 89–103.

Johnston, B. F. 1958. *The Staple Food Economies of Western Tropical Africa*. Stanford: Stanford University Press.

Jones, M. J. 1974. "Effects of Previous Crop on Yield and Nitrogen Response of Maize at Samaru, Nigeria." *Experimental Agriculture* 10: 278–279.

Jones, R. B., and Heisey, P. W. 1994. An Agronomic and Economic Analysis of the Results from the MOA/UNDP/FAO Fertilizer Demonstration Programme in 1989–1993. Lilongwe: Rockefeller Foundation.

Jones, R. B., and Wendt, J. W. 1995. "Contribution of Soil Fertility Research to Improved Maize Production by Smallholders in Eastern and Southern Africa," in D. C. Jewell, S. R. Waddington, J. K. Ransom, and K. V. Pixley (eds.), *Maize Research for Stress Environments: Proceedings of the Fourth Eastern and Southern Africa Regional Maize Conference, Harare, Zimbabwe, 28 March–1 April 1994*. Harare: International Maize and Wheat Improvement Center (CIMMYT).

Jones, S. P. 1994. *Privatisation and Policy Reform: Agricultural Marketing in Africa*. Oxford: University of Oxford Food Studies Group.

Judd, M. A.; Boyce, J. K.; and Evenson, R. E. 1987. "Investment in Agricultural Research and Extension," in V. Ruttan and C. Pray (eds.), *Policy for Agricultural Research*. Boulder, Colorado: Westview.

Kabaghe, C. L. 1992. "Zambia Seed Company Limited." First National Workshop on Seed Technology Education, Siavonga, Zambia, 14–16 July.

Kabambe, V. H., and Kumwenda, J. D. T. 1995. "Weed Management and Nitrogen Rate Effects on Maize Grain Yield and Yield Components," in D. C. Jewell, S. R. Waddington, J. K. Ransom, and K. V. Pixley (eds.), *Maize Research for Stress Environments: Proceedings of the Fourth Eastern and Southern Africa Regional Maize Conference, Harare, Zimbabwe, 28 March–1 April 1994*. Harare: International Maize and Wheat Improvement Center (CIMMYT).

Kaluwa, B. 1992. "Malawi Food Marketing: Private Trader Operation and State Intervention," in J. B. Wyckoff and M. Rukuni (eds.), *Food Security Research in*

Southern Africa: Policy Implications. Harare: University of Zimbabwe, Department of Agricultural Economics and Extension.

Kandoole, B. F., and Msukwa, L. 1992. "Household Food and Income Security Under Market Liberalization: Experience from Malawi," in J. B. Wyckoff and M. Rukuni (eds.), *Food Security Research in Southern Africa: Policy Implications.* Harare: University of Zimbabwe, Department of Agricultural Economics and Extension.

Kang, B. T.; Reynolds, L.; and Atta-Krah, A. N. 1990. "Alley Farming." *Agronomy Abstracts* 43: 315–359.

Karanja, D. D. 1990. "The Rate of Return to Maize Research in Kenya: 1955–88." M. Sc. thesis, Department of Agricultural Economics, Michigan State University, East Lansing, Michigan.

———. 1996. *An Economic and Institutional Analysis of Maize Research in Kenya.* MSU International Development Working Paper No. 57. East Lansing: Michigan State University, Department of Agricultural Economics.

KARI (Kenya Agricultural Research Institute). 1990. *Final Report, Fertilizer Use and Recommendations Project.* Nairobi: National Agricultural Research Laboratories (NARL), KARI.

Kassam, A. H.; Kowal, J.; Dagg, M.; and Harrison, M. N. 1975. "Maize in West Africa and Its Potential in the Savanna Area." *World Crops* 27: 73–78.

Keating, B. A.; Wafula, B. M.; and Watiki, J. M. 1992. "Development of a Modeling Capability for Maize in Semi-Arid Eastern Kenya," in M. E. Probert (ed.), *A Search for Strategies for Sustainable Dryland Cropping in Semi-Arid Eastern Kenya.* ACIAR Proceedings No. 41. Canberra: Australian Centre for International Agricultural Research (ACIAR).

Kelly, V.; Diagana, B.; Reardon, T.; Gaye, M.; and Crawford, E. 1996. *Cash Crop and Food Grain Productivity in Senegal: Historical View, New Survey Evidence, and Policy Implications.* MSU International Development Paper No. 20. East Lansing: Michigan State University.

Kennedy, E., and Reardon, T. 1994. "Shift to Non-Traditional Grains in the Diets of East and West Africa: Role of Women's Opportunity Cost of Time." *Food Policy* 19(1): 45–56.

Kenya (Government of), Central Bureau of Statistics, Ministry of Planning and National Development. 1966. *Statistical Digest.* Nairobi: Government Printers.

———, Ministry of Agriculture and Livestock Development. 1995. "Analysis of the Fertilizer Situation in Kenya." Unpublished report. Nairobi: Ministry of Agriculture and Livestock Development.

Keyter, C. 1975. *Maize Control in Southern Rhodesia, 1931–1941: The African Contribution Toward White Survival.* Local Series 34. Salisbury: Central African Historical Association.

Killick, T. 1966. "Agriculture and Forestry," in W. Birmingham, I. Neustadt, and E. Omaboe (eds.), *A Study of Contemporary Ghana. Volume 1. The Economy of Ghana.* London: Allen and Unwin.

Kim, S. K., and Winslow, M. D. 1991. "Progress in Breeding Maize for Striga Tolerance/Resistance at IITA," in J. K. Ransom, L. J. Musselman, A. D. Worsham, and C. Parker (eds.), *Proceedings, Fifth International Symposium on Parasitic Weeds, Nairobi.* Mexico City: International Maize and Wheat Improvement Center (CIMMYT).

Kirsten, J. F., and Sartoius von Bach, H. J. 1992. "Surplus Producers and the Food Price Dilemma in Traditional Agriculture in Southern Africa: Empirical Evidence from the Farmer Support Programme." *Agrekon* 31(4).

Koester, U. 1986. *Regional Cooperation to Improve Food Security in Southern and Eastern African Countries.* Research Report No. 53. Washington, D.C.: International Food Policy Research Institute (IFPRI).

Kumar, S. 1994. *Adoption of Hybrid Maize in Zambia: Effects on Gender Roles, Food Consumption, and Nutrition.* IFPRI Research Report 100. Washington, D.C.: International Food Policy Research Institute (IFPRI).

Kumwenda, J. D. T. 1995. "Soybean Spatial Arrangement in a Maize/Soybean Intercrop in Malawi." Second African Crop Science Conference for Eastern and Central Africa, Blantyre, Malawi, 19–24 February.

Kumwenda, J. D. T.; Kabambe, V. H. K.; and Sakala, W. D. M. 1993. "Maize-Soybean and Maize-Bean Intercropping Experiments," in *Maize Agronomy Annual Report for 1992/93.* Lilongwe: Chitedze Agricultural Research Station.

Kupfuma, B. 1994. "The Payoffs to Hybrid Maize Research and Extension in Zimbabwe: An Economic and Institutional Analysis." M. S. thesis, Michigan State University, East Lansing, Michigan.

Kydd, J. 1989. "Maize Research in Malawi: Lessons from Failure." *Journal of International Development* 1(1): 112–144.

LACE (Lusaka Agricultural Commodity Exchange). 1995. *Monthly Bulletins.* Lusaka: LACE.

Ladd, J. N., and Amato, M. 1985. "Nitrogen Cycling in Legume-Cereal Rotations," in B. T. Kang and J. van der Heide (eds.), *Nitrogen Management in Farming Systems in Humid and Sub-Humid Tropics.* Haren, The Netherlands: Institute for Soil Fertility; and Ibadan, Nigeria: International Institute for Tropical Agriculture.

Lafitte, H. R., and Edmeades, G. O. 1994. "Improvement for Tolerance to Low Soil Nitrogen in Tropical Maize. II. Grain Yield, Biomass Production, and N Accumulation." *Field Crops Research* 39: 15–25.

Laker-Ojok, R. 1994. *The Rate of Return to Agricultural Research in Uganda: The Case of Oilseeds and Maize.* MSU International Development Working Paper No. 42. East Lansing: Michigan State University.

Lele, U. 1971. *Foodgrain Marketing in India, Private Performance, and Public Policy.* Ithaca: Cornell University Press.

———. 1992. "Can Technology Transfer and Macroeconomic Adjustment Sustain Africa's Agricultural Revolution Without an Agricultural Sector Strategy? The Case of Sasakawa-Global 2000 Program in Tanzania." Report prepared for the Sasakawa-Global 2000 Program. Atlanta: Carter Center. Mimeo.

Lele, U.; Christiansen, R. E.; and Kadiresan, K. 1989. *Issues in Fertilizer Policy in Africa: Lessons from Development Programs and Adjustment Lending, 1970–87.* MADIA Discussion Paper No. 5. Washington, D.C.: World Bank.

Lele, U.; Oyejide, A.; Bindlish, V.; and Bumb, B. 1989. *Nigeria's Economic Development, Agriculture's Role and World Bank's Assistance, 1961–88: Lessons for the Future.* Washington, D.C.: World Bank.

Lele, U., and Bumb, B. 1995. "The Food Crisis in South Asia: The Case of India," in K. Sarwar Lateef (ed.), *The Evolving Role of the World Bank: Helping Meet the Challenge of Development.* Washington, D.C.: World Bank.

Lele, U. (ed.). 1991. *Aid to African Agriculture: Lessons from Two Decades of Donors' Experience.* Baltimore: Johns Hopkins.

Lipton, M. 1991. "Market Relaxation and Agricultural Development," in C. Colclough and J. Manor (eds.), *States or Markets.* Oxford: Clarendon Press.

Lipton, M., with Longhurst, R. 1989. *New Seeds and Poor People.* Baltimore: Johns Hopkins.

López-Pereira, M. A., and Filippello, M. P. 1995. *Emerging Roles of the Public and Private Sectors of Maize Seed Industries in the Developing World.* Economics Working Paper 95-01. Mexico City: International Maize and Wheat Improvement Center (CIMMYT).

Low, A. R. C. 1988. "Farm Household-Economics and the Design and Impact of Biological Research in Southern Africa." *Agricultural Administration and Extension* 29: 23–34.

———. 1993. "The Low-Input, Sustainable Agriculture (LISA) Prescription: A Bitter Pill for Farm-Households in Southern Africa." *Project Appraisal* 8(2): 97–101.

———. 1994. "Reorientation of Research and Extension for Small Farmers: Requirements for Effective Implementation." 32nd Annual Conference of the Agricultural Economics Association of Southern Africa, 19–20 September, Pretoria, South Africa.

Low, A. R. C., and Waddington, S. R. 1990. "On-Farm Research on Maize Production Technologies for Smallholder Farmers in Southern Africa: Current Achievements and Future Prospects," in B. Gebrekidan (ed.), *Maize Improvement, Production, and Protection in Eastern and Southern Africa: Proceedings of the Third Eastern and Southern Africa Regional Maize Workshop.* Nairobi: International Maize and Wheat Improvement Center (CIMMYT).

———. 1991. "Farming Systems Adaptive Research: Achievements and Prospects in Southern Africa." *Experimental Agriculture* 27: 115–125.

Lukanty, J., and Wood, A. P. 1990. "Agricultural Policy in the Colonial Period," in A. P. Wood, S. A. Kean, J. T. Milmo, and D. M. Warren (eds.), *The Dynamics of Agricultural Policy and Reform in Zambia.* Ames: Iowa State University Press.

Lynam, J. K., and Blackie, M. J. 1994. "Building Effective Agricultural Research Capacity: The African Challenge," in J. R. Anderson (ed.), *Agricultural Technology: Policy Issues for the International Community.* Wallingford: Commonwealth Agricultural Bureau (CAB) International.

MacColl, D. 1989. "Studies on Maize (*Zea mays* L.) at Bunda, Malawi. II. Yield in Short Rotation with Legumes." *Experimental Agriculture* 25: 367–374.

Mainwaring, C. 1922. "Seed Supply in Rhodesia." *Rhodesia Agricultural Journal* 19(5): 548–550.

Maize Board (South Africa). Various years. *Report on Maize.* Pretoria: Maize Board of South Africa.

Makings, S. 1966. "Agricultural Change in Northern Rhodesia/Zambia, 1945–1965." *Food Research Institute Studies* 6(2): 195–247.

Makken, F. 1993. "Case Studies of Malawi and Ethiopia," in H. van Reuler and W. H. Prins (eds.), *The Role of Plant Nutrients for Sustainable Food Crop Production in Sub-Saharan Africa.* Leidschendam: Dutch Association of Fertilizer Producers (VKP).

Manson, S. C.; Leighner, D. E.; and Vorst, J. J. 1986. "Cassava-Cowpea and Cassava-Peanut Intercropping. III. Nutrient Concentration and Removal." *Agronomy Journal* 78: 441–444.

Manyong, M. V.; Smith, J.; and Baker, D. 1995. *Addressing Heterogeneity of Environments Using Spatial Mapping and Statistical Analysis of Evolving Production Systems.* Ibadan: International Institute of Tropical Agriculture (IITA).

Manyong, M. V.; Smith, J.; Weber, G. K.; Jagtap, S. S.; and Oyewole, B. 1996. *Macro Characterization of Agricultural Systems in West Africa: An Overview.* Resource and Crop Management Research Monograph No. 21. Ibadan: International Institute of Tropical Agriculture (IITA).

Maredia, M. K., and Eicher, C. K. 1995. "The Economics of Wheat Research in Developing Countries: The One Hundred Million Dollar Puzzle." *World Development* 23(3): 401–412.

Marion, B. W. 1986. *The Organization and Performance of the U. S. Food System.* Lexington: Lexington Books.
Martin, M., and Lele, U. 1992. "Sensitivity Analysis of the Kilimo-Sasakawa-Global 2000 Farm Budget Under Alternative Policy Scenarios." Mimeo.
Mashingaidze, K. 1994. "Maize Research and Development," in M. Rukuni and C. K. Eicher (eds.), *Zimbabwe's Agricultural Revolution.* Harare: University of Zimbabwe Publications.
Mashiringwani, N. A. 1983. "The Present Nutrient Status of the Soils in the Communal Farming Areas of Zimbabwe." *Zimbabwe Agricultural Journal* 80: 73–75.
Masters, W. A., and Nuppenau, E. A. 1993. "Pan-Territorial Versus Regional Pricing for Maize in Zimbabwe." *World Development* 21(10): 1647–58.
Masters, W. A. 1994. *Government and Agriculture in Zimbabwe.* Westport, Connecticut: Praeger.
Mbekeani, Y. 1991. "Effects of Spacing on Water Relations, Light Interception and Biomass Partitioning in a *Leucaena*-Maize Alley Cropping System." M. Sc. thesis, Washington State University, Pullman, Washington.
McCalla, A. F. 1994. *Agriculture and Food Needs to 2025: Why We Should Be Concerned.* Sir John Crawford Memorial Lecture. Washington, D.C.: Consultative Group on International Agricultural Research (CGIAR).
McCown, R. L.; Keating, B. A.; Probert, M. E.; and Jones, R. K. 1992. "Strategies for Sustainable Crop Production in Semi-Arid Africa." *Outlook on Agriculture* 21(1): 21–31.
McIntire, J.; Bourzat, D.; and Pingali, P. 1992. *Crop-Livestock Interaction in Sub-Saharan Africa.* Washington, D.C.: World Bank.
McKenzie, J., and Chenoweth, F. 1992. "Zambia's Maize Policies: Consequences and Needed Reforms," in J. B. Wyckoff and M. Rukuni (eds.), *Food Security Research in Southern Africa: Policy Implications.* Harare: University of Zimbabwe.
Meinertzhagen, R. 1957. *Kenya Diary: 1902–1906.* Edinburgh and London: Oliver and Boyd.
Mellor, J. W. 1976. *The New Economics of Growth: A Strategy for India and the Developing World.* Ithaca: Cornell University Press.
Mellor, J. W.; Delgado, C.; and Blackie, M. J. (eds.). 1987. *Accelerating Food Production in Sub-Saharan Africa.* Baltimore: Johns Hopkins.
Metelerkamp, H. R. R. 1988. "Review of Crop Research Relevant to the Semiarid Areas of Zimbabwe," in *Proceedings of a Workshop on Cropping in the Semiarid Areas of Zimbabwe.* Harare: Agritex/Department of Research and Specialist Services (DR&SS)/Gemeinschaft für Technische Zusammenarbeit (GTZ).
Miller, T., and Tolley, G. 1989. "Technology Adoption and Agricultural Price Policy." *American Journal of Agricultural Economics* 71(4): 847–857.
Mills, B. F.; Hassan, R. M.; and Mwangi, P. 1995. *Maize Program Priorities in Kenya: Measuring Research-Induced Benefits with Multiple, Spatially Linked Production Zones.* Nairobi: Kenya Agricultural Research Institute (KARI), Priority-Setting Project.
Ministry of Agriculture (Malawi). 1984–1996. National Crop Estimates. Unpublished data.
Miracle, M. 1966. *Maize in Tropical Africa.* Madison: University of Wisconsin Press.
Mkandawire, R. K.; Jaffee, S.; and Bertoli, S. 1990. "Beyond Dualism: The Changing Face of the Leasehold Estate Sub-Sector in Malawi." Report prepared for

USAID/Malawi and USAID/Regional Economic Development Support Office (REDSO), East Africa.
MOA (Ministry of Agriculture, Policy Planning, Monitoring and Evaluation Department, Ghana). 1991. *Agriculture in Ghana: Facts and Figures*. Accra: MOA.
Moock, P. R. 1981. "Education and Technical Efficiency in Small-Farm Production." *Economic Development and Cultural Change* 29(4): 723–740.
Morris, M. L.; Clancy, C.; and López-Pereira, M. A. 1992. "Maize Research Investment and Impacts in Developing Countries." Part I of *1991–92 CIMMYT World Maize Facts and Trends: Maize Research Investment and Impacts in Developing Countries*. Mexico City: International Maize and Wheat Improvement Center (CIMMYT).
Mosley, P. 1975. *Maize Control in Kenya, 1920–1970*. Bath: University of Bath.
———. 1983. *The Settler Economies: Studies in the Economic History of Kenya and Southern Rhodesia 1900–63*. Cambridge: Cambridge University Press.
———. 1994. "Policy and Capital Market Constraints to the African Green Revolution: A Study of Maize and Sorghum Yields in Kenya, Malawi, and Zimbabwe, 1960–91," in G. A. Cornia and G. Helleiner (eds.), *From Adjustment to Development in Africa: Conflict, Controversy, Convergence, and Consensus?* New York: Macmillan.
Mudahar, M. S. 1986. "Fertilizer Problems and Policies in Sub-Saharan Africa," in A. U. Mokwunye and P. L. G. Vlek (eds.), *Management of Nitrogen and Phosphorus Fertilizers in Sub-Saharan Africa*. Dordrecht: Martinus Nijhoff.
Mughogho, S. K.; Bationo, A.; Christianson, B.; and Vlek, P. L. G. 1986. "Management of Nitrogen Fertilizers for Tropical African Soils," in A. U. Mokwunye and P. L. G. Vlek (eds.), *Management of Nitrogen and Phosphorus Fertilizers in Sub-Saharan Africa*. Dordrecht: Martinus Nijhoff.
Mugwira, L. M., and Mukurumbira, L. M. 1984. "Comparative Effectiveness of Manures from the Communal Area and Commercial Feedlots as Plant Nutrient Sources." *Zimbabwe Agricultural Journal* 81: 241–250.
Mugwira, L. M., and Shumba, E. M. 1986. "Rate of Manure Supplied in Some Communal Areas and Their Effect on Plant Growth and Maize Grain Yields." *Zimbabwe Agricultural Journal* 83: 99–104.
Muir, K. 1994. "Agriculture in Zimbabwe," in M. Rukuni and C. K. Eicher (eds.), *Zimbabwe's Agricultural Revolution*. Harare: University of Zimbabwe Publications.
Muir, K., and Takavarasha, T. 1989. "Pan-Territorial and Pan-Seasonal Pricing for Maize in Zimbabwe," in G. Mudimu and R. Bernsten (eds.), *Household and National Food Security in Southern Africa*. Harare: University of Zimbabwe.
Mukumbu, M. 1992. "The Effects of Market Liberalization on the Maize Milling Industry in Kenya," in *Proceedings of a Conference on Maize Supply and Marketing Under Market Liberalization, Egerton University, Policy Analysis Matrix Project, 18–19 June 1992*. Nairobi: Egerton University.
Mukumbu, M., and Jayne, T. S. 1994. "Urban Maize Meal Consumption Patterns: Strategies for Improving Food Access for Vulnerable Urban Households in Kenya." Symposium on Agricultural Policies and Food Security in Eastern Africa, 19–20 May, Nairobi, Kenya.
Mukurumbira, L. M. 1985. "Effects of Rate of Fertilizer Nitrogen and Previous Grain Legume Crop on Maize Yields." *Zimbabwe Agricultural Journal* 82: 177–179.
Mwangi, W. 1995. "Low Use of Fertilizers and Low Productivity in Sub-Saharan Africa." International Food Policy Research Institute (IFPRI)/Food and Agriculture Organization (FAO) workshop on Plant Nutrition Management, Food

Security and Sustainable Agriculture, and Poverty Alleviation in Developing Countries, 16–17 May, Viterbo, Italy.
Myers, R. J. K.; Palm, C. A.; Cuevas, E.; Gunatilleke, I. U. N.; and Brossard, M. 1994. "The Synchronization of Nutrient Mineralization and Plant Nutrient Demand," in P. I. Woomer and M. J. Swift (eds.), *The Biological Management of Tropical Soil Fertility.* Chichester: John Wiley.
Nandwa, S. M.; Anderson, J. M.; and Seward, P. D. 1995. "The Effect of Placement of Maize Stover and N Fertilization on Maize Productivity and N Use Efficiency in the Semi-Arid and Sub-Humid Agro-Ecological Zones of Kenya," in D. C. Jewell, S. R. Waddington, J. K. Ransom, and K. V. Pixley (eds.), *Maize Research for Stress Environments: Proceedings of the Fourth Eastern and Southern Africa Regional Maize Conference, Harare, Zimbabwe, 28 March–1 April 1994.* Harare: International Maize and Wheat Improvement Center (CIMMYT).
NARP (National Agricultural Research Project). 1994. *Ghana National Agricultural Research Strategic Plan: Final Report.* Accra: NARP.
Natarajan, M., and Shumba, E. M. 1990. "Intercropping Research in Zimbabwe: Current Status and Outlook for the Future," in S. R. Waddington, A. F. E. Palmer, and O. T. Edje (eds.), *Research Methods for Cereal/Legume Intercropping.* Mexico City: International Maize and Wheat Improvement Center (CIMMYT).
Ndambuki, F.; Kiplagat, J.; and Rubui, A. 1992. "A Review of the Maize Seed Industry in Kenya," in *Proceedings of a Workshop on the Review of the National Maize Research Program, Kakamega, Kenya, November 1990.* Nairobi: Kenya Agricultural Research Institute (KARI).
Ndayisenga, F., and Schuh, G. E. 1995. "Fertilizer Policy in Sub-Saharan Africa: Recurring Issues and Recommendations." Sasakawa-Global 2000 project. Atlanta: Carter Center.
Ninje, T., and Weaver, F. J. 1986. "Protein, Energy, and Carotene Content of Maize at Six Stages in the Preparation of *Ufa* by Village Processing Methods." *Ecology of Food and Nutrition* 15: 273–280.
Njoroge, K.; Kanampiu, N.; Otsyula, R.; Muthamia, Z.; Gathuri, C.; and Chivatsi, W. 1992. "The High-Altitude Maize Breeding Program," in *Proceedings of a Workshop on the Review of the National Maize Research Program, Kakamega, Kenya, November 1990.* Nairobi: Kenya Agricultural Research Institute (KARI).
Norman, D. W.; Simmons, E. B.; and Hays, H. M. 1982. *Farming Systems in the Nigerian Savanna: Research Strategies for Development.* Boulder, Colorado: Westview.
North, D. C. 1990. *Institutions, Institutional Change, and Economic Performance.* New York: Cambridge University Press.
Nwosu, A. C. 1995. "Fertilizer Supply and Distribution Policy in Nigeria," in A. E. Ikpi and J. K. Olayemi (eds.), *Sustainable Agriculture and Economic Development in Nigeria.* Morrilton: Winrock.
Nyoro, J. 1992. "Competitiveness of Maize Production Systems in Kenya," in *Proceedings of the Conference on Maize Supply and Marketing Under Market Liberalization.* Njoro: Egerton University, Policy Analysis Matrix (PAM).
Odhiambo, G., and Ransom, J. K. 1995. "Long Term Strategies for *Striga* Control," in D. C. Jewell, S. R. Waddington, J. K. Ransom, and K. V. Pixley (eds.), *Maize Research for Stress Environments: Proceedings of the Fourth Eastern and Southern Africa Regional Maize Conference, Harare, Zimbabwe, 28 March–1 April 1994.* Harare: International Maize and Wheat Improvement Center (CIMMYT).

Odhiambo, M., and Wilcock, D. 1989. "Reform of Maize Marketing in Kenya," in M. Rukuni, G. Mudimu, and T. S. Jayne (eds.), *Food Security Policies in the SADCC Region*. Harare: University of Zimbabwe.

Oehmke, J. F., and Crawford, E. W. 1996. "The Impact of Agricultural Technology in Sub-Saharan Africa." *Journal of African Economies* 5(2): 271–292.

Ofori, C. S. 1963. *The Response of Maize to N, P, and K Fertilizers on Peasant Farms in Ghana*. Kwadaso: Soil Research Institute.

Olver, R. C. 1988. "Zimbabwe Maize Breeding Program," in B. Gelaw (ed.), *Towards Self Sufficiency: A Proceedings of the Second Eastern, Central and Southern Africa Regional Maize Workshop, March 15–21, 1987*. Harare: International Maize and Wheat Improvement Center (CIMMYT).

Ong, C. 1994. "Alley Cropping—Ecological Pie in the Sky?" *Agroforestry Today* 6: 8–10.

Opoku-Apau, A.; Anchirinah, M. V.; Read, M.; and Tripp, R. 1987. "The Maize-Cassava Intercropping System of the Central Region, Ghana." Mimeo. Kumasi: Ghana Grains Development Project (GGDP).

Oputa, C. O.; Olunga, B. A.; Vaughan, O.; Farinde, O.; and Fapohunda, T. C. 1983. *On-Farm Adaptive Research (OFAR) Diagnostic Survey, Ifedapo West Local Government Area, Oyo State*. Ibadan: Federal Agricultural Coordinating Unit.

Otichillo, W., and Sinange, R. 1991. *Long Rains Maize and Wheat Production in Kenya in 1990*. Technical Report No. 140. Nairobi: Ministry of Planning and National Development, Department of Remote Sensing and Resource Surveys.

Pachai, B. 1973. "Land Policies in Malawi: An Examination of the Colonial Legacy." *Journal of Modern African History* 14(4): 681–698.

Page, S., and Chonyera, P. 1994. "The Promotion of Maize Fertiliser Packages: A Cause of Household Food Insecurity and Peasant Impoverishment in High Rainfall Areas of Zimbabwe." *Development Southern Africa* 11: 301–320.

Pardey, P. G., and Roseboom, J. 1991. *National Agricultural Research from a Regional and Agroecological Perspective*. Working Paper No. 40. The Hague: International Service for National Agricultural Research (ISNAR).

Pardey, P. G.; Roseboom, J.; and Anderson, J. 1991. "Regional Perspectives on Agricultural Research," in P. G. Pardey, J. Roseboom, and J. R. Anderson (eds.), *Agricultural Research Policy: International Quantitative Perspectives*. New York: Cambridge University Press.

Pardey, P. G.; Roseboom, J.; and Beintema, N. 1997. "Agricultural Research in Africa: Three Decades of Development." *World Development* 25 (3): 409–423.

Pearce, R. 1992. "Ghana," in A. Duncan and J. Howell (eds.), *Structural Adjustment and the African Farmer*. London: Overseas Development Institute (ODI).

Peters, P. E., and Herrera, M. G. 1989. "Cash Cropping, Food Security, and Nutrition: The Effects of Agricultural Commercialization Among Smallholders in Malawi." Mimeo. Cambridge: Harvard Institute for International Development (HIID).

Piha, M. I. 1993. "Optimizing Fertilizer Use and Practical Rainfall Capture in a Semi-Arid Environment with Variable Rainfall." *Experimental Agriculture* 29: 405–415.

Pinckney, T. C. 1988. *Storage, Trade, and Price Policy Under Production Instability: Maize in Kenya*. Research Report 71. Washington, D.C.: International Food Policy Research Institute (IFPRI).

———. 1993. "Is Market Liberalization Compatible with Food Security? Storage, Trade and Price Policies for Maize in Southern Africa," in A. Váldes and K. Muir-Leresche (eds.), *Agricultural Policy Reforms and Regional Market Integration in Malawi, Zambia, and Zimbabwe*. Washington, D.C.: International Food Policy Research Institute.

Pinstrup-Andersen, P. 1993. "Fertilizer Subsidies: Balancing Short-Term Responses with Long-Term Imperatives," in N. C. Russell and C. R. Dowswell (eds.), *Policy Options for Agricultural Development in Sub-Saharan Africa.* Mexico City: Centre for Applied Studies in International Negotiations (CASIN)/Sasakawa Africa Association (SAA)/Global 2000.

Pioneer Hi-Bred International. 1993. *1993 Annual Report.* Des Moines: Pioneer Hi-Bred International.

Pixley, K. V. 1995. "CIMMYT Mid-Altitude Maize Breeding Program: Report of Activities During 1993/94," in *Annual Research Report CIMMYT-Zimbabwe, November 1993 to October 1994.* Harare: International Maize and Wheat Improvement Center (CIMMYT).

Place, F.; Mwanza, S.; and Kwesiga, F. 1995. "A Cost-Benefit Analysis of Improved Fallows in Eastern Province, Zambia." Mimeo.

Plahar, W. A.; Osei-Yaw, A.; Galiba, M.; and Dake, F. B. 1987. "Properties and Consumer Acceptability of Selected Varieties of Maize Grown in Ghana." Mimeo. Accra: Global 2000.

Pray, C. E., and Ramaswami, B. 1993. "A Framework for Seed Policy Analysis in Developing Countries," in J. Abbott (ed.), *Agricultural and Food Marketing in Developing Countries: Selected Readings.* Wallingford: Commonwealth Agricultural Bureau (CAB) International.

Productive Farming. October 1995. Lusaka, Zambia.

Putterman, L. 1995. "Economic Reform and Smallholder Agriculture in Tanzania: A Discussion of Recent Market Liberalization, Road Rehabilitation, and Technology Dissemination Efforts." *World Development* 23(2): 311–326.

Quinten, K., and Sterkenburg, J. 1975. "Marketing of Smallholder Agricultural Produce in Malawi: The Production, Consumption and Marketing of Maize." AES No. 17, Vol. 2. Lilongwe: Ministry of Agriculture.

Quizon, J. B. 1985. *An Economic Appraisal of Withdrawing Fertilizer Subsidies in India.* Discussion Paper No. ARU 33. Washington, D.C.: World Bank, Research Unit, Agriculture and Rural Development Department.

Rattray, A. G. H. 1969. "Advances and Achievements in Crop Research," in *Proceedings of the Conference on Research and the Farmer, Salisbury, Rhodesia, September 18–19, 1969.* Harare: Department of Research and Specialist Services (DR&SS).

———. 1988. "Maize Breeding and Seed Production in Zimbabwe up to 1970," in *Proceedings of the Eighth South African Maize Breeding Symposium, March 15–17, 1988.* Pretoria: Department of Agriculture and Water Supply.

Ristanovic, D.; Gibson, P.; and Rao, K. N. 1986. "Development and Evaluation of Maize Hybrids in Zambia," in B. Gelaw (ed.), *To Feed Ourselves: Proceedings of the First Eastern, Central and Southern Africa Regional Maize Workshop.* Harare: International Maize and Wheat Improvement Center (CIMMYT).

Robertson, A. F. 1987. *The Dynamics of Productive Relationships: African Share Contracts in Comparative Perspective.* Cambridge: Cambridge University Press.

Rohrbach, D. D. 1989. *The Economics of Smallholder Maize Production in Zimbabwe: Implications for Food Security.* MSU International Development Paper No. 11. East Lansing: Michigan State University.

Roseboom, J.; Pardey, P. G.; Beintema, N. M.; and Mudimu, G. D. 1995. *Statistical Brief on the National Agricultural Research System of Zimbabwe.* The Hague: International Service for National Agricultural Research (ISNAR).

Rosegrant, M. W.; Agcaoili, M. C.; and Perez, N. D. 1995. *Global Food Projections to 2020: Implications for Investment.* Washington, D.C.: International Food Policy Research Institute (IFPRI).

RSA (Republic of South Africa). 1994. *Abstract of Agricultural Statistics.* Pretoria: Directorate of Agricultural Information.

Rubey, L. 1993. "Consumer Maize Meal Preferences in Zimbabwe: Survey Results and Policy Implications." Report prepared for the Ministry of Lands, Agriculture and Water Development, Harare, Zimbabwe.

———. 1995. "Maize Market Reform in Zimbabwe: Interactions Between Consumer Preferences, Small-Scale Enterprise Development and Alternative Marketing Channels." Ph. D. thesis, Department of Agricultural Economics, Michigan State University, East Lansing, Michigan.

Rubey, L., and Lupi, F. 1995. *Predicting the Effects of Market Reform in Zimbabwe: A Stated Preference Approach.* Staff Paper 95-16. East Lansing: Michigan State University, Department of Agricultural Economics.

Rukuni, M. 1996. *A Framework for Crafting Demand-Driven National Agricultural Research Institutions in Southern Africa.* Staff Paper No. 96-76. East Lansing: Michigan State University, Department of Agricultural Economics.

Rukuni, M., and Eicher, C. K. (eds.). 1994. *Zimbabwe's Agricultural Revolution.* Harare: University of Zimbabwe Publications.

Rusike, J. 1995. "An Institutional Analysis of the Maize Seed Industry in Southern Africa." Ph. D. thesis, Department of Agricultural Economics, Michigan State University, East Lansing, Michigan.

———. 1996a. "Constraints, Opportunities and Strategies for Increasing Seed Sector Performance in Zimbabwe and Zambia." Mimeo. East Lansing: Michigan State University.

———. 1996b. "Zimbabwe and Zambia Soil Fertility Case Study." Mimeo. East Lansing: Michigan State University.

Sahn, D., and Delgado, C. 1987. "The Nature and Implications for Market Interventions of Seasonal Food Price Variability," in D. Sahn (ed.), *Seasonal Variability in Third World Agriculture: The Consequences for Food Security.* Baltimore: Johns Hopkins.

Sahn, D., and Arulpragasam, J. 1991. "The Stagnation of Smallholder Agriculture in Malawi: A Decade of Structural Adjustment." *Food Policy* 16: 219–234.

Saka, A. R.; Bunderson, W. T.; Itimu, O. A.; Phombeya, H. S. K.; and Mbekeani, Y. 1994. "The Effects of *Acacia albida* on Soils and Maize Grain Yields Under Smallholder Farm Conditions in Malawi." *Forest Ecology and Management* 64: 217–230.

Sakala, W. D. M. 1994. "Crop Management Interventions in Traditional Maize Pigeonpea Intercropping Systems in Malawi." M. Sc. thesis, Bunda College of Agriculture, University of Malawi, Lilongwe, Malawi.

Sanders, J.; Bezuneh, T.; and Shroeder, A. C. 1994. *Impact Assessment of the SAFRAD Commodity Networks.* Washington, D.C.: U. S. Agency for International Development, Africa Bureau, Office of Analysis, Research and Technical Support, Division of Food, Agriculture and Resource Analysis (AFR/ARTS/FARA).

Sarris, A. 1992. *Options for Public Intervention to Enhance Food Security in Ghana.* Cornell Food and Nutrition Policy Program Monograph No. 14. Ithaca: Cornell Food and Nutrition Policy Program (CFNPP).

Saunders, A. R. Undated. "Maize in South Africa." Johannesburg: Central News Agency.

Scarborough, V. 1990. *Agricultural Policy Reforms Under Structural Adjustment in Malawi.* Occasional Paper 12. London: Department of Agricultural Economics, Wye College.

———. 1996. *Farmer-Led Approaches to Extension: Papers Presented at a Workshop in the Philippines, July 1995.* Agricultural Research and Extension

Network, Network Papers 59a, 59b, and 59c. London: Overseas Development Institute (ODI).

Schiff, M., and Valdés, A. 1992. *The Political Economy of Agricultural Pricing Policy: A Synthesis of the Economics in Developing Countries*. Baltimore: Johns Hopkins.

Schultz, T. W. 1964. *Transforming Traditional Agriculture*. New Haven: Yale University.

Scott, G. 1995. "Agricultural Transformation in Zambia: Past Experience and Future Prospects." USAID/Africa Bureau/Food Security and Productivity Unit Workshop, Agricultural Transformation in Africa, 26–29 September, Abidjan.

Segura, E. L.; Shetty, Y. T.; and Nishimizu, M. (eds.). 1986. *Fertilizer Producer Pricing in Developing Countries: Issues and Approaches*. Industry and Finance Series, Vol. 11. Washington, D.C.: World Bank.

Shaffer, J. D.; Weber, M.; Riley, H.; and Staatz, J. 1985. "Influencing the Design of Marketing Systems to Promote Development in Third World Countries," in *Agricultural Markets in the Semi-Arid Tropics: Proceedings of the International Workshop, 24–28 October 1983*. Pantacheru: International Center for Research in the Semi-Arid Tropics (ICRISAT).

Shalit, H., and Binswanger, H. P. 1985. *Is There a Theoretical Case for Fertilizer Subsidies?* Discussion Paper No. ARU 27. Washington, D.C.: World Bank.

Shepherd, A. 1989. "Approaches to the Privatization of Fertilizer Marketing in Africa." *Food Policy* 14(2): 143–154.

Shepherd, A., and Coster, R. 1987. *Fertilizer Marketing Costs and Margins in Developing Countries*. Rome: Food and Agriculture Organization (FAO) and Fertilizer Industry Advisory Council (FIAC).

Short, K. E., and Edmeades, G. O. 1991. "Maize Improvement for Water and Nitrogen Deficient Environments," in J. F. MacRobert (ed.), *Proceedings of the Crop Science Society of Zimbabwe Twenty-First Anniversary Crop Production Congress*. Harare: Crop Science Society of Zimbabwe.

Shumba, E. 1989. "Maize Technology Research in Mangwende, a High Potential Communal Area Environment in Zimbabwe. Part 2: The On-Farm Experimental Program." *Farming Systems Bulletin* 1: 1–13.

Shumba, E. M.; Waddington, S. R.; and Rukuni, M. 1992. "Use of Tine-Tillage, with Atrazine Weed Control, to Permit Earlier Planting of Maize by Small-Holder Farmers in Zimbabwe." *Experimental Agriculture* 28(4): 443–452.

Sibale, E. M. 1988. "Pounding Experiment on Maize," in "Maize Breeding Annual Report for the 1987/88 Season." Mimeo. Lilongwe: Maize Research Program.

Sipula, K. 1993. "Reforms of the Maize Market System in Zambia: Issues of Price and Market Policies, Cooperatives and Interprovincial Transportation." Ph. D. thesis, Department of Agricultural Economics, Michigan State University, East Lansing, Michigan.

Smale, M. 1995. "'Maize Is Life': Malawi's Delayed Green Revolution." *World Development* 23(5): 819–831.

Smale, M.; with Kaunda, Z. H. W.; Makina, H. L.; Mkandawire, M. M. M. K.; Msowoya, M. N. S.; Mwale, D. J. E. K.; and Heisey, P. W. 1991. "Chimanga Cha Makolo," in *Hybrids and Composites: An Analysis of Farmers' Adoption of Maize Technology in Malawi, 1989–91*. CIMMYT Economics Working Paper 91/04. Mexico City: International Maize and Wheat Improvement Center (CIMMYT).

Smale, M.; Kaunda, Z. H. W.; Makina, H. L.; and Mkandawire, M. M. M. K. 1993. *Farmers' Evaluation of Newly Released Maize Cultivars in Malawi: A Comparison of Local Maize, Semi-Flint and Dent Hybrids*. Lilongwe and Harare: International Maize and Wheat Improvement Center (CIMMYT).

Smale, M., and Heisey, P. W. 1994. "Maize Research in Malawi Revisited: An Emerging Success Story?" *Journal of International Development* 6(6): 689–706.

Smale, M.; Just, R.; and Leathers, H. D. 1994. "Land Allocation Decisions in HYV Adoption Models." *American Journal of Agricultural Economics* 76(3): 535–545.

Smaling, E. 1993. "Soil Nutrient Depletion in Sub-Saharan Africa," in H. Van Reuler and W. Prins (eds.), *The Role of Plant Nutrients for Sustainable Food Crop Production in Sub-Saharan Africa*. Leidschendam: Dutch Association of Fertilizer Producers.

Smith, J.; Woodworth, J. B.; and Dashiell, K. E. 1993. "Government Policy and Farm Level Technologies: The Expansion of Soyabean in Nigeria." *Agricultural Systems in Africa* 3: 20–32.

Smith, J.; Barau, A. D.; Goldman, A.; and Mareck, J. H. 1994a. "The Role of Technology in Agricultural Intensification: The Evolution of Maize Production in the Northern Guinea Savanna of Nigeria." *Economic Development and Cultural Change* 42(3): 537–554.

Smith, J.; Ogungbile, A. O.; Kling, J. G.; and Kim, S. K. 1994b. *Sustaining West Africa's Maize Revolution: Can Hybrid Maize Make a Contribution?* Ibadan: International Institute of Tropical Agriculture (IITA).

Smith, J., and Weber, G. 1994. "Strategic Research in Heterogeneous Mandate Areas," in J. R. Anderson (ed.), *Agricultural Technology: Policy Issues for the International Community*. Wallingford: Commonwealth Agricultural Bureau (CAB) International.

Smith, J., and Goldsmith, W. D. 1995. "Agricultural Transformation in India and Northern Nigeria: Exploring the Nature of Green Revolutions." *World Development* 23(2): 243–264.

Smith, L. D. 1995. "Malawi: Reforming the State's Role in Agricultural Marketing." *Food Policy* 20(6): 561–571.

Smith, R. C. 1979. *The Story of Maize and the Farmers' Co-op Limited*. Harare: Farmers' Co-op.

Snapp, S. S. 1995. "Improving Fertilizer Efficiency with Small Additions of High Quality Organic Inputs," in S. R. Waddington (ed.), *Report on the First Meeting of the Network Working Group. Soil Fertility Research Network for Maize-Based Farming Systems in Selected Countries of Southern Africa*. Lilongwe and Harare: Rockefeller Foundation Southern Africa Agricultural Sciences Program and the International Maize and Wheat Improvement Center (CIMMYT) Maize Program.

Sodhi, A. J. S. 1993. "Reforming the Agribusiness Markets: Case of Fertilizers from East Africa." International Training Workshop on Deregulation and Privatization Policies to Reform Agribusiness Markets, 19–30 April, Muscle Shoals, Alabama, and Washington, D.C.

Southworth, V. R.; Jones, W. O.; and Pearson, S. R. 1979. "Food Crop Marketing in Atebubu District, Ghana." *Food Research Institute Studies* 17(2): 157–195.

SPAAR (Special Program for African Agricultural Research). 1995. *Lessons Learnt from Implementation of the Frameworks for Action (FFAs)*. Washington, D.C.: World Bank.

Spencer, D. S. C., and Badiane, O. 1995. "Agriculture and Economic Recovery in African Countries," in G. H. Peters and D. D. Hedley (eds.), *Agricultural Competitiveness: Market Forces and Policy Choices. Proceedings of the Twenty-Second International Conference of Agricultural Economists, Held in Harare, Zimbabwe, August 22–29, 1994*. Aldershot: Darthmouth.

Sperling, D.; Sitch, L.; White, J.; and Zandamela, C. B. 1995. "The Evaluation of Open Pollinated Maize Varieties Under a Range of Zero Input Farming Systems in Central Mozambique," in *Maize Research for Stress Environments.* Proceedings of the Fourth Eastern and Southern Africa Regional Maize Conference, Mexico D. F., International Maize and Wheat Improvement Center (CIMMYT).

Spurling, A. T.; Pee, T.; Mkamanga, G.; and Nkwanyana, C. 1992. *Agricultural Research in Southern Africa: A Framework for Action.* Discussion Paper No. 184. Washington, D.C.: World Bank.

Stack, J. L. 1994. "The Distributional Consequences of the Smallholder Maize Revolution," in M. Rukuni and C. K. Eicher (eds.), *Zimbabwe's Agricultural Revolution.* Harare: University of Zimbabwe Publications.

Steffen, P. 1995. "The Roles and Limits of the Cereals Market in Assuring Food Security in Northeastern Mali." Ph. D. thesis, Department of Agricultural Economics, Michigan State University, East Lansing, Michigan.

Swift, M. J., and Woomer, P. 1993. "Organic Matter and the Sustainability of Agricultural Systems: Definition and Measurement," in K. Mulongoy and R. Merckx (eds.), *Soil Organic Matter Dynamics and Sustainability of Tropical Agriculture.* Chichester: Wiley-Sayce.

Tattersfield, R. 1982. "The Role of Research in Increasing Food Crop Potential in Zimbabwe." *Zimbabwe Science News* 16(1): 6–10.

Tattersfield, R., and Havazvidi, E. K. 1994. "The Development of the Seed Industry," in M. Rukuni and C. K. Eicher (eds.), *Zimbabwe's Agricultural Revolution.* Harare: University of Zimbabwe Publications.

Taylor, D. R. F. 1969. "Agricultural Change in Kikuyaland," in M. F. Thomas and G. W. Whittington (eds.), *Environment and Land Use in Africa.* London: Methuen.

Tegemeo Institute. 1996. Unpublished database on maize milling charges for whole meal. Nairobi, Egerton University/Tegemeo Institute for Agricultural Policy and Development.

Temu, A. E. M. 1982. "Rotational Green Manuring with *Crotalaria* in the Southern Highlands of Tanzania: Southern Highlands Maize Improvement Program Progress Report." Mimeo. Iringa, Tanzania: Southern Highlands Maize Improvement Program.

Thornton, D. 1973. *Agriculture in South East Ghana, Volume 1.* Development Study No. 12. Reading: Department of Agricultural Economics and Management, University of Reading.

Tomek, W. G. 1980. "Price Behavior on a Declining Terminal Market." *American Journal of Agricultural Economics* 62(3): 433–444.

Tomich, T. P.; Kilby, P.; and Johnston, B. F. 1995. *Transforming Agrarian Economies: Opportunities Seized, Opportunities Missed.* Ithaca: Cornell University Press.

Tripp, R. 1992. "Expectations and Realities in On-Farm Research." *Farming Systems Bulletin* 11: 1–13.

———. 1993. "Invisible Hands, Indigenous Knowledge and Inevitable Fads: Challenges to Public Sector Agricultural Research in Ghana." *World Development* 21(12): 2003–2016.

Tripp, R.; Marfo, K.; Dankyi, A. A.; and Read, M. 1987. *Changing Maize Production Practices of Small-Scale Farmers in the Brong-Ahafo Region, Ghana.* Mexico City: Ghana Grains Development Project (GGDP).

Truong, T. V., and Walker, S. T. 1990. "Policy Reform as Institutional Change: Privatizing the Fertilizer Subsector in Cameroon," in D. Brinkerhoff and A. Goldsmith (eds.), *Institutional Sustainability in Agriculture and Rural Development: A Global Perspective.* New York: Praeger.

Tshibaka, T. B., and Baanante, C. A. (eds.). 1988. *Fertilizer Policy in Tropical Africa.* Lomé: International Fertilizer Development Center (IFDC) and International Food Policy Research Institute (IFPRI).

Urban, F., and Nightingale, R. 1993. *World Population by Country and Region, 1950–1990 and Projections to 2050.* Staff Report No. AGES 9306. Washington, D.C.: U. S. Department of Agriculture.

USAID (United States Agency for International Development). 1985. *Tanzania Seed Multiplication.* USAID Project Impact Evaluation Report No. 55. Washington, D.C.: Government Printing Office.

———. 1995. *Market Price Survey.* Nairobi: USAID/Kenya.

USDA (U. S. Department of Agriculture). 1981. *Food Problems and Prospects in Sub-Saharan Africa: The Decade of the 1980s.* Washington, D.C.: USDA.

———. 1992. *World Indices of Agricultural and Food Production, 1980–89.* Statistical Bulletin No. 759. Washington, D.C.: USDA, Economic Research Service, Agricultural Trade Analysis Division.

UZ/RF LTSP (University of Zimbabwe/Rockefeller Foundation Land Tenure Studies Project). 1995. *Proceedings of the Workshop "Agricultural Service Institutions and Smallholder Development," Leopard Rock Motel, Mutare, March 1995.* Harare: University of Zimbabwe, Department of Agricultural Economics and Extension.

van der Bijl, G. 1987. *Farming Systems in a Changing Policy Environment: A Study of the Cassava and Maize Agricultural Economy in Kaoma District, Western Province.* Mongu and Lusaka: Provincial Planning Unit and University of Zambia.

Van der Walt, W. J. 1990. "The Role of Sansor in the Seed Trade and Research," in *Proceedings of the Ninth South African Maize Breeding Symposium.* Pretoria: Department of Agricultural Development.

Venkatesan, V. 1994. *Seed Systems in Sub-Saharan Africa: Issues and Options.* Washington, D.C.: World Bank.

Vercruijsse, E. 1988. *The Political Economy of Peasant Farming in Ghana.* Occasional Paper No. 106. The Hague: Institute of Social Studies.

Versteeg, M., and Koudokpon, V. 1993. "Participative Farmer Testing of Four Low External Input Technologies to Address Soil Fertility Decline in Mono Province (Benin)." *Agricultural Systems* 42: 265–276.

Viyas, V. S. 1983. "Asian Agriculture: Achievements and Challenges." *Asian Development Review* 1: 27–44.

Vlek, P. L. G. 1990. "The Role of Fertilizers in Sustaining Agriculture in Sub-Saharan Africa." *Fertilizer Research* 26: 327–339.

von Braun, J., and Puetz, D. 1987. "An African Fertilizer Crisis: Origin and Economic Effects in the Gambia." *Food Policy* 12(4): 337–348.

Waddington, S. R., and Ransom, J. K. 1995. "Linking Soils, Agronomy, and Crops Research for Maize: CIMMYT Initiatives in Southern and Eastern Africa." Conference on Soils Management in Eastern and Southern Africa, 30 January–2 February, Bellagio, Italy.

Weber, G.; Elemo, K.; and Lagoke, S. T. O. 1995. "Weed Communities in Intensified Cropping Systems of the Northern Guinea Savanna." *Weed Research* 35: 167–178.

Weber, G.; Elemo, K.; Lagoke, S. T. O.; Awad, A.; and Oikeh, S. 1995. "Population Dynamics and Determinants of *Striga hermonthica* on Maize and Sorghum in Savanna Farming Systems." *Crop Protection* 14: 283–290.

Weber, G., and Smith, J. 1995. *Targeting Research on Soil-Borne Constraints to Resource and Farming Domains in the Northern Guinea Savanna of West Africa.* Ibadan: International Institute of Tropical Agriculture (IITA).

Weber, G.; Smith, J.; and Manyong, V. M. 1996. "System Dynamics and the Definition of Research Domains for the Northern Guinea Savanna of West Africa." *Agriculture, Ecosystems and Environment* 57: 133–148.

Weber, M. T.; Staatz, J. M.; Holtzman, J. S.; Crawford, E. W.; and Bernsten, R. H. 1988. "Informing Food Security Decisions in Africa: Empirical Analysis and Policy Dialogue." *American Journal of Agricultural Economics* 70(5): 1044–52.

Weber, M. T.; Tschirley, D.; Varela, R.; Santos, A. P.; and F. De Marrule, H. 1992. "Reflections on Relationships Between Food Aid and Maize Pricing/Marketing in Maputo, Mozambique." Mimeo. Ministry of Agriculture, Michigan State University (MOA/MSU) Food Security Project.

Weinmann, H. 1972. *Agricultural Research and Development in Southern Rhodesia, 1890–1923.* Department of Agriculture, Occasional Paper No. 4. Salisbury: University of Rhodesia.

———. 1975. *Agricultural Research and Development in Southern Rhodesia: 1924–1950.* Series in Science, No. 2. Salisbury: University of Rhodesia.

Wendt, J. W.; Jones, R. B.; and Itimu, O. A. 1994. "An Integrated Approach to Soil Fertility Improvement in Malawi, Including Agroforestry," in E. T. Craswell and J. Simpson (eds.), *Soil Fertility and Climatic Constraints in Dryland Agriculture.* ACIAR Proceedings No. 54. Canberra: Australian Centre for International Agricultural Research (ACIAR).

Westlake, M. 1994. *The Impact of Deficient Commodity Pricing and Payment Systems and Delay in the Payment of Farmers: Lessons from Kenya.* Development Discussion Paper No. 475. Cambridge: Harvard Institute for International Development (HIID).

Wiggins, S., and Cromwell, E. 1995. "NGOs and Seed Provision to Smallholders in Developing Countries." *World Development* 23(3): 413–422.

Willey, R. W. 1979. "Intercropping—Its Importance and Research Needs. Part I. Competition and Yield Advantages." *Field Crops Abstracts* 32: 1–10.

Williams, L. B., and Allgood, J. H. 1990. *Fertilizer Situation and Markets in Malawi.* Muscle Shoals, Alabama: International Fertilizer Development Center (IFDC).

Williamson, O. E. 1990. "Political Institutions: The Neglected Side of the Story: Comment." *Journal of Law, Economics and Organization* 6: 263–266.

Woomer, P. L.; Martin, A.; Albrecht, A.; Resck, D. V. S.; and Scharpenseel, H. W. 1994. "The Importance and Management of Soil Organic Matter in the Tropics," in P. L. Woomer and M. J. Swift (eds.), *The Biological Management of Tropical Soil Fertility.* Chichester: Wiley-Sayce.

World Bank. 1981. *Accelerated Development in Sub-Saharan Africa.* Washington, D.C.: World Bank.

———. 1994a. *Adjustment in Africa: Reforms, Results and the Road Ahead.* New York: Oxford University Press.

———. 1994b. *Agricultural Extension: Lessons from Completed Projects.* Washington, D.C.: World Bank.

———. 1995. *Kenya Poverty Assessment.* Report No. 13152-KE. Washington, D.C.: World Bank.

———. 1996a. *World Development Report, 1996.* New York: Oxford University Press.

———. 1996b. *Taking Action for Poverty Reduction in Sub-Saharan Africa: Report of an Africa Region Task Force.* Report No. 15575-AFR. Washington, D.C.: World Bank.

Wright, P. D., and Nieuwoudt, W. L. 1993. "Price Distortions in the South African Maize Economy: A Comparative Political Analysis." *Agrekon* 32(2): 51–59.

Young, A. 1989. *Agroforestry for Soil Conservation.* Wallingford: Commonwealth Agricultural Bureau (CAB) International.

Yudelman, M.; Coulter, J.; Goffin, P.; McCune, D.; and Ocloo, E. 1991. "An Evaluation of the Sasakawa-Global 2000 Project in Ghana," in N. C. Russell and C. R. Dowswell (eds.), *Africa's Agricultural Development in the 1990s: Can It Be Sustained?* Mexico City: Centre for Applied Studies in International Negotiations (CASIN)/Saskawa African Association (SAA)/Global 2000.

Zamseed. 1982–1994. "Sales Data." Mimeo. Lusaka: Zamseed.

Zimbabwe, Government of. 1991. *Estimates of Expenditures for the Year Ending June 30, 1992.* Harare: Government Printer.

———. 1995. *The Agricultural Sector of Zimbabwe.* Harare: Ministry of Agriculture.

Contributors

Malcolm J. Blackie is a senior scientist with the Rockefeller Foundation Southern Africa Agricultural Sciences Program, Malawi.

Derek Byerlee is principal economist in the Agricultural and Natural Resources Department, World Bank, Washington, D.C.

Carl K. Eicher is University Distinguished Professor in the department of Agricultural Economics at Michigan State University, East Lansing, Michigan.

M. A. B. Fakorede is a maize breeder at Obafemi Awolowo University, Ile-Ife, Nigeria.

Rashid M. Hassan is a professor in the Department of Agricultural Economics, University of Pretoria, South Africa.

Paul W. Heisey is an economist with the International Maize and Wheat Improvement Center (CIMMYT) and is stationed in Mexico.

Julie A. Howard is an assistant professor in the Department of Agricultural Economics, Michigan State University, East Lansing, Michigan.

Thomas S. Jayne is an associate professor in the Department of Agricultural Economics, Michigan State University, East Lansing, Michigan.

David Jewell is a maize breeder with the International Maize and Wheat Improvement Center (CIMMYT) and is stationed in Harare, Zimbabwe.

Share Jiriyengwa is deceased. He was formerly the head of the Planning Unit, Grain Marketing Board, Zimbabwe.

Richard B. Jones is an agronomist with the International Crops Research Institute for Semi-Arid Tropics (ICRISAT), Nairobi.

Stephen Jones is a senior research officer with the Food Studies Group, Oxford University.

Daniel D. Karanja is an agricultural economist with the Kenya Agricultural Research Institute (KARI), Kenya.

John D. T. Kumwenda is a senior agronomist with the Maize Commodity Team, Ministry of Agriculture and Livestock Development, Malawi.

Bernard Kupfuma is an agricultural economist with the Department of Research and Specialist Services, Ministry of Agriculture, Harare, Zimbabwe.

M. V. Manyong is an agricultural economist with the International Institute of Tropical Agriculture (IITA) and is stationed in Ibadan, Nigeria.

Kofi Marfo is an economist with the Crops Research Institute, Kumasi, Ghana.

Mulinge Mukumbu is a research scholar with Egerton University/Policy Analysis Matrix Project, Nairobi, Kenya.

Catherine Mungoma is principal agricultural research officer in the Research Branch of the Department of Agriculture, Ministry of Agriculture, Food, and Fisheries, Lusaka, Zambia.

Wilfred Mwangi is an economist with the International Maize and Wheat Improvement Center (CIMMYT) and is stationed in Addis Ababa, Ethiopia.

Lawrence Rubey is an agricultural economist with the U.S. Agency for International Development.

Joseph Rusike is a lecturer in the Department of Agricultural Economics and Extension, University of Zimbabwe, Harare.

Melinda Smale is an economist with the International Maize and Wheat Improvement Center (CIMMYT) and is stationed in Mexico.

Joyotee Smith is an economist with the Center for International Forestry Research (CIFOR) and is stationed in Bogor, Indonesia. She conducted the research described in this book under a previous appointment with the International Institute for Tropical Agriculture (IITA), Ibadan, Nigeria.

Sieglinde S. Snapp is a soil scientist with the International Crops Research Institute for Semi-Arid Tropics (ICRISAT), Malawi.

Robert Tripp is a research fellow with the Overseas Development Institute, London, England.

David Tschirley is an associate professor in the Department of Agricultural Economics, Michigan State University, East Lansing, Michigan,

Stephen R. Waddington is an agronomist with the International Maize and Wheat Improvement Center (CIMMYT) and coordinator of the Soil Fertility Network for Maize-Based Farming Systems in Countries of Southern Africa and is stationed in Zimbabwe.

Richard W. Ward is an associate professor of Crop and Soil Sciences, Michigan State University, East Lansing, Michigan.

Georg Weber is with PASOLAC (Programa de Agricultura Sostenible en Laderas de América Central), Nicaragua.

Index

Africa, 3–22, 247–262; agricultural research in, 6, 254, 256; colonial governments in, 19–20; extension in, 137–141, 257; fertilizer marketing, 258; fertilizer use, 6, 12, 193–211, 249; food consumption, 3, 10; food crisis in, 3–4, 21, 122, 247, 260; food imports, 11; food policy debates, 5–7; food production, 3–4, 5, 7, 10, 12; improved maize varieties, 249; infrastructure, 203; maize consumption, 16–17; maize economy, 4–5, 9–22; maize markets, 19–20; maize production, 16, 17–19, 247–262; maize revolution, 248–251; maize research, 127–137, 142; marketing reforms, 4, 7, 20–21, 259–260; producers in, 15; seed industry, 176–177; soil fertility management, 157–172; yields, 17, 18, 250. *See also* Ghana; Kenya; Malawi; Nigeria; Zambia; Zimbabwe
Africa Pacific Seeds, 181, 183
Agricultural Development and Marketing Corporation (ADMARC), 69, 70–71, 72, 76
Agricultural research. *See* Maize research; Soil fertility, research
Agricultural systems: evolutionary paths, 108–111; market-driven, 108–111, 114, 116, 118, 121, 122; population-driven, 108–111, 114, 117, 122; in western Africa, 107–111, 118
Agroecological regions, 13–15, 103–104, 111–118; forest areas, 96, 103, 104, 108, 111, 112, 116; in Ghana, 96–97, 103; in Kenya, 82, 84; savanna areas, 16, 97, 103, 104, 108, 111–118, 121–123, 251; transition zone, 96, 97, 103, 104; in western Africa, 107–111; in Zambia, 46; in Zimbabwe, 26–27
Agroforestry systems, 164–165
Angola, 18
Animal traction, 132
Arnold, H. C., 29
Asia: green revolution in, 3, 4, 25, 157–158, 204

Berg Report, 213, 225
Burundi, 13

Cameroon, 207–208
Cargill Hybrid Seeds, 70, 77, 183, 184, 185, 186, 187
Carnia-Asgrow Seed, 181, 187
Carter Center, 100
Cassava, 96, 115
Chiluba, Frederick, 55
CIMMYT. *See* International Maize and Wheat Improvement Center
Cocoa, 97
Commercial maize producers, 27–30, 240–241; access to credit, 89–90; in Ghana, 104; implicit taxation of, 223–224; in South Africa, 181; technology adoption, 82, 84; in Zambia, 46, 59, 185; in Zimbabwe, 26, 27–30, 33, 35, 41, 148–149
Consumer maize preferences, 145–155, 259; eliciting information on,

295

150–153, 154; in Mozambique, 152–153; in Zimbabwe, 150, 152
Contingent valuation techniques, 150–153, 154
Credit policies, 207; in Kenya, 85, 86, 89–90; in Malawi, 69, 202; in Zambia, 52, 56–57, 59; in Zimbabwe, 32–33, 223
Crop management, 131–132, 142
Crop rotations, 163–165

Demand. *See* Maize consumption
Donor policies, 4, 192, 206–207, 213, 224–225, 253, 254, 256, 261
Dualistic agricultural structure, 26, 58, 215, 238
Drought, 147, 152, 234

Eastern Africa. *See* Kenya; Tanzania
Ellis, R. T., 68
Ethiopia, 11, 129, 157, 195, 207
European farmers, 215, 216, 220, 240, 241–242
Exchange rate overvaluation, 203, 232
Extension programs, 137–141; farming systems' approach, 138; funding, 141; in Ghana, 99–100; institutional issues, 140; in Kenya, 89, 91, 139; Sasakawa-Global 2000 (SG 2000), 34, 100, 104, 140; Training and Visit (T&V), 85, 89, 99, 138–140, 141, 257

FAO. *See* Food and Agriculture Organization
Farmer support organizations, 41–42, 59, 92
Fertilizer: adoption, 194, 199, 204–205; availability, 162, 202; chemical, 6, 131, 160–161, 169, 194; distribution systems, 69, 202–203, 206–207, 208, 258–259; donor-financed, 202–203, 206–207; market reforms, 56, 207–210; organic, 162–165, 170; prices, 72, 199–201, 202–205, 258–259; production, 71, 194, 208; recommendations, 70, 131, 160–161, 167–168; research priorities, 208–210; supply, 202–204, 206–207; technical response to, 74–75, 101, 109, 118, 161, 193, 197–199, 205–206
Fertilizer subsidies, 53, 204–208, 209, 259; in Ghana, 100–101; in Malawi, 69, 72; in Nigeria, 113, 114, 117–118, 123
Fertilizer use, 160–165, 170, 193–211, 249; application rates, 195, 197, 201; disequilibrium in, 195; efficiency, 160–162; in Ghana, 100–101, 103, 105, 205; in Kenya, 82, 84, 86–87, 197; in Malawi, 70, 75, 160, 197, 205; in Nigeria, 115, 116–117, 197, 205; profitability, 169, 199; risk aversion, 201–202; by smallholders, 33–34; in Zambia, 53–54, 57–58, 197, 205, 234; in Zimbabwe, 33–34, 160–161, 197, 234. *See also* Soil Fertility; Yields
Food and Agriculture Organization (FAO), 49, 167
Food-deficit households, 74–75, 236
Food security, 21, 130, 154–155, 236, 238, 259; in Kenya, 93; urban, 230; in western Africa, 107, 122–123

Ghana, 95–106; agroecological regions in, 96–97, 104; extension programs, 99–100, 104; fertilizer subsidies, 100–101; fertilizer use, 100–101, 103, 105, 205; improved maize varieties, 99, 101, 103; increasing maize productivity, 104–106; land issues, 104, 105; maize consumption, 95–96, 104; maize economy, 95–97; maize marketing, 95, 101–102; maize prices, 21, 101–102; maize production, 96–97, 99, 250; maize research, 97–99, 105–106; maize technology adoption, 102–104; producers in, 104; reform in, 21, 105; seed distribution network, 100; soil fertility, 197; yields, 103
Ghana Grains Development Project (GGDP), 97–98
Green revolution, 25, 27–34, 37, 41–42, 183; in Asia, 3, 4, 157–158, 204; fiscal sustainability, 37; preconditions for, 35–36; role of state, 34–35

Hand-hoe agriculture, 15, 132, 255
Harrison, M. N., 82
Hoyle, S., 68
Hybrid maize varieties, 112, 129, 148, 149, 176, 181, 253; adoption patterns, 75–76; fertilizer responsiveness, 74–75; in Kenya, 82–87; in Malawi,

67, 68–69, 70, 72–78; profitability of, 75; smallholders, 134; yields, 67–68, 73–74, 75, 77–78, 118; in Zambia, 50; in Zimbabwe, 25, 29–32, 40. *See also* Improved maize varieties; Maize research; Maize technology adoption, OPVs; SR 52

Improved maize varieties, 129–131, 249; demand, 189; in Kenya, 183–184; in Malawi, 185; in Nigeria, 112–113; producer preferences, 131; resistance of, 135; in South Africa, 181; yields, 103, 133, 142. *See also* Hybrid maize varieties; Maize research; Maize technology adoption; SR 52; Yields
Infrastructure, 113, 123, 208, 203, 249; improvement, 240; investment in 20, 260; in Kenya, 87–88
India, 197, 247
Intercropping, 164
International agricultural research centers (IARCs), 128, 129, 135, 136
International Institute of Tropical Agriculture (IITA), 112, 113, 116, 128, 165
International Maize and Wheat Improvement Center (CIMMYT), 15, 49, 68, 128, 166
International Monetary Fund (IMF), 54
Investment, 20, 60, 205, 240

Kaunda, Kenneth, 51, 54
Kenya, 9–10, 81–93; agroecological regions, 82, 84; credit policies, 85, 86; extension services, 85, 89, 137, 139; fertilizer use, 82, 84, 86–87, 197; hybrid maize varieties, 82–87; improved maize varieties, 183–184; infrastructure, 87–88; maize consumption, 16, 81; maize production, 81–82, 87, 90–93, 184, 232–233, 249; maize research, 82–84, 87, 91–92, 133, 134, 184; maize technology adoption, 82, 84, 85; marketing policies, 89–90, 216, 217, 220, 221, 223, 225, 226; market reforms, 90, 91, 93; population growth, 91; public investment, 91; seed industry in, 87–88, 183; soil fertility, 92; yields, 82–87, 184

Kenya Agricultural Research Institute (KARI), 84, 85, 87, 89, 92
Kenya Seed Company, 88, 183, 184

Labor issues, 132, 159, 255
Land use, 64–66, 104, 247; expansion phase, 108–109; intensification phase, 11–12, 108–109
Land tenure, 105
Latin America, 130, 204
Legumes, 116, 121, 163–165, 169–170
Lesotho, 129
Lever Brothers, 70, 86

Maize, area planted to. *See* Maize production
Maize attributes, 67, 146–150, 154
Maize consumption, 16–17; demand, 17, 151, 239; in Ghana, 95–96, 104; in Kenya, 16, 81; in Malawi, 63, 66–67; regional, 16; urban, 150–153, 230; in Zambia, 46, 51. *See also* Consumer maize preferences
Maize economy, 4–5, 9–22; evolution of, 9–10; in Ghana, 95–97; in Nigeria, 111–114; in Zimbabwe, 26–27
Maize prices, 230–232; in Ghana, 21, 101–102; in Malawi, 72, 74; in Mozambique, 152–153; in Zimbabwe, 151. *See also* Price instability; Price policies
Maize processing, 67, 153, 219, 229–230
Maize production, 13–19, 21, 193–211, 247–262; in Ghana, 96–97, 99, 104; in Kenya, 81–93, 184, 232–233, 249; in Malawi, 10, 64, 186; in Nigeria, 111, 113–117, 123, 250; in rain-fed areas, 197–198; smallholder, 30–34, 147, 181, 183, 238, 251–252; strategies to increase, 91–93, 123, 251–253; sustaining, 116–118, 123–124, 250; in Tanzania, 187; in western Africa, 107–111, 122–123; in Zambia, 45–46, 185, 232–233, 234; in Zimbabwe, 27, 31–32, 216, 219, 232–233, 249. *See also* Agroecological regions
Maize research, 127–137, 141–143, 145–155, 248–249, 256–257; capacity, 135–137; consumer preferences, 145–155; continuity, 135–136; evolution of, 127–128; funding issues, 20, 38–39, 77, 84, 87, 106, 136, 137,

256; in Ghana, 97–99, 105–106; impacts of, 129–135; intensity of, 128; in Kenya, 81, 82–84, 87, 91–92, 133, 134, 184; in Malawi, 68–69, 77, 136; management, 136; in Nigeria, 112–113, 118–122; private sector, 56, 137; public, 188, 189; returns to investment, 31, 133–134, 256; smallholder-oriented, 134, 248–249; in South Africa, 178; strategic issues, 134–135; in Zambia, 48–50, 54, 59; in Zimbabwe, 29, 133, 134, 136, 148, 160, 183. *See also* Maize technology development

Maize research priorities, 92–93, 105, 118, 121–122, 142, 153–155, 255

Maize technology adoption, 75–76, 84–87, 118–121, 141–142, 187–190, 248–251; in Ghana, 102–104; hybrid maize, 25, 29–32, 40, 48–50, 72–73, 82–84, 92, 118, 129; improved maize varieties, 114, 116, 129–131, 141, 147–149; in Kenya, 82–87, 92; in Malawi, 72–73, 75–76, 158; in Nigeria, 114, 116, 118, 119–120; OPVs, 102–103; in Zambia, 48–50; in Zimbabwe, 25, 29–32, 40, 147–149. *See also* Fertilizer, adoption

Maize technology development, 95–106, 127–142, 176, 252, 253; approaches to, 145–146; breeding programs, 68–69, 130, 145–147, 150, 154, 181, 183, 185–186; path dependence, 148; in Tanzania, 187–188; time horizon, 38, 145, 147, 189. *See also* Improved maize varieties; Maize research

Malawi, 63–79; credit policies, 69, 202; cultural significance of maize, 66–67; farm size, 65, 67; fertilizer distribution networks, 69–71; fertilizer use, 70, 75, 160, 197, 205; food-deficit households, 74–75, 236; improved maize technologies, 63, 72–76, 185–186, 249; land use, 64–66; local maize varieties, 67, 74; maize consumption, 63, 66–67; maize processing, 67; maize production, 10, 64, 186; maize research, 68–69, 77, 150; marketing policies, 71–72, 217, 220, 221, 225, 226, 228; population growth, 64–66; price policies, 72, 74; rural incomes, 66; seed industry, 69–71, 77, 185–186; smallholder agriculture, 77, 159, 186; soil fertility, 64, 65–66, 167–168; yields, 64, 66, 67, 68–70, 77–78, 165

Mali, 60, 133

Marketing boards, 7, 20, 214, 219, 224–225, 228, 230; costs, 54, 223; deficits, 37, 54, 236–237; in Kenya, 89, 90, 91; in Malawi, 70, 71–72; in Zambia, 51–52, 54, 185; in Zimbabwe, 36, 37

Marketing costs, 228–230

Marketing policies, 213–228; colonial, 9, 17, 27, 51, 89, 215–220; post-independence, 220–224; reform, 213, 214, 225, 228, 259–260. *See also* Ghana; Kenya; Malawi; Market liberalization; Nigeria; Price policies; South Africa; Tanzania; Zambia; Zimbabwe

Market liberalization, 201, 224–237, 240–241; assessment of, 228–237; challenges, 238–239; complementary investments, 240; fertilizer, 207–210; food deficits, 235–236; grain flows, 228–229; limitations of, 238; maize production, 234–235; marketing board deficits, 236–237; supply response to, 230–235; sustainability of, 240–241; welfare effects, 234–235. *See also* Marketing policies, reform

Markets, government-controlled, 19–20, 215–224, 238

Millet, 111

Mozambique, 12, 18, 60, 135, 148, 149, 152–154

Mugabe, Robert, 31

Multinational companies: fertilizer, 56, 57; seed, 40, 173, 181–187, 258

National agricultural research systems (NARSs), 37–39, 128, 136, 137, 256–257. *See also* Ghana; Kenya; Maize research; Malawi; Nigeria; Zambia; Zimbabwe

National Seed Company of Malawi, 70, 77

Nigeria, 107, 111–124; fertilizer subsidies, 20, 113, 114, 117–118, 205; fertilizer use, 115, 116–117, 197, 205; improved maize varieties, 112–113, 114, 116; infrastructure, 113, 114, 116;

maize economy in, 111–114; maize production, 111, 113–116, 117, 123, 250; maize research, 112–113, 118–122; maize technology adoption, 114, 116, 205
Nongovernmental organizations (NGOs), 141, 142, 254, 256

Open-pollinated maize varieties (OPVs), 48, 50, 51, 68, 72, 103, 129, 134, 176, 181, 185, 248, 249, 253

Pannar Seed Company, 181, 183, 184, 185, 186, 187
Pan-territorial prices, 203, 234
Pioneer Hi-Bred International, 173, 181, 183, 184, 185, 187
Price instability, 215, 239–240, 260
Price policies, 20, 51–52, 101–102, 213–241. *See also* Fertilizer subsidies; Maize prices; Marketing policies
Policy reform. *See* Market liberalization; Marketing policies, reform
Population growth, 3–4, 64–66, 91, 234, 235, 250, 253. *See also* Agricultural systems, population-driven

Research. *See* Maize research; Soil fertility, research
Research scientists, 38–39, 127, 128
Ristanovic, Dusan, 49–50
Rwanda, 16, 157

Sasakawa Foundation, 100
Sasakawa-Global 2000 (SG 2000). *See* Extension programs
Seed industry, 20, 173–192, 257–258; distribution networks, 69, 87–88, 100, 137, 188; evolution of, 177–187; in Ghana, 100; in Kenya, 87–88, 183–184; life cycle model of, 174–176, 187–191; in Malawi, 69, 185–186; in South Africa, 178, 179, 181; in Tanzania, 186–187, 189, 192; in Zambia, 182–183; in Zimbabwe, 182–183. *See also* Multinational companies, seed
Seed quality, 189, 190
Seed supply, 173–176
Soil fertility, 131, 157–172, 255; improvement strategies, 159–166, 253; inorganic sources of, 160–161; in Kenya, 92; in Malawi, 64, 65–66, 78; nutrient depletion, 109, 116–117, 157, 166, 169, 250; organic sources of, 162–165, 168, 169; research, 166–168, 206, 255. *See also* Fertilizer use
Sorghum, 111, 113, 114
South Africa: marketing policies, 213, 217, 225, 226, 237; seed industry in, 178, 179, 181, 191
Southern Africa: consumer maize preferences, 145–155; marketing and pricing policies, 213–241; seed industry in, 173, 176–187. *See also* Malawi; Zambia; Zimbabwe
Soybeans, 121, 163
Spot markets, 239
SR 52, 25, 29, 48–49, 52, 147, 182, 185
Striga hermonthica, 117, 121, 135, 166, 250, 254, 256
Structural adjustment programs, 20, 55, 57, 58, 105, 173, 186, 191, 192, 213, 225–227, 235, 238, 250. *See also* Market liberalization; Marketing policies, reform
Sudan, 195
Swaziland, 129
Swedish International Development Authority (SIDA), 49, 52

Tanzania, 10, 140; fertilizer use, 205; marketing and pricing policies, 213, 217, 220, 221, 225, 226, 228, 236; seed industry in, 186–187, 189, 192
Training and Visit System (T&V). *See* Extension programs

Uganda, 16, 134
United National Independence Party (UNIP), 51, 54
U.S. Agency for International Development (USAID), 49, 184, 186

Weed control, 99, 105, 117, 255
Western Africa: agricultural systems in, 16, 107–111, 122–123; food security in, 107, 122–123; maize consumption, 16; maize production, 10, 15, 247, 249; maize research, 129, 248, 253. *See also* Ghana; Nigeria
World Bank, 54, 58, 68, 89, 106, 111, 138, 139

Yields, 17–18, 21, 41, 139, 165, 193, 250, 252–253; in Ghana, 103; instability, 19, 134–135; in Kenya, 82–87; in Malawi, 64, 66, 67, 70, 77–78, 165; smallholder, 27, 183; in South Africa, 181; in Zambia, 48, 165, 185; in Zimbabwe, 27, 149, 250. *See also* Fertilizer, technical response to; Fertilizer use; Hybrid maize; Maize production; Open-pollinated maize varieties

Zaire, 12, 134, 250
Zambia, 45–61; agroecological characteristics, 46; commercial farmers in, 46, 59, 184, 223–224; credit system, 52, 56, 58, 59; fertilizer use, 53–54, 57–58, 197, 205, 234; maize consumption, 46, 51; maize prices, 51–52, 54, 55, 236; maize production, 45–46, 185, 232–233, 234; maize research, 48–50, 54, 59; maize technology development, 48–50, 58; marketing board costs, 54, 223; marketing policies, 51–52, 54, 216, 218, 220, 222, 223, 225, 227, 228; marketing reforms, 54–58, 185, 259; seed industry in, 52–53, 184–185; smallholder production, 46, 48–50, 149; socioeconomic characteristics, 45–46, 58; yields, 48, 165, 185
Zamseed, 52, 56, 184, 185
Zimbabwe, 25–43; agroecological regions, 26–27; colonial policies, 27–30; commercial farmers in, 26, 27–30, 33, 35, 41, 223–224; consumer maize preferences, 147–148, 150–152; credit system, 32–33, 223, 234; crop buying stations, 37, 220, 223, 234; Department of Research and Specialist Services (DR&SS), 38–39; farmer support organizations, 29–30, 36–37; fertilizer use, 33–34, 160–161, 197, 234; green revolution in, 25, 27–34, 37, 41–42, 183; hybrid maize in, 25, 29–32, 40; infrastructure, 30, 35; institutional innovation, 40–41; maize economy in, 26–27; maize production, 27, 31–32, 216, 219, 232–233, 249; maize research, 29, 31–34, 37–40, 133, 134, 136, 148, 160, 183; marketing policies, 35, 216, 218, 220, 222, 225, 227, 228, 236; political leadership, 34–37; research scientists in, 38–39; seed industry in, 25, 40, 182–183; smallholder production, 18–19, 25, 26, 30–34, 41, 146; technological innovation, 37–39; yields, 27, 149, 250

About the Book

Although relatively new to Africa, maize has recently replaced cassava as the continent's most important food crop, and increased maize production has the potential of helping to reverse Africa's food crisis. This book presents the results of extensive field research on the maize economy in six African countries, as well as broader-based studies of maize research and extension (R&E), soil fertility, seed distribution, fertilizer, and marketing and processing.

The main finding of the research is that R&E and associated input and marketing interventions have, as intended, produced rapid increases in maize production. Nevertheless, the results tell only a qualified success story—crop yields are still low, yield gains are threatened by a loss of soil fertility, and many farmer-support services require government subsidies. The authors outline the technical, institutional, and policy reforms needed to significantly accelerate maize production in Africa.

Derek Byerlee is principal economist in the Agricultural and Natural Resources Department at the World Bank. Formerly, he served as director of the Economics Program at the International Maize and Wheat Improvement Center (CIMMYT). His most recent book is *International Spillovers and Economies of Size in Agricultural Research: Wheat Research in Developing Countries* (coedited with M. K. Maredia). **Carl K. Eicher** is University Distinguished Professor in the agricultural economics department at Michigan State University. He has been a visiting professor at the University of Nigeria, Stanford University, and the University of Zimbabwe. His numerous publications include "Facing Up to Africa's Food Crisis," *Foreign Affairs,* and *Agricultural Development in the Third World* (coedited with John Staatz).